Lecture Notes in Computer Science 10336

Commenced Publication in 1973
Founding and Former Series Editors:
Gerhard Goos, Juris Hartmanis, and Jan van Leeuwen

More information about this series at http://www.springer.com/series/7407

Mingyu Xiao · Frances Rosamond (Eds.)

Frontiers in Algorithmics

11th International Workshop, FAW 2017
Chengdu, China, June 23–25, 2017
Proceedings

 Springer

Editors
Mingyu Xiao 🆔
University of Electronic Science
and Technology of China
Chengdu
China

Frances Rosamond
University of Bergen
Bergen
Norway

ISSN 0302-9743   ISSN 1611-3349 (electronic)
Lecture Notes in Computer Science
ISBN 978-3-319-59604-4   ISBN 978-3-319-59605-1 (eBook)
DOI 10.1007/978-3-319-59605-1

Library of Congress Control Number: 2017941503

LNCS Sublibrary: SL1 – Theoretical Computer Science and General Issues

Printed on acid-free paper

This Springer imprint is published by Springer Nature
The registered company is Springer International Publishing AG
The registered company address is: Gewerbestrasse 11, 6330 Cham, Switzerland

Editors
Mingyu Xiao (iD)
University of Electronic Science
  and Technology of China
Chengdu
China

Frances Rosamond
University of Bergen
Bergen
Norway

ISSN 0302-9743   ISSN 1611-3349 (electronic)
Lecture Notes in Computer Science
ISBN 978-3-319-59604-4   ISBN 978-3-319-59605-1 (eBook)
DOI 10.1007/978-3-319-59605-1

Library of Congress Control Number: 2017941503

LNCS Sublibrary: SL1 – Theoretical Computer Science and General Issues

Printed on acid-free paper

This Springer imprint is published by Springer Nature
The registered company is Springer International Publishing AG
The registered company address is: Gewerbestrasse 11, 6330 Cham, Switzerland

Mingyu Xiao · Frances Rosamond (Eds.)

Frontiers in Algorithmics

11th International Workshop, FAW 2017
Chengdu, China, June 23–25, 2017
Proceedings

 Springer

Preface

This volume contains the papers presented at FAW 2017: the 11th International Frontiers of Algorithmics Workshop, held during June 23–25, 2017, in Chengdu, China. The workshop brings together researchers working on all aspects of theoretical computer science and algorithms.

FAW 2017 was the 11th conference in the series. The previous ten meetings were held during August 1–3, 2007, in Lanzhou, June 19–21, 2008, in Changsha, June 20–23, 2009, in Hefei, August 11–13, 2010, in Wuhan, May 28–31, 2011, in Jinhua, May 14–16, 2012, in Beijing, June 26–28, 2013, in Dalian, June 28–30, 2014, in Zhangjiajie, July 3–5, 2017, in Guilin, and June 30–July 2, in Qingdao. FAW is already playing an important regional and international role, promising to become a focused forum on current research trends on algorithms seen throughout China and other parts of Asia.

In all, 61 submissions were received from 18 countries and regions: Canada, China, Colombia, France, Germany, Greece, Hong Kong, India, Israel, Japan, Korea, Lebanon, Macao, Malaysia, Norway, Poland, Taiwan, and USA. The FAW 2017 Program Committee selected 24 papers for presentation at the conference. In addition, we had three plenary speakers: John Hopcroft (Cornell University, USA), Michael R. Fellows (University of Bergen, Norway), and Matthew Katz (Ben-Gurion University of The Negev, Israel). We express our sincere thanks to them for their contributions to the conference and proceedings.

We would like to thank the Program Committee members and external reviewers for their hard work in reviewing and selecting papers. We are also very grateful to all the authors who submitted their work to FAW 2017.

Finally, we wish to thank the editors at Springer and the local organizer, University of Electronic Science and Technology of China, for their support and hard work in the preparation of this conference.

June 2017

Mingyu Xiao
Frances Rosamond

Organization

Program Co-chairs

Frances A. Rosamond   University of Bergen, Norway
Mingyu Xiao   University of Electronic Science and Technology of China,
       China

Program Committee

Leizhen Cai	The Chinese University of Hong Kong, SAR China
Jianer Chen	Texas A&M University, USA
Yijia Chen	Fudan University, China
Zhi-Zhong Chen	Tokyo Denki University, Japan
Wenqiang Dai	University of Electronic Science and Technology of China
Fei Deng	University of California, Davis, USA
Qilong Feng	Central South University, China
Henning Fernau	University of Trier, Germany
Rudolf Fleischer	German University of Technology in Oman
Yuxi Fu	Shanghai Jiao Tong University, China
Takuro Fukunaga	National Institute of Informatics, Japan
Jiong Guo	Shandong University, China
Kun He	Huazhong University of Science and Technology, China
Giuseppe F. Italiano	University of Rome, Italy
Ragesh Jaiswal	Indian Institute of Technology Delhi, India
Eun Jung Kim	University of Paris Dauphine, France
Christian Komusiewicz	Friedrich Schiller University Jena, Germany
Michael Lampis	Université Paris Dauphine, France
Van Bang Le	University of Rostock, Germany
Minming Li	City University of Hong Kong, SAR China
Guohui Lin	University of Alberta, Canada
Chunmei Liu	Howard University, USA
Kazuhisa Makino	Kyoto University, Japan
Sheung-Hung Poon	Brunei Technological University, Brunei
Md Saidur Rahman	Bangladesh University of Engineering and Technology
Frances A. Rosamond	University of Bergen, Norway
Hao Wang	Norwegian University of Science and Technology, Norway
Jianxin Wang	Central South University, China
Gerhard J. Woeginger	Eindhoven University of Technology, The Netherlands
Jigang Wu	Guangdong University of Technology, China
Mingyu Xiao	University of Electronic Science and Technology of China
Ke Xu	Beihang University, China

Yinfeng Xu	Xi'an Jiaotong University, China
Boting Yang	University of Regina, Canada
Guang Yang	Chinese Academy of Sciences, China
Guochuan Zhang	Zhejiang University, China
Jialin Zhang	Chinese Academy of Sciences, China
Shengyu Zhang	The Chinese University of Hong Kong, SAR China
Binhai Zhu	Montana State University, USA
Daming Zhu	Shandong University, China

Organizing Committee

Mingyu Xiao	University of Electronic Science and Technology of China
Wenqiang Dai	University of Electronic Science and Technology of China
Shaowei Kou	University of Electronic Science and Technology of China
Yuqing Wang	University of Electronic Science and Technology of China
Jianan Zhang	University of Electronic Science and Technology of China

Additional Reviewers

Ali, Mohsen	Huang, Mingzhang	Stege, Ulrike
Brauße, Franz	Istrate, Gabriel	Wang, Jingbo
Casel, Katrin	Kuinke, Philipp	Wang, Jiongbo
Chau, Vincent	Li, Fagen	Wu, Chao
Chen, Zhihuai	Li, Minming	Xu, Yao
Chini, Peter	Mnich, Matthias	Yang, Feidiao
Dong, Xiaoju	Nishat, Rahnuma Islam	Yang, Qizhe
Epstein, Leah	Paulusma, Daniel	Ye, Deshi
Gao, Xiaofeng	Poon, Sheung-Hung	Zhang, Chihao
Gentner, Michael	Rushdi, Muhammad	Zhang, Jia
Han, Xin	Shen, Weiran	Zheng, Feifeng
Hartline, Jason	Sikora, Florian	

Contents

On the Complexity of Minimizing the Total Calibration Cost

Eric Angel[1], Evripidis Bampis[2], Vincent Chau[3(✉)],
and Vassilis Zissimopoulos[4]

[1] IBISC, Université d'Évry Val d'Essonne, Évry, France
angel@ibisc.fr
[2] Sorbonne Universités, UPMC Univ. Paris 06, UMR 7606, LIP6, Paris, France
Evripidis.Bampis@lip6.fr
[3] Shenzhen Institutes of Advanced Technology, Academy of Sciences,
Shenzhen, China
vincentchau@siat.ac.cn
[4] Department of Informatics and Telecommunications,
National and Kapodistrian University of Athens, Athens, Greece
vassilis@di.uoa.gr

Abstract. Bender et al. (SPAA 2013) proposed a theoretical framework for testing in contexts where safety mistakes must be avoided. Testing in such a context is made by machines that need to be often calibrated. Since calibrations have non negligible cost, it is important to study policies minimizing the calibration cost while performing all the necessary tests. We focus on the single-machine setting and we study the complexity status of different variants of the problem. First, we extend the model by considering that the jobs have arbitrary processing times and that the preemption of jobs is allowed. For this case, we propose an optimal polynomial time algorithm. Then, we study the case where there is many types of calibrations with different lengths and costs. We prove that the problem becomes NP-hard for arbitrary processing times even when the preemption of the jobs is allowed. Finally, we focus on the case of unit-time jobs and we show that a more general problem, where the recalibration of the machine is not instantaneous, can be solved in polynomial time.

1 Introduction

The scheduling problem of minimizing the number of calibrations has been recently introduced by Bender et al. in [2]. It is motivated by the Integrated Stockpile Evaluation (ISE) program [1] at Sandia National Laboratories for testing in contexts where safety mistakes may have serious consequences. Formally,

This work has been supported by the ALGONOW project of the THALES program and the Special Account for Research Grants of National and Kapodistrian University of Athens, by NSFC (Nos. 61433012, U1435215), and by Shenzhen basic research grant JCYJ20160229195940462.

© Springer International Publishing AG 2017
M. Xiao and F. Rosamond (Eds.): FAW 2017, LNCS 10336, pp. 1–12, 2017.
DOI: 10.1007/978-3-319-59605-1_1

the problem can be stated as follows: we are given a set $\mathcal{J}$ of n jobs (tests), where each job j is characterized by its release date r_j, its deadline d_j and its processing time p_j. We are also given a (resp. a set of) testing machine(s) that must be calibrated on a regular basis. The calibration of a machine has a unit cost and it is instantaneous, i.e., a machine can be calibrated between the execution of two jobs that are processed in consecutive time-units. A machine stays calibrated for T time-units and a job can only be processed during an interval where the machine is calibrated. The goal is to find a feasible schedule performing all the tests (jobs) between their release dates and deadlines and minimizing the number of calibrations. Using the classical three-field notation in scheduling [4], the problem can be denoted as $P|\ r_j, d_j, T\ |\ (\# \ calibrations)$. Bender et al. [2] studied the case of *unit-time* jobs. They considered both the single-machine and multiple-machine problems. For the single-machine case, they showed that there is a polynomial-time algorithm, called the Lazy Binning algorithm that solves the problem optimally. For the multiple-machine case, they proposed a 2-approximation algorithm. However, the complexity status of the multiple-machine case with unit-time jobs remained open. Bender et al. [2] stated, *"As a next step we hope to generalize our model to capture more aspects of the actual ISE problem. For example, machines may not be identical, and calibrations may require machine time. Moreover, some jobs may not have unit size"*.

Fineman and Sheridan [3] studied a first generalization of the problem by considering that the jobs have arbitrary processing times. They considered the multiple-machine case where the execution of a job is not allowed to be interrupted once it has been started. Since the feasibility problem is NP-hard, they considered a *resource-augmentation* [5] version of the problem. They were able to relate this version with the classical *machine-minimization* problem [8] in the following way: suppose there is an s-speed α-approximation algorithm for the machine-minimization problem, then there is an $O(\alpha)$-machine s-speed $O(\alpha)$-approximation for the resource-augmentation version of the problem of minimizing the number of calibrations.

In this paper, we focus on the single-machine case without resource augmentation and we study the complexity status of different variants of the problem. In Sect. 2, we study the problem when the jobs may have arbitrary processing times and the preemption of the jobs is allowed: the processing of any job may be interrupted and resumed at a later time. We denote this variant of the problem as $1\ |\ r_j, d_j, pmtn, T\ |\ (\# \ calibrations)$. Clearly, by using the optimal algorithm of Bender et al. for unit-time jobs, we can directly obtain a pseudopolynomial-time algorithm by just replacing every job by a set of unit-time jobs with cardinality equal to the processing time of the job. We propose a polynomial time algorithm for this variant of the problem. Then, in Sect. 3, we study the case of scheduling a set of jobs when K different types of calibrations are available. Each calibration type is associated with a length T_i and a cost f_i. The objective is to find a feasible schedule minimizing the total calibration cost. We show that the problem, denoted as $1\ |\ r_j, d_j, pmtn, \{T_1, \ldots, T_K\}\ |\ cost(calibrations)$ for arbitrary processing times is NP-hard, even when the preemption of the jobs is allowed.

Given the NP-hardness of the problem for arbitrary processing times, in Sect. 4, we study the case of unit-time jobs. We propose a polynomial time algorithm based on dynamic programming. We present the algorithm for a more general setting where each calibration takes λ units of time during which the machine cannot be used. We denote this variant as $1 \mid \tilde{r_j}, d_j, p_j = 1, \lambda + \{T_1, \ldots, T_K\} \mid cost(calibrations)$.

2   Arbitrary Processing Times and Preemption

We suppose here that the jobs have arbitrary processing times and that the preemption of the jobs is allowed. An obvious approach in order to obtain an optimal preemptive schedule is to divide each job j into p_j unit-time jobs with the same release date and deadline as job j and then apply the Lazy Binning (LB) algorithm of [2] that optimally solves the problem for instances with unit-time jobs. However, this idea leads to a pseudopolynomial-time algorithm. Here, we propose a more efficient way for solving the problem. Our method is based on the idea of Lazy Binning. Before introducing our algorithm, we briefly recall LB: at each iteration, a time t is fixed and the (remaining) jobs are scheduled, starting at time $t+1$ using the Earliest Deadline First (EDF) policy[1]. If a feasible schedule exists (for the remaining jobs), t is updated to $t+1$, otherwise the next calibration is set to start at time t which is called the current *latest-starting-time* of the calibration. Then, the jobs that are scheduled during this calibration interval are removed and this process is iterated after updating t to $t+T$, where T is the calibration length. The polynomiality of the algorithm for unit-time jobs comes from the observation that the starting time of any calibration is at a distance of no more than n time-units before any deadline. In our case however, i.e. when the jobs have arbitrary processing times, a calibration may start at a distance of at most $P = \sum_{j=1}^{n} p_j$ time-units before any deadline.

Definition 1. *Let* $\Psi := \bigcup_i \{d_i - P, d_i - P + 1, \ldots, d_i\}$ *where* $P = \sum_{j=1}^{n} p_j$.

Proposition 1. *There exists an optimal solution in which each calibration starts at a time in* Ψ.

Proof. Let σ be an optimal solution in which there is at least one calibration that does not start at a time in Ψ. We show how to transform the schedule σ into another optimal schedule that satisfies the statement of the proposition.

Let $c_{i'}$ be the first calibration of σ that starts at time $t' \notin \Psi$. Let $c_{i'}, \ldots, c_i$ be the maximum set of consecutive calibrations such that when a calibration finishes another starts immediately. We denote by c_{i+1} the next calibration that is not adjacent to calibration c_i. We can push the set of calibrations $c_{i'}, \ldots, c_i$ to the right (we delay the calibrations) until:

---

[1] An EDF policy is a schedule in which at any time, the job with the smallest deadline among the available jobs is scheduled first.

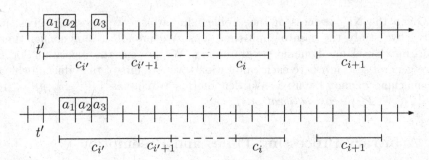

Fig. 1. Illustration of Proposition 1. The first schedule is an optimal schedule. The second one is obtained after pushing the continuous block of calibrations $c_{i'}, \ldots, c_i$ to the right.

- either we reach the next calibration c_{i+1},
  or $c_{i'}$ starts at a time in Ψ (Fig. 1).

Note that this transformation is always possible. Indeed, since $c_{i'}$ starts at a time that is in a distance more than P from a deadline, it is always possible to push the scheduled jobs to the right. In particular, if there are no jobs scheduled when calibration $c_{i'}$ starts, then there are no modifications for the execution of jobs. Otherwise, there is at least one job scheduled when calibration $c_{i'}$ starts. Let $a_1, \ldots, a_e$ be the continuous block of jobs. Since the starting time of job a_1 is at a distance (to the left hand side) more than P from a deadline, then all these jobs can be pushed to the right by one unit. This transformation is possible because no job of this block finishes at its deadline. Note that after this modification, jobs can be assigned to another calibration.

We can repeat the above transformation until we get a schedule satisfying the statement of the proposition. □

For jobs with arbitrary processing times when the preemption of the jobs is allowed, we propose the following algorithm whose idea is based on the Lazy Binning algorithm: we first compute the current latest-starting-time of the calibration such that no job misses its deadline (this avoids to consider every time in Ψ). This calibration time depends on some deadline d_k. At each iteration, among the remaining jobs, we compute for every deadline the sum of the processing times of all these jobs (or of their remaining parts) having a smaller than or equal deadline and we subtract it from the current deadline. The current latest-starting-time of the calibration is obtained by choosing the smallest computed value. Once the calibration starting time is set, we schedule the remaining jobs in the EDF order until reaching d_k and we continue to schedule the available jobs until the calibration interval finishes. In the next step, we update the processing time of the jobs that have been processed. We repeat this computation until there is no processing time left. A formal description of the algorithm, that we call the Preemptive Lazy Binning (PLB) algorithm, is given below (Algorithm 1).

Algorithm 1. Preemptive Lazy Binning (PLB)

1: Jobs in $\mathcal{J}$ are sorted in non-decreasing order of deadline
2: **while** $\mathcal{J} \neq \emptyset$ **do**
3:     $t \leftarrow \max_{i \in \mathcal{J}} d_i,\ k \leftarrow 0$
4:     **for** $i \in \mathcal{J}$ **do**
5:       **if** $t > d_i - \sum_{j \leq i, j \in \mathcal{J}} p_j$ **then**
6:         $t \leftarrow d_i - \sum_{j \leq i, j \in \mathcal{J}} p_j$
7:         $k \leftarrow i$
8:       **end if**
9:     **end for**
10:    $u \leftarrow t + \left\lceil \frac{d_k - t}{T} \right\rceil \times T$
11:    Calibrate the machine at time $t, t+T, t+2T, \ldots, u-T$
12:    Schedule jobs $\{j \leq k \mid j \in \mathcal{J}\}$ from t to d_k by applying the EDF policy and remove them from $\mathcal{J}$.
13:    Schedule fragment of jobs from $k+1, \ldots, n$ in $[d_k, u)$ in EDF order
14:    Let q_j for $j = k+1, \ldots, n$ be the processed quantity in $[d_k, u)$
15:    //Update processing time of jobs
16:    **for** $i = k+1, \ldots, n$ **do**
17:       $p_i \leftarrow p_i - q_i$
18:       **if** $p_i = 0$ **then**
19:         $\mathcal{J} \leftarrow \mathcal{J} \setminus i$
20:       **end if**
21:    **end for**
22: **end while**

We can prove the optimality of this algorithm using a similar analysis as the one for the Lazy Binning algorithm in [2].

Proposition 2. *The schedule returned by Algorithm PLB is a feasible schedule in which the starting time of each calibration is maximum.*

Proof. The condition in line 5 in Algorithm PLB ensures that we always obtain a feasible schedule. In fact, we compute the latest-starting-time at each step and this time is exactly the latest time of the first calibration.

By fixing a deadline d_i, we know that jobs that have a deadline earlier than d_i have to be scheduled before d_i, while the other jobs are scheduled after d_i. When we update t for every deadline d_i in the algorithm, we assume that there is no idle time between $d_i - \sum_{j \leq i, j \in \mathcal{J}} p_j$ and d_i. Note that if $d_i - \sum_{j \leq i, j \in \mathcal{J}} p_j < 0$, then the schedule is not feasible. For the sake of contradiction, suppose that a feasible schedule exists in which some calibration is not started at a time computed by the algorithm. We will show that the starting time of this calibration is not maximum. Denote this time by t'. Since, the starting time of the calibration is not one of $d_i - \sum_{j \leq i, j \in \mathcal{J}} p_j\ \forall i$, then there is at least one unit of idle time between the starting time of the calibration and some deadline d_i. Hence, it is possible to delay all calibrations starting at t' or after, as well as the execution of the jobs inside these calibrations by keeping the EDF order. This can be done in a similar way as in the proof of Proposition 1. $\qquad \square$

Proposition 3. *Algorithm PLB is optimal.*

Proof. It is sufficient to prove that Algorithm PLB returns the same schedule as Lazy Binning after splitting all jobs to unit-time jobs. We denote respectively PLB and LB the schedules returned by these algorithms.

Let t' be the first time at which the two schedules differ. The jobs executed before t' are the same in both schedules since the jobs are scheduled in the EDF order. Given that the schedules are the same before t', the remaining jobs are the same after t'. Two cases may occur:

- a job is scheduled in $[t', t' + 1)$ in PLB but not in LB. This means that the machine is not calibrated at this time slot in the schedule produced by LB. Since the calibrations are the same before t' in both schedules, then a calibration starting at t' is necessary in PLB. Thanks to Proposition 2, we have a contradiction to the fact that we were looking for the latest-starting-time of the calibration.
- a job is scheduled in $[t', t' + 1)$ in LB but not in PLB. This means that there does not exist a feasible schedule starting at $t' + 1$ with the remaining jobs. Hence, PLB is not feasible. This case cannot happen thanks to Proposition 2. □

Proposition 4. *Algorithm PLB has a time complexity in $O(n^2)$.*

Proof. We first sort jobs in the non-decreasing order of their deadlines in $O(n \log n)$ time. At each step, we compute the first latest-starting-time of the calibration in $O(n)$ time. Then the scheduling of jobs in the EDF order takes $O(n)$ time. We need also to update the processing times of the jobs whose execution has been started. This can be done in $O(n)$ time. At each step, we schedule at least one job. Hence, there are at most n steps. □

3 Arbitrary Processing Times, Preemption and Many Calibration Types

In this section, we consider a generalization of the model of Bender et al. in which there are more than one types of calibration. Every calibration type is associated with a length T_i and a cost f_i. We are also given a set of jobs, each one characterized by its processing time p_j, its release time r_j and its deadline d_j. Each job can be scheduled only when the machine is calibrated regardless of the calibration type. Our objective is to find a feasible preemptive schedule minimizing the total calibration cost. We prove that the problem is NP-hard.

Proposition 5. *The problem of minimizing the calibration cost is NP-hard for jobs with arbitrary processing times and many types of calibration, even when the preemption is allowed.*

In order to prove the NP-hardness, we use a reduction from the UNBOUNDED SUBSET SUM problem (which is NP-hard) [6,7]. In an instance of the

UNBOUNDED SUBSET SUM problem, we are given a set of n items where each item j is associated to a value κ_j. We are also given a value V. We aim to find a subset of the items that sums to V under the assumption that an item may be used more than once.

Proof. Let Π be the preemptive scheduling problem of minimizing the total calibration cost for a set of n jobs that have arbitrary processing times in the presence of a set of K calibration types.

Given an instance of the UNBOUNDED SUBSET SUM problem, we construct an instance of problem Π as follows. For each item j, create a calibration length $T_j = \kappa_j$ and of cost $f_j = \kappa_j$. Moreover, we create n jobs with positive arbitrary processing times such that $\sum_i p_i = V$ with $r_i = 0$ and $d_i = V \ \forall i$.

We claim that the instance of the UNBOUNDED SUBSET SUM problem is feasible if and only if there is a feasible schedule for problem Π of cost V.

Assume that the instance of the UNBOUNDED SUBSET SUM problem is feasible. Therefore, there exists a subset of items C' such that $\sum_{j \in C'} \kappa_j = V$. Note that the same item may appear several times. Then we can schedule all jobs, and calibrate the machine according to the items in C' in any arbitrary order. Since the calibrations allow all the jobs to be scheduled in $[0, V)$, then we get a feasible schedule of cost V for Π.

For the opposite direction of our claim, assume that there is a feasible schedule for problem Π of cost V. Let C be the set of calibrations that have been used in the schedule. Then $\sum_{j \in C} T_j = V$. Therefore, the items which correspond to the calibrations in C form a feasible solution for the UNBOUNDED SUBSET SUM problem. □

4 Unit-Time Jobs, Many Calibration Types and Activation Length

Since the problem is NP-hard when many calibration types are considered even in the case where the calibrations are instantaneous, we focus in this section on the case where the jobs have unit processing times. We also assume that there is an activation length, that we denote by λ. This means that in this section, the calibrations are no more instantaneous, but each of them takes λ units of time during which no job can be processed. For feasibility reasons, we allow to recalibrate the machine at any time point, even when it is already calibrated. To see this, consider the instance given in Fig. 2. The machine has to be calibrated at time 0 and requires $\lambda = 3$ units of time for being available for the execution of jobs. At time 3 the machine is ready to execute job 1 and it remains calibrated for $T = 4$ time units. If we do not have the possibility to recalibrate an already calibrated machine then the earliest time at which we can start calibrating the machine is at time 7. This would lead to the impossibility of executing job 2. However, a recalibration at time 4 would lead to a feasible schedule.

It is easy to see that the introduction of the activation length into the model makes necessary the extension of the set of "important" dates that we have used

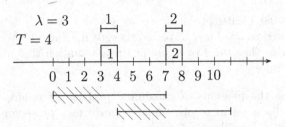

Fig. 2. An infeasible instance if we do not have the possibility to recalibrate at any time. We have a single machine, two unit-time jobs and a single type of calibration of length $T = 4$. The activation length, i.e. the time that is required in order for the calibration to be effective is $\lambda = 3$. Job 1 is released at time 3 and its deadline is 4. Job 2 is released at time 7 and its deadline is 8.

in Sect. 2 (Definition 1). Indeed, jobs can be scheduled at a distance bigger than n from a release date or a deadline. However, as we prove below, it is still possible to define a polynomial-size time-set.

In the worst case, we have to calibrate n times and schedule n jobs. Thus the calibration can start at a time at most $n(\lambda + 1)$ time units before a deadline. Note that it is not necessary to consider every date in $[d_i - n(\lambda + 1), d_i]$ for a fixed i. In the sequel, we suppose without loss of generality that jobs are sorted in non-decreasing order of their deadline, $d_1 \leq d_2 \leq \ldots \leq d_n$.

Definition 2. *Let* $\Theta := \bigcup_i \{d_i - j\lambda - h, \ j = 0, \ldots, n, \ h = 0, \ldots, n\}$.

Proposition 6. *There exists an optimal solution in which each calibration starts at a time in Θ.*

Proof. We show how to transform an optimal schedule into another schedule satisfying the statement of the proposition without increasing the total calibration cost. Let c_j be the last calibration that does not start at a date in Θ. We can shift this calibration to the right until:

- one job of this calibration finishes at its deadline and hence, it is no more possible to push this calibration to the right anymore. This means that there is no idle time between the starting time of this calibration and this deadline. Thus the starting time of this calibration is in Θ.
- the current calibration meets another calibration. In this case, we continue to shift the current calibration to the right while this is possible. Perhaps, there will be an overlap between calibration intervals, but as we said before, we allow to recalibrate the machine at any time. If we cannot shift to the right anymore, either a job ends at its deadline (and we are in the first case), or there is no idle time between the current calibration and the next one. Since there is at most n jobs and the next calibration starts at a time $d_i - j\lambda - h$ for some i, j, h, then the current calibration starts at a time $d_i - (j+1)\lambda - (h+h')$ where h' is the number of jobs scheduled in the current calibration with $h' + h \leq n$ and $j \leq n - 1$. □

Moreover the set of starting times of jobs has also to be extended by considering the activation length λ.

Definition 3. *Let* $\Phi := \{t + a \mid t \in \Theta,\ a = 0, \ldots, n\} \cup \bigcup_i \{r_i, r_i + 1, \ldots, r_i + n\}.$

As for the starting time of calibrations, the worst case happens when we have to recalibrate after the execution of every job.

Proposition 7. *There exists an optimal solution in which the starting times and completion times of jobs belong to Φ.*

Proof. The first part of the proof comes from Proposition 1. Indeed, jobs can only be scheduled when the machine is calibrated. Let i be the first job that is not scheduled at a time in Φ in an optimal solution. Thanks to Proposition 1, we know that a calibration occurs before a deadline. Job i belongs to some calibration that starts at time $t \le d_j$ for some other job j. By moving job i to the left, the cost of the schedule does not increase, since this job belongs to the same calibration. Two cases may occur:

- job i meets another job i' (Fig. 3(a)). In this case, we consider the continuous block of jobs $i'', \ldots, i', i$. We assume that at least one job in this block is scheduled at its release date and job i is at a distance at most n of this release date (because there is at most n jobs). Otherwise, we can shift this block of jobs to the left by one time unit (Fig. 3(b)). Indeed, this shifting is possible because no job in $\{i'', \ldots, i'\}$ is executed at a starting time of a calibration (if it is the case, job i is in Φ by definition). Since job i' was in Φ, by moving this block, job i will be scheduled at a time in Φ.
- job i meets its release date, thus its starting time is in Φ.   □

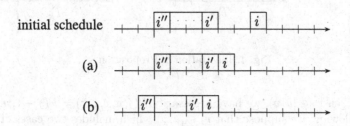

Fig. 3. Illustration of Proposition 7

Definition 4. *Let* $S(j, u, v) = \{i \mid i \le j \text{ and } u \le r_i < v\}$. *We define* $F(j, u, v, t, k)$ *as the minimum cost of a schedule of the jobs in* $S(j, u, v)$ *such that:*

- *all these jobs are scheduled during the time-interval* $[u, v)$
- *the last calibration of the machine is at time t for a length of $\lambda + T_k$ (where the time-interval $[t, t + \lambda)$ corresponds to the activation length)*
- *the first calibration is not before u.*

We are now ready to give our dynamic programming algorithm. We examine two cases depending on whether r_j belongs to the interval $[u, v)$. Otherwise, there is two subcases: whether job j is scheduled in the last calibration or not.

Proposition 8. *One has* $F(j, u, v, t, k) = F'$

$$
F' := \begin{cases}
F(j-1, u, v, t, k) & \text{if } r_j \notin [u, v) \\[2mm]
\displaystyle\min_{\substack{u' \in \Phi,\ r_j \leq u' < t + T_k + \lambda \\ t' + \lambda \leq u' < v \\ u' < d_j}} F(j-1, u, u', t, k) & \text{last calibration} \\[6mm]
\displaystyle\min_{\substack{u' \in \Phi,\ r_j \leq u' < t' + T_{k'} + \lambda \\ t' \in \Theta,\ t' + \lambda \leq u' < v \\ u' < d_j \\ 1 \leq k' \leq K}} \left\{ \begin{array}{c} F(j-1, u, u', t', k') \\ +F(j-1, u'+1, v, t, k) \end{array} \right\} & \text{otherwise}
\end{cases}
$$

with $F(0, u, v, t, k) := f_k,\ \forall t + \lambda \leq v\ \&\ t \geq u$.
$F(0, u, v, t, k) := +\infty$ *otherwise*.

The objective function for our problem is $\min_{t \in \Theta, 1 \leq k \leq K} F(n, \min_i r_i, \max_i d_i, t, k)$ (Fig. 4).

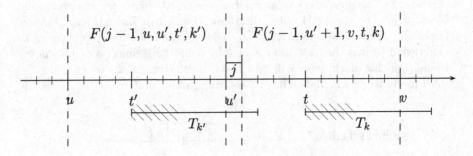

Fig. 4. Illustration of Proposition 8

Proof. When $r_j \notin [u, v)$, we have necessarily $F(j, u, v, t, k) = F(j-1, u, v, t, k)$. In the following, we suppose that $r_j \in [u, v)$ which includes two cases. The first one is when job j is scheduled in the last calibration.
We first prove that $F(j, u, v, t, k) \leq F'$.

We consider a schedule S_1 that realizes $F(j-1, u, u', t', k')$ and a schedule S_2 that realizes $F(j-1, u'+1, v, t, k)$. We build a schedule as follows: from time u to time u' use S_1, then execute job j in $[u', u'+1)$, and finally from $u'+1$ to time v use S_2. Moreover, it contains all jobs in $\{i \mid i \leq j \text{ and } u \leq r_i < v\}$. Since the first calibration in S_2 does not begin before $u'+1$, then we have a feasible schedule.

So $F(j, u, v, t, k) \leq F'$.

We now prove that $F(j, u, v, t, k) \geq F'$.

Since $j \in \{i \mid i \leq j \text{ and } u \leq r_i < v\}$, job j is scheduled in all schedules that realize $F(j, u, v, t, k)$.

Among such schedules, let $\mathcal{X}$ denote the schedule of $F(j, u, v, t, k)$ in which the starting time of job j is maximal. We claim that all jobs in $\{i \leq j, u \leq r_i < v\}$ that are released before u' are completed at u'. If it is not the case, we could swap the execution of such a job with job j, getting in this way a feasible schedule with the same cost as before. Formally, let i be a job with $\{i \leq j, u \leq r_i < u'\}$ that is scheduled after $u' + 1$. We can swap the execution of job i with job j, the resulting schedule is feasible since job j has larger deadline than job i, and job i is released before u'. This will contradict the fact that the starting time of job j is maximal.

We consider a schedule S_1 that realizes $F(j-1, u, u', t', k')$ and a schedule S_2 that realizes $F(j-1, u'+1, v, t, k)$. Then, the restriction of S_1 in the schedule $\mathcal{X}$ to $[u, u')$ will be a schedule that meets all constraints related to $F(j-1, u, u', t', k')$. Hence its cost is greater than $F(j-1, u, u', t', k')$. Similarly, the restriction of S_2 in the schedule $\mathcal{X}$ to $[u' + 1, v)$ is a schedule that meets all constraints related to $F(j-1, u'+1, v, t, k)$.

Finally, $F(j, u, v, t, k) \geq F'$. $\square$

Proposition 9. *The problem of minimizing the total calibration cost with arbitrary calibration lengths, activation length and unit-time jobs can be solved in time* $O(n^{16} K^2)$.

Proof. This problem can be solved with the dynamic program in Proposition 8. Recall that the objective function is $\min_{t \in \Theta, 1 \leq k \leq K} F(n, \min_i r_i, \max_i d_i, t, k)$. The size of both sets Θ and Φ is $O(n^3)$. Indeed, by rewriting the set Φ, we have

$$\Phi = \bigcup_i \{r_i, r_i + 1 \ldots, r_i + n\} \cup \{t + a \mid t \in \Theta, \ a = 0, \ldots, n\}$$

$$= \bigcup_i \{r_i, r_i + 1 \ldots, r_i + n\} \bigcup_i \left\{ \begin{array}{c} d_i - j\lambda - k + a, \ j = 0, \ldots, n \\ k = 0, \ldots, n, \ a = 0, \ldots, n \end{array} \right\}$$

$$= \bigcup_i \{r_i, r_i + 1 \ldots, r_i + n\} \bigcup_i \{d_i - j\lambda + k, \ j = 0, \ldots, n, \ k = -n, \ldots, n\}$$

The size of the table is $O(n^{10} K)$. When each value of the table is fixed, the minimization is over the values u', t' and k', so the time complexity is $O(n^6 K)$. Therefore the overall complexity time is $O(n^{16} K^2)$. $\square$

Note that when there is no feasible schedule, the objective function $\min_{t \in \Theta, 1 \leq k \leq K} F(n, \min_i r_i, \max_i d_i, t, k)$ will return $+\infty$.

5   Conclusion

We considered different extensions of the model introduced by Bender et al. in [2]. We proved that the problem of minimizing the total calibration-cost on a single machine can be solved in polynomial time for the case of jobs with arbitrary processing times when the preemption is allowed. Then we proved that the problem becomes NP-hard for arbitrary processing times when there are many calibration types, even if the preemption of jobs is authorized. Finally, we considered the case with many calibration types, where the calibrations are not instantaneous but take machine time, and we proved that the problem can be solved in polynomial time using dynamic programming for unit-time jobs. An interesting question is whether it is possible to find a lower time-complexity algorithm for solving this version of the problem, either optimally, or in approximation. Of course, it would be of great interest to study the case where more than one machines are available. Recall that the complexity of the simple variant studied by Bender et al. remains unknown for the multiple machines problem.

References

1. New integrated stockpile evaluation program to better ensure weapons stockpile safety, security, reliability (2006). http://www.sandia.gov/LabNews/060331.html
2. Bender, M.A., Bunde, D.P., Leung, V.J., McCauley, S., Phillips, C.A.: Efficient scheduling to minimize calibrations. In: 25th ACM Symposium on Parallelism in Algorithms and Architectures, SPAA 2013, pp. 280–287. ACM (2013). http://doi.acm.org/10.1145/2486159.2486193
3. Fineman, J.T., Sheridan, B.: Scheduling non-unit jobs to minimize calibrations. In: Blelloch, G.E., Agrawal, K. (eds.) Proceedings of the 27th ACM on Symposium on Parallelism in Algorithms and Architectures, SPAA 2015, Portland, OR, USA, 13–15 June 2015, pp. 161–170. ACM (2015). http://doi.acm.org/10.1145/2755573.2755605
4. Graham, R.L., Lawler, E.L., Lenstra, J.K., Kan, A.: Optimization and approximation in deterministic sequencing and scheduling: a survey. Ann. Discret. Math. **5**, 287–326 (1979)
5. Kalyanasundaram, B., Pruhs, K.: Speed is as powerful as clairvoyance. J. ACM **47**(4), 617–643 (2000). http://doi.acm.org/10.1145/347476.347479
6. Lai, T.: Worst-case analysis of greedy algorithms for the unbounded knapsack, subset-sum and partition problems. Oper. Res. Lett. **14**(4), 215–220 (1993). http://dx.doi.org/10.1016/0167-6377(93)90072-O
7. Lueker, G.: Two NP-complete problems in nonnegative integer programming. Report No. 178, Computer Science Laboratory, Princeton University (1975)
8. Phillips, C.A., Stein, C., Torng, E., Wein, J.: Optimal time-critical scheduling via resource augmentation. Algorithmica **32**(2), 163–200 (2002). http://dx.doi.org/10.1007/s00453-001-0068-9

On the Fixed-Parameter Tractability of Some Matching Problems Under the Color-Spanning Model

Sergey Bereg[1], Feifei Ma[2], Wencheng Wang[2], Jian Zhang[2], and Binhai Zhu[3]([✉])

[1] Department of Computer Science, University of Texas-Dallas,
Richardson, TX 75080, USA
besp@utdallas.edu

[2] State Key Laboratory of Computer Science, Institute of Software,
Chinese Academy of Sciences, Beijing 100080, China
{maff,whn,zj}@ios.ac.cn

[3] Gianforte School of Computing, Montana State University,
Bozeman, MT 59717, USA
bhz@montana.edu

Abstract. Given a set of n points P in the plane, each colored with one of the t given colors, a color-spanning set $S \subset P$ is a subset of t points with distinct colors. The minimum diameter color-spanning set (MDCS) is a color-spanning set whose diameter is minimum (among all color-spanning sets of P). Somehow symmetrically, the largest closest pair color-spanning set (LCPCS) is a color-spanning set whose closest pair is the largest (among all color-spanning sets of P). Both MDCS and LCPCS have been shown to be NP-complete, but whether they are fixed-parameter tractable (FPT) when t is a parameter are still open. (Formally, the problem whether MDCS is FPT was posed by Fleischer and Xu in 2010.) Motivated by this question, we consider the FPT tractability of some matching problems under this color-spanning model, where $t = 2k$ is the parameter. The results are summarized as follows: (1) Min-Sum Matching Color-Spanning Set, namely, computing a matching of $2k$ points with distinct colors such that their total edge length is minimized, is FPT; (2) MaxMin Matching Color-Spanning Set, namely, computing a matching of $2k$ points with distinct colors such that the minimum edge length is maximized, is FPT; (3) MinMax Matching Color-Spanning Set, namely, computing a matching of $2k$ points with distinct colors such that the maximum edge length is minimized, is FPT; and (4) k-Multicolored Independent Matching, namely, computing a matching of $2k$ vertices in a graph such that the vertices of the edges in the matching do not share common edges in the graph, is W[1]-hard. With (2), we show that LCPCS is in fact FPT.

1 Introduction

Given a set of n points Q with all points colored in one of the t given colors, a *color-spanning* set (sometimes also called a *rainbow* set) is a subset of t points

© Springer International Publishing AG 2017
M. Xiao and F. Rosamond (Eds.): FAW 2017, LNCS 10336, pp. 13–21, 2017.
DOI: 10.1007/978-3-319-59605-1_2

with distinct colors. (In this paper, as we focus on matching problems, we set $t = 2k$. Of course, in general t does not always have to be even.) In practice, many problems require us to find a specific color-spanning set with certain property due to the large size of the color-spanning sets. For instance, in data mining a problem arises where one wants to find a color-spanning set whose diameter is minimized (over all color-spanning sets), which can be solved in $O(n^t)$ time using a brute-force method [2,14]. (Unfortunately, this is still the best bound to this date.)

Since the color-spanning set problems were initiated in 2001 [1], quite some related problems have been investigated. Many of the traditional problems which are polynomially solvable, like Minimum Spanning Tree, Diameter, Closest Pair, Convex Hull, etc., become NP-hard under the color-spanning model [6,7,9]. Note that for the hardness results the objective functions are usually slightly changed. For instance, in the color-spanning model, we would like to maximize the closest pair and minimize the diameter (among all color-spanning sets). On the other hand, some problems, like the Maximum Diameter Color-Spanning Set, remain to be polynomially solvable [4].

In [6,7], an interesting question was raised. Namely, if t is a parameter, is the NP-complete Minimum Diameter Color-Spanning Set (MDCS) problem fixed-parameter tractable? This question is still open. In this paper, we try to investigate some related questions along this line. The base problem we target at is the matching problem, both under the geometric model and the graph model. We show that an important graph version is W[1]-hard while all other versions in consideration are fixed-parameter tractable (FPT). With that, we show that the symmetric version of MDCS, the Largest Closest Pair Color-Spanning Set is in fact FPT.

This paper is organized as follows. In Sect. 2, we define the basics regarding FPT algorithms and the problems we will investigate. In Sect. 3, we illustrate the positive FPT results on the geometric version MinSum Matching (and a related graph version). In Sect. 4, we show the positive results on the MaxMin Matching and MinMax Matching under the color-spanning model. In Sect. 5, we show that a special graph version is W[1]-hard. In Sect. 5, we conclude the paper.

2   Preliminaries

We make the following definitions regarding this paper. An Fixed-Parameter Tractable (FPT) algorithm is an algorithm for a decision problem with input size n and parameter k whose running time is $O(f(t)n^c) = O^*(f(t))$, where $f(-)$ is any computable function on t and c is a constant. FPT algorithms are efficient tools for handling some NP-complete problems as they introduce an extra dimension t. If an NP-complete problem, like Vertex Cover, admits an FPT algorithm, then it is basically polynomially solvable when the parameter t is a small constant [3,8].

Of course, it is well conceived that not all NP-hard problems admit FPT algorithms. It has been established that

$$FPT \subseteq W[1] \subseteq W[2] \subseteq \cdots W[z] \subseteq XP,$$

where XP represents the set of problem which must take $O(n^t)$ time to solve (i.e., not FPT), with t being the parameter. Typical problems in W[1] include Independent Set and Clique, etc. For the formal definition and foundation, readers are referred to [3,8].

Given a set Q of n points in the plane with t colors, a *color-spanning* set $S \subset Q$ is a subset of t points with distinct colors. If S satisfies a property Π among all color-spanning sets of Q, we call the corresponding problem of computing S the Property-Π Color-Spanning Set. For instance, the Minimum Diameter Color-Spanning Set (MDCS) is one where the diameter of S is minimized (among all color-spanning sets of Q) and the Largest Close Pair Color-Spanning Set (LCPCS) is one where the closest pair of S is maximized (among all color-spanning sets of Q). All the distances between two points in the plane are Euclidean (or L_2). We next define the matching problems we will investigate in this paper.

Given a set P of n points in the plane with $2k$ colors, a *color-spanning* set $S \subset P$ is a subset of $2k$ points with distinct colors. The points in S always form a perfect matching, i.e., a set M of k edges connecting the $2k$ points in S. Among all these matchings (over all color-spanning sets), if a matching M satisfies a property Π, we call the problem the *Property-Π Matching Color-Spanning Set* or *Property-Π Color-Spanning Matching*. The three properties we focus on are MinSum, MinMax and MaxMin.

MinSum means that the sum of edge lengths in M is minimized, MinMax means that the maximum edge length in M is minimized, and MaxMin means that the minimum edge length in M is maximized. The main purpose of this paper is to investigate the FPT tractability of the three problems: MinSum Matching Color-Spanning Set, MinMax Matching Color-Spanning Set, and MaxMin Matching Color-Spanning Set. We show that all these problems are in fact FPT.

We also briefly mention some of the related problems on graphs, where we are given a general graph G whose vertices are colored with $2k$ colors, the problem is to determine whether a perfect matching M exists such that M contains exactly $2k$ vertices of distinct colors. We call this problem k-Multicolored Matching, and we will show that this problem is FPT.

Finally, we will study a special version on graphs where the (vertices of the) edges in M cannot share edges in G. We call the problem k-Multicolored Independent Matching, and we will show that this problem is W[1]-hard.

3   MinSum Matching Color-Spanning Set Is FPT

In this section, we consider the MinSum Matching Color-Spanning Set (MSMCS) problem, namely, given a set P of n points in the plane, each colored with one of the $2k$ colors, identify $2k$ points with distinct colors such that they induce a matching with the minimum total weight (among all feasible color-spanning matchings). Recall that the weight of an edge (p_i, p_j) is the Euclidean distance between p_i and p_j. For a point p_i, let $color(p_i)$ be the color of p_i. For this problem, we in fact have a good property of the optimal solution which is stated as follows.

Lemma 1. *In an optimal solution of MSMCS, let p_i and p_j be a matched edge in the optimal matching, then (p_i, p_j) must be the closest pair between points of $color(p_i)$ and $color(p_j)$.*

With this property, among $\binom{2k}{2}$ pairs of colors, we need to select k disjoint pairs and for each feasible solution compute the matching (by computing the bichromatic closest pair between points in each of the k paired colors) in $O(kn \log n)$ time [13]. This gives an FPT algorithm running in $k^{O(k)} O(n \log n)$ time. We hence have

Theorem 1. *MinSum Matching Color-Spanning Set (MSMCS) is FPT.*

We next consider the graph version of the MSMCS problem, or, the k-Multicolored Matching problem, which is formally defined as follows.

INSTANCE: An undirected graph $G = (V, E)$ with each vertex colored with one of the $2k$ given colors.

QUESTION: Is there a matching $E' \subseteq E$ including all the $2k$ colors? That is, are there k disjoint edges in E', and all the vertices of the edges in E' have different colors.

The following theorem shows that k-Multicolored Matching is also FPT.

Theorem 2. *k-Multicolored Matching is FPT.*

Proof. We could simulate the method for Theorem 1 as follows. First, make G into a complete weighted graph (K_n, w). For an edge $e \in E(K_n)$, if $e \in E(G)$ then set $w(e) = 1$; if $e \notin E(G)$, then set $w(e) = 2$. Then, once the k pairs of colors are given, we could compute the minimum weight matching (by computing the bichromatic closest pair between points in each of the k paired colors) in $O(kn^2)$ time. Then, there is a solution for k-Multicolored Matching if and only if the minimum color-spanning matching, over all possible pairs of $2k$ colors, has a weight k. Similar to Theorem 1, the running time is $k^{O(k)} O(n^2)$. The theorem is hence proved. $\square$

In the next section, we investigate the MaxMin and MinMax Matching Color-Spanning Sets problems.

4   MaxMin and MinMax Matching Color-Spanning Sets Are FPT

We first study the MaxMin Matching Color-Spanning Set problem. The first attempt is to try to see whether a property similar to Lemma 1 holds or not.

In Fig. 1, we show an example where such a property does not hold for MaxMin Matching Color-Spanning Set, i.e., the minimum edge length is maximized among all feasible color-spanning matchings. In Fig. 1, the MinSum Color-Spanning Matching is $\{(a,c),(b,d)\}$, with a total weight of $3 - 2\epsilon$. The Max-Sum Color-Spanning Matching is $\{(c,d),(e,f)\}$, with a total weight of $3 + 6\epsilon$. The optimal solution for MaxMin Matching Color-Spanning Set is $\{(c,e),(d,f)\}$, with a solution value of 1.5 (while the total weight is 3). Again, note that (c,e) and (d,f) do not form the closest pairs among the subsets of respective colors.

Nonetheless, it is easy to see that MaxMin Matching Color-Spanning Set is NP-complete. The reduction is from LCPCS: just compute an optimal solution for the MaxMin Color-Spanning Matching on set P, the edge with the minimum weight in the matching must be a solution for LCPCS. We next show that Maxmin Matching Color-Spanning Set is FPT.

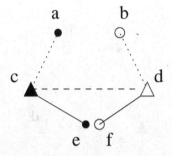

Fig. 1. A simple multicolored point set, the dotted, dashed and solid segments have lengths $1.5 - \epsilon$, 3 and 1.5 respectively; moreover, $d(e,f) = 6\epsilon$.

Theorem 3. *MaxMin Matching Color-Spanning Set is NP-complete and is FPT.*

Proof. We first enumerate all possible ways to pair colors. Following Theorem 1, there are an $k^{O(k)}$ number of such valid pairs. Let C_i and C_j be the set of points with $color_i$ and $color_j$ respectively. If in the optimal solution for MaxMin Matching Color-Spanning Set $color_i$ and $color_j$ are paired together, then the edge connecting two points $p \in C_i$ and $q \in C_j$ must be the ℓ-th closest pair between C_i, C_j, with $\ell \leq k$. (The reason is that there are only k edges in the matching, the other $k - 1$ edges in the matching could make at most the first $k - 1$ closest pairs between C_i and C_j infeasible — or, are longer than these closest pairs.) Then our FPT algorithm for MaxMin Matching Color-Spanning set is straightforward: given each valid pairing of $2k$ colors, for each pair of colors $color_i$ and $color_j$, compute the $\leq k$ closest pairs between the corresponding point sets with $color_i$ and $color_j$ respectively, then enumerate k^k possible solutions for the matching and return the one whose minimum edge length is maximized. For a fixed pairing of $2k$ colors, this takes $O^*(k^k)$ time. As we have $k^{O(k)}$ number of such pairings, the total running time is still $O^*(k^{O(k)})$. $\qquad\square$

Corollary 1. *The parameterized version of LCPCS, where the number of colors k is the parameter, is FPT.*

Now we consider the MinMax Matching Color-Spanning Set, namely, the maximum edge length is minimized among all feasible color-spanning matchings. Not surprisingly, the property illustrated in Lemma 1 also does not hold. In fact, such a matching might have nothing to do with the MinSum Color-Spanning Matching or the MaxSum Color-Spanning Matching. In Fig. 2, the MinSum Color-Spanning Matching is $\{(a, b), (c, d)\}$, with a total weight of 3. The MaxSum Color-Spanning Matching is $\{(a, c), (b, d)\}$ or $\{(c, d), (e, f)\}$, with a total weight of $4 + 2\epsilon$. For the MinMax Color-Spanning Matching problem, all of the above matchings give a solution value of $2 + \epsilon$. The optimal solution is $\{(c, e), (d, f)\}$, with a solution value of $1.5 + \epsilon$ (while the total weight is $3 + 2\epsilon$). Also, note that (c, e) and (d, f) do not form the farthest pairs among the subsets of respective colors.

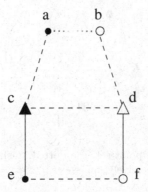

Fig. 2. A simple multicolored point set, the dotted, dashed and solid segments have lengths $1 - \epsilon$, $2 + \epsilon$ and $1.5 + \epsilon$ respectively.

It is unknown whether MinMax Color-Spanning Matching is NP-complete. Nonetheless, we show that it is FPT. The algorithm is almost identical to that for MaxMin Color-Spanning Matching in Theorem 4. The only difference is that when a pairing is fixed for point sets C_i and C_j, with colors $color_i$ and $color_j$ respectively, we enumerate the $\leq k$ farthest pairs between C_i and C_j.

Corollary 2. *MinMax Color-Spanning Matching is FPT.*

In the next section, we show that a special version on graphs is in fact W[1]-hard.

5 k-Multicolored Independent Matching Is W[1]-Hard

The k-Multicolored Independent Matching problem is defined as follows.

INSTANCE: An undirected graph $G = (V, E)$ with each vertex colored with one of the $2k$ given colors.

QUESTION: Is there an independent matching $E' \subseteq E$ including all the k colors? That is, are there k edges in E' such that any two edges in E' are not incident to any vertex in V, and all the vertices of the edges in E' have different colors.

The problem originates from an application in shortwave radio broadcast, where the matched nodes represent the shortwave channels which should not directly affect each other [12]. (We also comment that this problem seems to be related to the uncolored version of 'Induced Matching' which is known to be W[1]-hard as well [10,11].) We will show that this problem is not only NP-complete but also W[1]-hard. The problem to reduce from is the k-Multicolored Independent Set, which is defined as follows.

INSTANCE: An undirected graph $G = (V, E)$ with each vertex colored with one of the k given colors.

QUESTION: Is there an independent set $V' \subseteq V$ including all the k colors? That is, are there k vertices in V' incurring no edge in E, and all the vertices in V' have different colors.

When $U \subseteq V$ contains exactly k vertices of different colors, we also say that U is *colorful*.

For completeness, we first prove the following lemma, similar to what was done by Fellows *et al.* on k-Multicolored Clique problem [5].

Lemma 2. *k-Multicolored Independent Set is W[1]-complete.*

Proof. The proof can be done through a reduction from k-Independent Set. Given an instance $(G = (V, E), k)$ for k-Independent Set, we first make k copies of G, G_i's, such that the vertices in each G_i are all colored with color i, for $i = 1..k$. For any $u \in V$, let u_i be the corresponding mirror vertex in G_i. Then, for each $(u, v) \in E$ and for each pair of i, j, with $1 \leq i \neq j \leq k$, we add four edges (u_i, u_j), (v_i, v_j), (u_i, v_j) and (u_j, v_i). Let the resulting graph be G'. It is easy to verify that G has a k-independent set if and only if G' has a k-multicolored independent set. As k-Independent Set is W[1]-complete [3], the lemma follows.
□

The following theorem shows that k-Multicolored Independent Matching is not only NP-complete but also W[1]-hard.

Theorem 4. *k-Multicolored Independent Matching is W[1]-hard, i.e., it does not admit any FPT algorithm unless FPT=W[1].*

Proof. We reduce k-Multicolored Independent Set (IS) to the k-Multicolored Independent Matching problem.

Given an instance of k-Multicolored IS problem, i.e., a graph $G = (V, E)$ with each vertex in $V = \{v_1, v_2, \ldots, v_n\}$ colored with one of the k colors $\{1, 2, \ldots, k\}$, the question is whether one could compute an IS of size k, each with a distinct color.

We construct an instance for the k-Multicolored Independent Matching as follows. First, make a copy of G (with the given coloring of k colors). Then, construct a set $U = \{u_1, u_2, \ldots, u_k\}$ such that u_i has color $k + i$. Finally, we connect each $u_i \in U$ to each $v_j \in V$, i.e., we construct a set $E' = \{(u_i, v_j)|u_i \in U, v_j \in V, 1 \le i \le k, 1 \le j \le n\}$. Let the resulting graph be $G' = (V \cup U, E \cup E')$, with each vertex in G' colored with one of the $2k$ colors. We claim that G has a colorful independent set of size k if and only if G' has a colorful independent matching of size k. The details are given as follows.

If G has a colorful independent set $V' \subseteq V$ of size k, we select the k vertices in V' and match them up with the k vertices in U. As the vertices in V' are independent and no two vertices in U share an edge (i.e., vertices in U are also independent), we have a colorful independent matching for G'.

If G' has a colorful independent matching of size k, then exactly k vertices of V must match up with the k vertices in U. (Otherwise, if two vertices v_i and v_j in V form an edge in the optimal colorful matching then we cannot have k edges in the matching. This is because at least two vertices in U cannot match up with the vertices in V of the same color as v_i and v_j. Then the colorful matching contains at most $k - 1$ edges, a contradiction.) By the definition of colorful independent matching, the k vertices from V cannot share any edge hence form a independent set for G.

As the reduction takes polynomial time, the theorem is proved.   □

We have the following corollary.

Corollary 3. *The optimization version of k-Multicolored Independent Matching (called Multicolored Maximum Independent Matching) does not admit a factor $n^{1-\epsilon}$ polynomial-time approximation, for some $\epsilon > 0$, unless $P = NP$.*

Proof. As the reductions in Lemma 2 and Theorem 4 are both L-reductions, the Multicolored Maximum Independent Matching problem is as hard to approximate as the Independent Set problem, which does not admit a factor $n^{1-\epsilon}$ polynomial-time approximation, for some $\epsilon > 0$, unless P = NP [15].   □

6   Closing Remarks

Motivated by the open question of Fleischer and Xu, we studied the FPT tractability of LCPCS and some related matching problems under the color-spanning model. We show in this paper that most of these problems are FPT, except one version on graphs which can be considered as a generalization of the multicolored independent set problem. The original question on the FPT tractability of Minimum Diameter Coloring-Spanning Set (MDCS), is, unfortunately, still open.

Acknowledgments. This research is partially supported by NSF of China under project 61628207. We also thank Ge Cunjing for pointing out some relevant reference.

References

1. Abellanas, M., Hurtado, F., Icking, C., Klein, R., Langetepe, E., Ma, L., Palop, B., Sacristán, V.: Smallest color-spanning objects. In: Heide, F.M. (ed.) ESA 2001. LNCS, vol. 2161, pp. 278–289. Springer, Heidelberg (2001). doi:10.1007/3-540-44676-1_23

2. Chen, Y., Shen, S., Gu, Y., Hui, M., Li, F., Liu, C., Liu, L., Ooi, B.C., Yang, X., Zhang, D., Zhou, Y.: MarcoPolo: a community system for sharing and integrating travel information on maps. In: Proceedings of the 12th International Conference on Extending Database Technology (EDBT 2009), pp. 1148–1151 (2009)
3. Downey, R., Fellows, M.: Parameterized Complexity. Springer, New York (1999)
4. Fan, C., Luo, J., Wang, W., Zhong, F., Zhu, B.: On some proximity problems of colored sets. J. Comput. Sci. Technol. **29**(5), 879–886 (2014)
5. Fellows, M., Hermelin, D., Rosamond, F., Vialette, S.: On the parameterized complexity of multiple-interval graph problems. Theoret. Comput. Sci. **410**(1), 53–61 (2009)
6. Fleischer, R., Xu, X.: Computing minimum diameter color-spanning sets. In: Lee, D.-T., Chen, D.Z., Ying, S. (eds.) FAW 2010. LNCS, vol. 6213, pp. 285–292. Springer, Heidelberg (2010). doi:10.1007/978-3-642-14553-7_27
7. Fleischer, R., Xu, X.: Computing minimum diameter color-spanning sets is hard. Info. Process. Lett. **111**(21–22), 1054–1056 (2011)
8. Flum, J., Grohe, M.: Parameterized Complexity Theory. Springer, Berlin (2006)
9. Ju, W., Fan, C., Luo, J., Zhu, B., Daescu, O.: On some geometric problems of color-spanning sets. J. Comb. Optim. **26**(2), 266–283 (2013)
10. Moser, H., Sikdar, S.: The parameterized complexity of the induced matching problem. Discret. Appl. Math. **157**(4), 715–727 (2009)
11. Moser, H., Thilikos, D.: Parameterized complexity of finding regular induced subgraphs. J. Discret. Algorithms **7**(2), 181–190 (2009)
12. Ma, F., Gao, X., Yin, M., Pan, L., Jin, J., Liu, H., Zhang, J.: Optimizing shortwave radio broadcast resource allocation via pseudo-boolean constraint solving and local search. In: Rueher, M. (ed.) CP 2016. LNCS, vol. 9892, pp. 650–665. Springer, Cham (2016). doi:10.1007/978-3-319-44953-1_41
13. Preparata, F.P., Shamos, M.I.: Computational Geometry: An Introduction. Springer, New York (1985)
14. Zhang, D., Chee, Y.M., Mondal, A., Tung, A.K.H., Kitsuregawa, M.: Keyword search in spatial databases: towards searching by document. In: Proceedings of the 25th IEEE International Conference on Data Engineering (ICDE 2009), pp. 688–699 (2009)
15. Zuckerman, D.: Linear degree extractors and the inapproximability of max clique and chromatic number. Theory Comput. **3**(1), 103–128 (2007)

The Complexity of Finding (Approximate Sized) Distance-d Dominating Set in Tournaments

Arindam Biswas[1], Varunkumar Jayapaul[2(✉)], Venkatesh Raman[1], and Srinivasa Rao Satti[3]

[1] The Institute of Mathematical Sciences, Chennai 600113, India
{barindam,vraman}@imsc.res.in
[2] Chennai Mathematical Institute, Chennai 603103, India
varunkumarj@cmi.ac.in
[3] Seoul National University, Seoul, Korea
ssrao@cse.snu.ac.kr

Abstract. A tournament is an orientation of a complete graph. For a positive integer d, a distance-d dominating set in a tournament is a subset S of vertices such that every vertex in $V \setminus S$ is reachable by a path of length at most d from one of the vertices in S. When $d = 1$, the set is simply called a dominating set. While the complexity of finding a k-sized dominating set is complete for the complexity class $LOGSNP$ and the parameterized complexity class $W[2]$, it is well-known that every tournament on n vertices has a dominating set of size $g(n) = \lg n - \lg \lg n + 2$ that can be found in $O(n^2)$ time. We first show that for any k, one can find a dominating set of size at most $k + g(n)$ in $O(n^2/k)$ time, and prove an (unconditional) lower bound of $\Omega(n^2/k)$ for any $k > \epsilon \lg n$ for any $\epsilon > 0$. Hence in particular, we can find a $(1 + \epsilon) \lg n$ sized dominating set in the optimal $\Theta(n^2 / \lg n)$ time.

For distance-d dominating sets, it is known that any tournament has a distance-2 dominating set consisting of a single vertex. Such a vertex is called a king or a 2-king and can be found in $O(n\sqrt{n})$ time. It follows that there is a vertex, from which every other vertex is reachable by a path of length at most d for any $d \geq 2$ and such a vertex is called a d-king. A d-king can be found in $O\left(n^{1+1/2^{d-1}}\right)$ for any $d \geq 2$ [3]. We generalize our result for dominating set to show that for $d \geq 2$,

- we can find a k-sized distance-d dominating set in a tournament in $O\left(k(n/k)^{1+1/2^{d-1}}\right)$ time for any $k \geq 1$, and
- we can find a $(g(n) + k)$-sized distance-d dominating set in a tournament using $O\left(k(n/k)^{1+1/(2^d-1)}\right)$ time for any $k \geq 1$.

The second algorithm provides a faster runtime for finding distance-d dominating sets that are larger than $g(n)$. We also show that the second algorithm is optimal whenever $k \geq \epsilon \lg n$ for any $\epsilon > 0$.

Thus our algorithms provide tradeoffs between the (additive) approximation factor and the complexity of finding distance-d dominating sets in tournaments. For the problem of finding a d-king, we show some additional results.

Keywords: Tournament · King · d-Cover · Dominating set

© Springer International Publishing AG 2017
M. Xiao and F. Rosamond (Eds.): FAW 2017, LNCS 10336, pp. 22–33, 2017.
DOI: 10.1007/978-3-319-59605-1_3

1   Introduction and Motivation

Tournaments, Dominating Sets and Guarantees. A *tournament* is a digraph in which there is exactly one directed edge between every pair of vertices. As they model many practical scenarios (game tournaments, voting strategies), tournaments are well-studied in structural and algorithmic graph theory. Various structural and algorithmic properties of tournaments are known and Moon's [17] early monograph on this subject lead to much subsequent work in the area.

It is well-known that every tournament has a dominating set of size at most $\lceil \lg n \rceil$ and can be found in $O(n^2)$ time. In fact, one can guarantee existence of a slightly smaller sized dominating set that can also be found in the same time. There exists a dominating set of size $g(n) = \log n - \lg \lg n + 2$ in a tournament on n vertices [12] (see also Sect. 2.2). Thus, the minimum dominating set can be found in $n^{O(\lg n)}$ time and hence it is unlikely to be NP-complete. Papadimitriou and Yannakakis [18] show that finding the smallest dominating set in a tournament is complete for a complexity class $LOGSNP$. It is also known that there are tournaments where the minimum dominating set size is $\Omega(\lg n)$ [4,9] (see also [16]) using which it has been shown that finding a k-sized dominating set is complete for the parameterized complexity class $W[2]$ (where the parameter is the solution size k) and hence, even an $f(k)n^{O(1)}$ time algorithm is unlikely for any function f of k [6].

It is also well-known that every tournament has a 'centre' or a *king* vertex, i.e. a vertex from which every other vertex can be reached by a path of length at most 2 (the maximum outdegree vertex of the tournament is one such vertex). A king is also called a 2-king. In general, for any $d \in \mathbb{N}$, a d-king is a vertex in the tournament from which every other vertex can be reached by a dipath of length at most d, and a d-cover is a set of vertices S such that every other vertex in the tournament can be reached from some vertex in S by a dipath of length at most d. A 1-cover is simply a *dominating set*, and d-covers are referred to as *distance-d* dominating sets. We consider the complexity of finding dominating sets of size more than $\lg n - \lg \lg n + 2$ and more generally the complexity of finding d-covers.

As is standard, we assume that the given tournament has the vertex set $V = \{1, 2, \ldots n\}$, and is represented by its adjacency matrix where the (i, j)-th entry denotes the direction of the edge between vertex i and vertex j. We call a query to the matrix entry an edge query or an edge probe.

Motivation. Above or below guarantee parameterization is a common well-studied paradigm in parameterized complexity [7,10,11,13,14]. Here, one studies the parameterized complexity of the problem where the parameter is a bound above or below the guarantee (based on whether the guarantee is an upper or lower bound, and whether the problem is a maximization or a minimization problem). For example, for the dominating set problem in tournaments, the natural 'below-guarantee' parameterization question is the complexity of finding a dominating set of size at most $g(n) - k$ where k is the parameter. In that context, it is not clear whether there is an infinite family of tournaments whose minimum dominating set size is $g(n)$, as we only know of infinite family of tournaments where the minimum dominating set size is $\lceil \lg n \rceil - 2\lceil \lg \lg n \rceil$ (roughly $g(n) - \lg \lg n$) [12]. Tightening the gap to find the "tight" bound seems open.

Here we depart from the usual below guarantee parameterization when we have an upper bound. We consider the complexity of finding a dominating set of size at most $g(n) + k$. Clearly such a dominating set exists, as a dominating set remains a dominating set even if we add a few more vertices. Our question is whether we can find such a (larger sized) dominating set faster, without necessarily finding a dominating set of size at most $g(n)$ first. We answer the question affirmatively and even prove adversarial lower bounds for finding such dominating sets (or more generally d-covers). Such parameterization can be considered as obtaining trade-offs between additive approximation and the running time of finding a parameter. This could trigger other such work on approximation, and can be considered as another variation of the recent popular approaches of 'FPT within P' [1,8] where one investigates the complexity of parameters that are already solvable in polynomial time, with a view to obtaining faster algorithms for some relaxed parameterized versions. In our case, the parameter in question is the additive approximation error.

Our Results

Dominating Sets and d-Covers. For dominating sets, our main result is that we can find a $\lg n - \lg \lg n + 2 + k$ sized dominating set in a tournament using only $O\left(n^2/k\right)$ time and that this bound is optimal as long as $k > \epsilon \log n$ for any $\epsilon > 0$. The lower bound is shown using adversary arguments. We generalize this to show that $\lg n - \lg \lg n + 2 + k$ sized d-covers can be found in $O\left(k(n/k)^{1+1/(2^d-1)}\right)$ time for any $d \geq 2$ and $k \geq 1$. We also prove a matching adversary based lower bound to show that this bound is optimal for $k > \epsilon \lg n$ for any $\epsilon > 0$.

Kings. While a king vertex forms a distance 2 dominating set by itself, motivation for finding kings comes from various practical scenarios. Acharya et al. [2] study kings to help find bounds for density estimation, where the objective is to approximate an unknown distribution based on its samples. Ajtai et al. [3] study kings to perform sorting and searching operations when the comparisons between two elements can be imprecise if the difference between the two vertices is less than a certain threshold. While the maximum outdegree vertex in a tournament is a king, it is known [5] that finding a maximum outdegree vertex in a tournament can require $\Omega\left(n^2\right)$ edge queries. Shen et al. [19] gave the first known non-trivial upper bound of $O\left(n^{3/2}\right)$ edge queries for finding a king in a tournament. They also gave the first lower bound of $\Omega\left(n^{4/3}\right)$ edge queries for finding a king in a tournament. Similar results were later proved independently by Ajtai et al. [3] who generalized the bounds for finding a d-king. Narrowing this gap between upper and lower bound has been mentioned as an open problem in the papers [3,19]. We prove the following results regarding the complexity of finding kings.

- If an adversary gives orientations based on a pro-low strategy used to show $\Omega\left(n^{4/3}\right)$ lower bound (see Sect. 3.3) [19], then there is a $O\left(n^{4/3}\right)$ algorithm for finding a king; this implies that we need a completely different adversary strategy to prove a better lower bound, or there exists a better algorithm to find a king;
- If a (2-)king can be found in $O\left(n^{4/3}\right)$ time, then a d-king, for any $d \geq 3$ can be found in the optimal $O\left(n^{1+1/(2^d-1)}\right)$ time. The (matching) lower bound for d-king using the pro-low adversary was proved by Ajtai et al. [3]. This result was conjectured in the conclusions of [3];
- While a (2-)king can be found in $O\left(n\sqrt{n}\right)$ time, verifying whether a vertex is a (2-)king may take $\Omega\left(n^2\right)$ edge-queries.

Organization of the Paper. In Sect. 2 we give the necessary terminologies and some basic results on tournaments that we use in the paper. Section 3 describes several upper and lower bounds for finding above-guarantee d-covers in a tournament. In Sect. 4, we describe algorithms (against a weak adversary) and lower bounds for verifying whether a vertex is a king. We conclude with pointers to some open problems in Sect. 5.

2   Preliminaries

2.1   Definitions and Notation

Let $T = (V, E)$ be a tournament. A *dominator* in T is a vertex $u \in V$ such that for any other vertex $v \in S, (u, v) \in E$. A *king* in T is a vertex $u \in V$ such that for any other vertex $v \in V$ there is a $u \to v$ dipath of length at most 2. We denote by $V(T)$ the vertex set of T, and by $E(T)$ the edge set of T. $|T|$ denotes the order of T, i.e., the size of $V(T)$ and for a subset $S \subseteq V(T)$, $T[S]$ denotes the subtournament of T induced by S.

Let $d \in \mathbb{N}$ and $S \subseteq V$. S is called a *d-cover* of T if for every vertex $v \in V \backslash S$, there is vertex $u \in S$ such that there is a $u \to v$ dipath of length at most d.

A 1-cover is also called a *dominating set*. A king by itself is a 2-cover. If a d-cover is of size 1, we call the unique element in the set a d-king. Let $u \in V$ be a vertex. An out-neighbour of u is a vertex $v \in V$ such that $(u, v) \in E$; u is said to *dominate* v. Similarly, an in-neighbour of u is a vertex $v \in V$ such that $(v, u) \in E$. For any $v \in V(T)$, $N^+(v)$ denotes the set of out-neighbours of v, and $N^-(v)$ denotes the set of its in-neighbours. We also define $\deg^+(v) = |N^+(v)|$, and $\deg^-(v) = |N^-(v)|$.

Given a subset S of vertices, *performing a round robin tournament* on S refers to the process of querying all possibles edges between every pair of vertices in S to know their orientations. Using the same tournament analogy, we sometimes refer to the out-neighbours of a vertex v as vertices that *lose* to v, and to its in-neighbours as vertices that *win* against it. In other words, if there is an edge from u to v, we say that u wins against v and v loses to u.

We assume that the vertex set of the tournament is $\{1, 2, \ldots n\}$ and is given in the form of an adjacency matrix where the (i, j)-th entry denotes the direction of the edge between i and j. By an edge query or an edge probe, we mean a probe to the adjacency matrix.

Remark 1. To keep the notation simple, we sometimes omit ceilings and floors on fractions when we actually mean integers, typically on the sizes of the vertex subsets we handle. This doesn't affect the asymptotic analysis.

2.2 Elementary Results

Lemma 1 [15]. *Let T be a tournament and $v \in V(T)$. If a vertex u in $N^-(v)$ is a king in $T[N^-(v)]$, then u is a king in T.*

Proof. The vertex u reaches every vertex in $N^+(v)$ through v. Every other vertex in T is reachable from u by a dipath of length at most 2 as it is a king in $T[N^-(v)]$. □

The following lemma follows from the fact that the sum of the indegrees, and the sum of the outdegrees of the vertices of a tournament are both $\binom{n}{2}$.

Lemma 2 [12]. *In any tournament T of order n, there is a vertex with outdegree at least $(n-1)/2$ and a vertex with indegree at least $(n-1)/2$ and such vertices can be found in $O(n^2)$ time.*

The following result is immediate from Lemmas 1 and 2. We simply find a vertex in the tournament that dominates at least $(n-1)/2$ other vertices, include it in the dominating set and recurse on the in-neighbors of the vertex.

Theorem 1 [16]. *For any tournament T of order n, a dominating set of size at most $\lceil \lg n \rceil$ can be found in time $O(n^2)$.*

The following tighter upper bound is known [12], we give the proof here for completion as it is not so well-known to the best of our knowledge.

Theorem 2. *In any tournament $T(V, E)$ of order n, a dominating set of size at most $\lg n - \lg \lg n + 2$ exists and can be found in time $O(n^2)$.*

Proof. We proceed as in the case of Theorem 1, but bail out the recursion after $\lceil \lg n - \lg \lg n + 1 \rceil$ steps at which point, the partial dominating set D we have found has at most $\lceil \lg n - \lg \lg n \rceil + 1$ vertices, and the resulting subtournament R (whose vertices are yet to be dominated) has at most $n/2^{\lg n - \lg \lg n + 1}$, i.e. at most $(\lg n)/2$ vertices. By the definition of R, the vertices of R dominate all of D. Now we check in $O(n^2)$ time by looking at the (out)neighborhood of $V(R)$ in $V \setminus D$ whether $V(R)$ dominates all vertices of $V \setminus D$. If so, we output $V(R)$ whose size is at most $(\lg n)/2$. Otherwise, there is some vertex x in $V \setminus (D \cup V(R))$ that dominates the vertices of R. In this case, we output $D \cup \{x\}$ as the dominating set. Either way, the size of the dominating set is at most $\lg n - \lg \lg n + 2$. □

3   Finding d-Covers

3.1   Finding a Dominating Set (1-Cover) of Size $(k+\lg N-\lg\lg n+2)$

We start with the following above-guarantee result for 1-cover (dominating set).

Theorem 3. *A $(k+\lg n-\lg\lg n+2)$ sized 1-cover in a tournament on n vertices can be found in $O\left(n^2/k\right)$ time.*

Proof. Let V be the input set of vertices. If $k \leq 2$ apply Theorem 2. Otherwise, pick a subset of vertices of size $2(n/k) + 1$ and find a vertex (say u) which dominates at least (n/k) vertices inside this subset using Lemma 2. This vertex can be found using $O\left(n^2/k^2\right)$ time. Delete u's out-neighbors add u to the dominating set. Now recurse on the remaining tournament as long as $|V| \geq 2n/k + 1$. Once $|V| \leq 2n/k$, apply Theorem 2 and find a 1-cover of size $\lg n - \lg\lg n + 2$ for the remaining tournament in $O\left((n/k)^2\right)$ time, and add these vertices to the output cover. The output cover is a 1-cover for the tournament, of size at most $k + \lg n - \lg\lg n + 2$. The total time taken is $O\left((k-2)(n/k)^2 + (n/k)^2\right) = O\left(n^2/k\right)$. □

3.2   Finding d-Covers ($d \geq 2$)

For $d \geq 2$, the lower bound on the size of the d-cover is 1. We first make the following simple generalization of the $O\left(n^{1+1/2^{d-1}}\right)$ algorithm [3] to find a d-king to show the following.

Theorem 4. *In any tournament of order n, a d-cover of size k can be found in $O\left(k(n/k)^{1+1/2^{d-1}}\right)$ time.*

Proof. Simply partition the vertex set into k parts to get k sub-tournaments on roughly n/k vertices each. Find a d-king in each and add them all to the k-sized d-cover. □

In what follows, we give an improved algorithm for finding d-covers of size more than $\lceil \lg n \rceil$. We start with a lemma similar to Lemma 2 for 2-covers.

Lemma 3. *In any tournament of order n, a vertex that reaches at least $(n-1)/2$ other vertices via dipaths of length at most 2 can be found in time $O\left(n^{4/3}\right)$.*

Proof. Let V be the vertex set of the tournament. Pick a subset of vertices of size $2n^{1/3} + 1$ and find a vertex (say u) which dominates at least $n^{1/3}$ of those vertices. Take exactly $n^{1/3}$ of u's losers and remove them from V. Also add u to a set of vertices called C. Repeat this process on the resulting V as long as $|V| \geq 2n^{1/3} + 1$. Once $|V| < 2n^{1/3} + 1$, perform a round-robin tournament on all the vertices in C and find a vertex k that dominates at least $(|C| - 1)/2$ vertices using Lemma 2. The size of C is at least $n^{2/3} - n^{1/3} - 1$.

Thus vertex k can reach at least $n^{1/3}(n^{2/3}-n^{1/3}-1)/2 = n/2-(n^{2/3}-n^{1/3})/2$ vertices with path of length exactly two, and can reach $(n^{2/3}-n^{1/3}-1)/2$ vertices with path of length 1. Thus the vertex k is king of at least $(n-1)/2$ vertices.

The total number of edge queries made is $O(n^{4/3})$, since round robin on set C costs $O(n^{4/3})$ edge queries, as well as the time taken to create the set C is $O(n^{2/3}(n^{1/3})^2) = O(n^{4/3})$. $\qquad\square$

Corollary 1. *A $(\lg n - \lg\lg n + 2)$ sized 2-cover in a tournament on n vertices can be found in $O(n^{4/3})$ time.*

Proof. Use Lemma 3 to find a vertex $v \in V$ which has a path of length at most 2, to at least $(n-1)/2$ vertices (say to the set C). Remove v and the vertices in C from V after adding v to the 2-cover set S and adding C to S', and repeat this process on the remaining vertices, till the number of vertices in S is $\lg n - \lg\lg n + 1$. At this point the number of vertices remaining in V is at most $(\lg n)/2$. Find the relation of all vertices with these $(\lg n)/2$ with the vertices in S' using $O(n\lg n)$ queries. One of the two cases happen. Either some vertex v' in S' dominates all vertices in V, in which case we add v' to S and output it as the 2-cover, or no vertex can dominate all vertices in V, in which case V forms a dominating set of size $(\lg n)/2$ and we output as 2-cover, since any 1-cover is a valid 2-cover. This would give a $\lg n - \lg\lg n + 2$ sized 2-cover. $\qquad\square$

Theorem 5. *A $(k+\lg n-\lg\lg n+2)$ sized 2-cover in a tournament on n vertices can be found in $O\left(k(n/k)^{4/3}\right)$ time.*

Proof. Take a subset of input vertices of size $2n/k$ from V. Using Lemma 3 find a 2-king u of at least half of these vertices using $O((n/k)^{4/3})$ edge queries. Remove u from V and add it to the output set C, and remove all vertices in V that can be reached by a path of length at most 2 from u. Recurse on the remaining tournament as long as $|V| \geq 2n/k$. When $|V| < 2n/k$, find a $\lg n - \lg\lg n + 2$ sized 2-cover, by using $O((n/k)^{4/3})$ edge queries using Corollary 1, and add this 2-cover to the set C. The set C is a 2-cover for all the vertices in the tournament of size at most $k + \lg n - \lg\lg n + 2$. The total time spent is $O\left(k(n/k)^{4/3} + (n/k)^{4/3}\right) = O\left(k(n/k)^{4/3}\right)$. $\qquad\square$

Now we generalize the above theorem for $d \geq 2$.

Lemma 4. *Let $d \geq 2$ and $k \geq 1$ be an integer. If a k-sized 2-cover can be found in $O\left(k(n/k)^{4/3}\right)$ time, then a k-sized d-cover can be found in $O\left(k(n/k)^{1+1/(2^d-1)}\right)$ time.*

Proof. We prove this by induction on d. For $d = 2$, there is nothing to prove. Let $d \geq 3$. Assume that the lemma is true for all integers from 2 to $d-1$.

Let V be the set of vertices and $s = (n/k)^{1/(2^d-1)}$. As $d \geq 3, n/s = n^{1-1/(2^d-1)}k^{1/(2^d-1)} > 3 + \lg n$ (for large enough n), and so using Theorem 3 find a (n/s)-sized 1-cover of V using $O\left(n^2/(n/s)\right) = O\left(ns\right)$ time. Now find a k-sized $(d-1)$-cover for these n/s vertices using $O\left(k(n/ks)^{1+1/(2^{d-1}-1)}\right) = O\left(ns\right)$ time using induction hypothesis. The resulting set is a k-sized d-cover of entire input, which is found in $O\left(ns\right) = O\left(k(n/k)^{1+1/(2^d-1)}\right)$ time.   □

Thus we have from Theorem 5, Lemma 4 and Theorem 3,

Theorem 6. *A $(k+\lg n-\lg\lg n+2)$-sized d-cover in a tournament on n vertices can be found in $O\left(k(n/k)^{1+1/(2^d-1)}\right)$ time for any $k \geq 1$ and $d \geq 1$.*

By setting $k = 1$ in Lemma 4, we have

Corollary 2. *If a 2-king can be found in $O\left(n^{4/3}\right)$ time, then we can find a d-king in $O\left(n^{1+1/(2^d-1)}\right)$ time.*

The above corollary was mentioned as a likely possibility (without proof) in the conclusion of [3].

3.3   Lower Bounds

In this section, we show that most of the algorithms in the previous section are optimal using an adversary argument. Central to the lower bound result in [19] and [3] for finding a king, is, the adversary that answers according to the pro-low strategy defined below.

Definition 1 (Pro-Low Strategy). *Let T be a tournament whose edge directions are determined by the following adversary strategy for an algorithm that queries edges: when the edge uv is queried, it is assigned the direction $u \rightarrow v$ if $\deg^+(u) \leq \deg^+(v)$, and $u \leftarrow v$ otherwise. Here $\deg^+(u)$ and $\deg^+(v)$ denote the outdegree of u and v before the edge query is made.*

Lemma 5 [3,19]. *Let T be a tournament and $S \subseteq V(T)$. Suppose an adversary answers edge queries involving vertices $u, v \in S$ using the pro-low strategy, i.e., it compares out-degrees within $T[S]$. If a vertex v in S achieves $\deg_S^+(v) = t$, at least $t(t+1)/2$ edge queries must have been made by the algorithm.*

Generalizing Lemma 5, Ajtai et al. [3] show that

Lemma 6 [3]. *When playing against a pro-low adversary, if an algorithm finds a vertex that reaches at least t vertices by a path of length at most d for $d \geq 1$, then at least $\Omega(t^{1+1/(2^d-1)})$ edge queries must have been made.*

Setting $t = n$ gives the lower bound for finding a d-king. We generalize the above lemma to show

Lemma 7. *If an algorithm playing against the pro-low adversary finds a k-sized d-cover for some $k, d \geq 1$, then it must have made at least $k(n/k)^{1+1/(2^d-1)}$ edge queries.*

Proof. Suppose the algorithm returns a set of vertices $C = \{v_1, v_2, v_3, \ldots, v_k\}$ as the k sized d-cover of the entire tournament. Perform a breadth-first search from the vertices of C with all vertices in C at level 0, those that have direct edges from them at level 1 etc. The breadth-first search tree has d levels. Now, for each vertex in level i (from d to 1) assign a unique vertex at level $i - 1$ from which there is a direct edge to it. This partitions the vertices into k sets, where the i-th set contains the vertex v_i, and all vertices assigned to it from level 1, and all vertices assigned to those level 1 vertices from level 2 and so on. Thus we have k subtournaments $T_1, T_2, \ldots, T_k$ induced on the partition, and let s_i be the number of vertices in the sub-tournament T_i. Note that v_i has an at most d-length path to every vertex in T_i and hence by Lemma 6, the number of edge queries made in the tournament T_i is at least $(s_i)^{1+1/(2^d-1)}$. Thus the total number of edge queries done by all the vertices is at least $\sum_{i=1}^{k}[(s_i)^{1+1/(2^d-1)}]$. As $\sum_{i=1}^{k} s_i = n$, we have that the number of edge queries made is at least $k(n/k)^{1+1/(2^d-1)}$ using Jensen's inequalities (as $\sum_{i=1}^{k}[(s_i)^{1+1/(2^d-1)}]$ attains its minimum when all the s_i's are the same as their average value which is n/k). $\qquad \square$

Note that this lower bound matches asymptotically the time bound in Theorem 6, though the upper bound in Theorem 6 is to find a d-cover of size at most $g(n)+k$. Hence it follows that the bound of Theorem 6 is optimal for $k \geq \Omega(g(n))$ or $k \geq \epsilon \lg n$ for any $\epsilon > 0$.

4   Finding Kings

As mentioned in the introduction, it is known that $O(n^{3/2})$ probes are sufficient and $\Omega(n^{4/3})$ probes are necessary to find a king in a tournament on n vertices. Narrowing this gap between upper and lower bound has been mentioned as an open problem in literature [3,19]. In this section, we prove improved upper bounds against weak adversaries.

4.1   Finding a King Against a Weak Adversary

Theorem 7. *If the adversary follows a pro-low strategy as described in Definition 1 then we can find a king in* $O(n^{4/3})$ *time.*

Proof. Take a sample of $2n^{1/3}$ from the input V and find a vertex v that dominates at least half of these vertices (that exists from Lemma 2) using a round-robin tournament on the sample. Remove v and all vertices that lose to v from V, and add v to the set F. Repeat this process of sampling and discarding elements, and adding elements to F, till there are at most $n^{2/3}$ elements left in V, which we call R. At this point F has $f = n^{2/3} - n^{1/3}$ vertices. Perform a round robin on all the vertices of F, and arrange vertices of F in non-decreasing order of their out-degrees. Let $v_1, v_2, \ldots v_f$ be the vertices of F in that order. Let $t_1, t_2, \ldots, t_{n^{2/3}}$ be the vertices in R. Now compare each t_i with all vertices in F in the same order. We first make the following claims.

Claim. If for some i, t_i wins the first q vertices of F and loses to v_{q+1}, then t_{i+1} will win at least the first $q+1$ vertices of F.

Proof. As t_{i+1} starts of exactly as t_i when t_i made no comparisons, it will also win the first q comparisons of F. As t_i lost to v_{q+1} at that time out-degree(v_{q+1}) $\geq$ out-degree(t_i) at that time, which is the same as out-degree(t_{i+1}) now. But after the win out-degree(v_{q+1}) is strictly more than out-degree(t_{i+1}) now, and so t_{i+1} will win v_{q+1} now, as the adversary plays the pro-low strategy.   □

The following claim follows from previous claim and the fact that $|R| > |F|$, and every vertex in F starts off with outdegree at least $n^{1/3}$ and every vertex in R starts with outdegree 0.

Claim. There exists a vertex in R that wins against all vertices of F.

Let F' be the set of all vertices in R that defeat everyone in F. Now the algorithm finds a king k in the sub-tournament induced by F', and declares that vertex as the king of the entire tournament.

Claim. The declared vertex k is a king of the entire tournament.

Proof. Clearly, k reaches all vertices of F' by a path of length at most 2. Let x be a vertex in $R \backslash F'$. Then x has lost to some vertex y in F. Then k can reach x through y as k has defeated all vertices of F. For the same reason, k can reach all vertices of F by a direct edge. The vertex k can reach a vertex in $V \backslash (R \cup F)$ through a vertex in F, as such vertices are dominated directly by a vertex in F.   □

The total number of comparisons made is clearly O $(n^{4/3})$.

The following corollary follows from Lemma 4 and Theorem 7.

Corollary 3. *If the adversary follows a pro-low strategy as described in Definition 1 then we can find a d-king in* O $\left(n^{1+1/(2^d-1)}\right)$ *time.*

The above lemma implies that we need a completely different adversary strategy to improve the lower bounds.

4.2 Verification of Kings

Lemma 8. *Let T be a tournament of order n. Given a vertex $v \in V(T)$, it takes* $\Omega\left(n^2\right)$ *edge queries to decide whether v is a king in T.*

Proof. Consider the following adversary strategy. The adversary arbitrarily fixes subsets $A, B \subseteq V(T) \backslash \{v\}$ such that $|A| = \lfloor (n-1)/2 \rfloor$ and $|B| = \lceil (n-1)/2 \rceil$, and then assigns edge directions such that $N^+(v) = B$ and $N^-(v) = A$. The remaining edge directions are assigned dynamically.

Suppose $u \in A, w \in B$, and the edge uw is queried. The following cases arise.

- There are other non-queried u–B edges: the adversary assigns the direction $u \rightarrow w$.
- All other u–B edges have been queried: the adversary assigns the direction $w \rightarrow u$.

For edges uw where $u \in A$ and $w \in A$ or $u \in B$ and $w \in B$, the adversary answers arbitrarily. Note that v directly reaches each vertex in $N^+(v)$ by a directed edge, and it can only reach vertices in $N^-(v)$ through vertices in $N^+(v)$. For any vertex $u \in A$, an algorithm cannot decide whether v can reach u until it queries all u–B edges, since the status of v depends on how the adversary answers when the last u–B edge query is made. By the same token, the algorithm cannot decide whether v can reach all vertices of A unless it asks all edges between A and B.

Thus for every $u \in A$, $\lceil (n-1)/2 \rceil$ queries must be made, i.e. a total of $\lfloor (n-1)/2 \rfloor \cdot \lceil (n-1)/2 \rceil = \Omega\left(n^2\right)$ edge queries must be made to determine whether v is a king.   □

5   Conclusions and Open Problems

We have investigated the complexity of finding above-guarantee sized d-dominating sets in tournaments. While our algorithms are not very involved, they are strengthened by the fact that the algorithms to find above $\Omega(\lg n)$ sized d-dominating sets are provably optimal. We have also provided some additional insights into the complexity of finding a d-king which may help narrow the gap between upper and lower bound for the complexity of the problem. We believe that our work will spur further work on developing above or below guarantee parameterizations in polynomially solvable problems. We end with some specific problems from the work on tournaments.

- There is still a gap for the complexity of finding a d-king for $d \geq 2$. We have shown that to improve the upper bound, it is sufficient to improve the upper bound for finding 2-kings. We don't know whether one can find d-kings (for $d > 3$) faster than finding 2-kings, which is the same as asking whether the converse of Corollary 2 is true.
- Are the bounds in Theorem 6 optimal for $k < \epsilon \lg n$, for $d \geq 2$? For $d = 1$, we know that finding a k-sized dominating set is $W[2]$-complete.
- Find the tight upper bound guarantee for the minimum number of vertices in a dominating set in a tournament, and find the parameterized complexity of parameterizing below the bound.

References

1. Abboud, A., Williams, V.V., Wang, J.R.: Approximation and fixed parameter sub-quadratic algorithms for radius and diameter in sparse graphs. In: Proceedings of the Twenty-Seventh Annual ACM-SIAM Symposium on Discrete Algorithms, SODA 2016, Arlington, VA, USA, 10–12 January 2016, pp. 377–391 (2016)

2. Acharya, J., Falahatgar, M., Jafarpour, A., Orlitksy, A., Suresh, A.T.: Maximum selection and sorting with adversarial comparators and an application to density estimation. Comput. Res. Repos. **abs/1606.02786**, 1–24 (2016)
3. Ajtai, M., Feldman, V., Hassidim, A., Nelson, J.: Sorting and selection with imprecise comparisons. ACM Trans. Algorithms **12**(2), 19:1–19:19 (2016)
4. Alon, N., Spencer, J.: The Probabilistic Method. Wiley, Hoboken (1992)
5. Balasubramanian, R., Raman, V., Srinivasaragavan, G.: Finding scores in tournaments. J. Algorithms **24**(2), 380–394 (1997)
6. Downey, R., Fellows, M.: Parameterized Complexity. Springer, New York (1999)
7. Garg, S., Philip, G.: Raising the bar for vertex cover: fixed-parameter tractability above a higher guarantee. In: Proceedings of the Twenty-Seventh Annual ACM-SIAM Symposium on Discrete Algorithms, SODA 2016, Arlington, VA, USA, 10–12 January 2016, pp. 1152–1166 (2016)
8. Giannopoulou, A.C., Mertzios, G.B., Niedermeier, R.: Polynomial fixed-parameter algorithms: a case study for longest path on interval graphs. In: Husfeldt, T., Kanj, I. (eds.) 10th International Symposium on Parameterized and Exact Computation (IPEC 2015), Leibniz International Proceedings in Informatics (LIPIcs), vol. 43, pp. 102–113. Schloss Dagstuhl-Leibniz-Zentrum fuer Informatik, Dagstuhl, Germany (2015)
9. Graham, R.L., Spencer, J.H.: A constructive solution to a tournament problem. Canad. Math. Bull. **14**, 45–48 (1971)
10. Gutin, G., Yeo, A.: Constraint satisfaction problems parameterized above or below tight bounds: a survey. In: Bodlaender, H.L., Downey, R., Fomin, F.V., Marx, D. (eds.) The Multivariate Algorithmic Revolution and Beyond. LNCS, vol. 7370, pp. 257–286. Springer, Heidelberg (2012). doi:10.1007/978-3-642-30891-8_14
11. Lokshtanov, D., Narayanaswamy, N.S., Raman, V., Ramanujan, M.S., Saurabh, S.: Faster parameterized algorithms using linear programming. ACM Trans. Algorithms **11**(2), 15:1–15:31 (2014)
12. Lu, X., Wang, D., Wong, C.K.: On the bounded domination number of tournaments. Discret. Math. **220**(1–3), 257–261 (2000)
13. Mahajan, M., Raman, V.: Parameterizing above guaranteed values: MaxSat and MaxCut. J. Algorithms **31**(2), 335–354 (1999)
14. Mahajan, M., Raman, V., Sikdar, S.: Parameterizing above or below guaranteed values. J. Comput. Syst. Sci. **75**(2), 137–153 (2009)
15. Maurer, S.B.: The king chicken theorems. Math. Mag. **53**(2), 67–80 (1980)
16. Megiddo, N., Vishkin, U.: On finding a minimum dominating set in a tournament. Theor. Comput. Sci. **61**, 307–316 (1988)
17. Moon, J.: Topics on tournaments. In: Selected Topics in Mathematics. Athena series. Holt, Rinehart and Winston (1968)
18. Papadimitriou, C.H., Yannakakis, M.: On limited nondeterminism and the complexity of the V-C dimension. J. Comput. Syst. Sci. **53**(2), 161–170 (1996). http://dx.doi.org/10.1006/jcss.1996.0058
19. Shen, J., Sheng, L., Wu, J.: Searching for sorted sequences of kings in tournaments. SIAM J. Comput. **32**(5), 1201–1209 (2003)

On Computational Aspects of Greedy Partitioning of Graphs

Piotr Borowiecki[✉]

Faculty of Electronics, Telecommunications and Informatics,
Gdańsk University of Technology, Gdańsk, Poland
pborowie@eti.pg.gda.pl

Abstract. In this paper we consider a problem of graph $\mathcal{P}$-coloring consisting in partitioning the vertex set of a graph such that each of the resulting sets induces a graph in a given additive, hereditary class of graphs $\mathcal{P}$. We focus on partitions generated by the greedy algorithm. In particular, we show that given a graph G and an integer k deciding if the greedy algorithm outputs a $\mathcal{P}$-coloring with a least k colors is $\mathbb{NP}$-complete for an infinite number of classes $\mathcal{P}$. On the other hand we get a polynomial-time certifying algorithm if k is fixed and the family of minimal forbidden graphs defining the class $\mathcal{P}$ is finite. We also prove co$\mathbb{NP}$-completeness of the problem of deciding whether for a given graph G the difference between the largest number of colors used by the greedy algorithm and the minimum number of colors required in any $\mathcal{P}$-coloring of G is bounded by a given constant. A new Brooks-type bound on the largest number of colors used by the greedy $\mathcal{P}$-coloring algorithm is given.

Keywords: Graph partitioning · Computational complexity · Graph coloring · Greedy algorithm · Grundy number · Minimal graphs

1 Introduction and Problem Statement

We are interested, specifically, in partitions of the vertex set of simple, finite and undirected graph in which each of the resulting sets induces a graph belonging to a given additive, hereditary class of graphs. A class $\mathcal{P}$ of graphs is called *hereditary* if for every graph G in the class all induced subgraphs of G belong to $\mathcal{P}$, and it is called *additive* if for each graph G all of whose components are in $\mathcal{P}$ it follows that also G is in $\mathcal{P}$. All classes of graphs considered in this paper are additive and hereditary. It is well known that if a class $\mathcal{P}$ is additive and hereditary, then it can be characterized by the family $\boldsymbol{F}(\mathcal{P})$ of connected *minimal forbidden graphs* consisting of graphs G such that $G \notin \mathcal{P}$ but each proper induced subgraph of G belongs to $\mathcal{P}$ (for graphs G and H we write $H \leq G$ if G contains H, that is, if there exists an induced subgraph of G isomorphic to H). The family $\boldsymbol{F}(\mathcal{P})$ should be distinguished from the family of *minimal graphs* in the class $\mathcal{P}$

Supported by the Polish National Science Center grant DEC-2011/02/A/ST6/00201.

M. Xiao and F. Rosamond (Eds.): FAW 2017, LNCS 10336, pp. 34–46, 2017.
DOI: 10.1007/978-3-319-59605-1_4

denoted by $\min_{\leq}\mathcal{P}$ which consists of all graphs in $\mathcal{P}$ that do not contain any graph from $\mathcal{P}$ as a proper induced subgraph.

Returning to the notion of partitions, a $\mathcal{P}$-*coloring* of a graph $G = (V, E)$ is a partition $(V_1, \ldots, V_k)$ of its vertex set $V(G)$ such that each graph $G[V_i]$ induced by a *color class* V_i with $i \in \{1, \ldots, k\}$ belongs to the class $\mathcal{P}$. Equivalently, every $\mathcal{P}$-coloring can be seen as a partition into $\mathcal{P}$-independent sets, where a set of vertices is $\mathcal{P}$-*independent* if the graph induced by the vertices of that set belongs to $\mathcal{P}$. For example, if $F(\mathcal{P}) = \{K_2\}$, then we deal with the classical *proper coloring* in which we simply partition the vertex set into independent sets. The smallest k for which there exists some $\mathcal{P}$-coloring of a graph G with k colors is called the $\mathcal{P}$-*chromatic number* of G and it is denoted by $\chi_{\mathcal{P}}(G)$.

The greedy $\mathcal{P}$-coloring algorithm (for brevity, we say *the greedy algorithm*) colors the vertices of a graph G, one by one, in some order $(v_1, \ldots, v_{n(G)})$ that is independent of the algorithm. Following the order, the algorithm colors each vertex v_i using the smallest color such that its assignment to v_i results in a $\mathcal{P}$-coloring of the graph induced by $\{v_1, \ldots, v_i\}$ (throughout the paper we assume that $\mathcal{P}$ is a class for which deciding the membership in $\mathcal{P}$ is polynomial). A $\mathcal{P}$-coloring produced by the greedy algorithm is called a *Grundy $\mathcal{P}$-coloring*. Note that every Grundy $\mathcal{P}$-coloring of a graph G with k colors can be seen as a surjection $\varphi : V(G) \to \{1, \ldots, k\}$. The number of colors used by the greedy algorithm strongly depends on vertex ordering; with the largest number of colors denoted by $\Gamma_{\mathcal{P}}(G)$ and called the $\mathcal{P}$-*Grundy number* of a graph G. By interpolation theorems of Cockayne et al. [7] the number of colors used by the greedy algorithm can take any value from $\chi_{\mathcal{P}}(G)$ to $\Gamma_{\mathcal{P}}(G)$. For proper coloring, the notion of the Grundy number is usually attributed to Christen and Selkow [6]. Considering generalization of this notion we focus on the following problems.

GRUNDY $\mathcal{P}$-COLORING

    Input: A graph G and positive integer k.

    Question: Does G have a Grundy $\mathcal{P}$-coloring with at least k colors?

GRUNDY $(\mathcal{P}, k)$-COLORING

    Input: A graph G.

    Question: Does G have a Grundy $\mathcal{P}$-coloring with at least k colors?

Both problems have been intensively studied in context of proper coloring. Goyal and Vishwanathan [10] and Zaker [17] proved that for proper coloring the former problem is NP-complete. Does it also hold for *every nontrivial fixed $\mathcal{P}$*? On the other hand, for proper coloring, by finite basis theorem of Gyárfás et al. [11] the latter problem admits a polynomial-time solution (for related concepts see Zaker [17], Borowiecki and Sidorowicz [3]). Determining the Grundy number is known to be polynomial, e.g., for trees [13] and P_4-laden graphs [1]. For some classes of graphs, e.g., interval graphs [16], complements of as well bipartite as chordal graphs [12] and $\{P_5, K_4 - e\}$-free graphs [5], polynomial-time constant-factor approximation algorithms are known. For a deeper discussion on approximability and other aspects of the Grundy number we refer to [10,14,15].

In Sect. 3 of this paper we give a polynomial-time certifying algorithm for GRUNDY $(\mathcal{P}, k)$-COLORING when $F(\mathcal{P})$ is finite (this strongly relies on structural properties related to critical partitions introduced in Sect. 2). In Sect. 4 we show that GRUNDY $\mathcal{P}$-COLORING is $\mathbb{NP}$-complete for every class $\mathcal{P}$ such that $\boldsymbol{F}(\mathcal{P}) = \{K_p\}$ with $p \geq 3$, while in Sect. 5 we prove that for every integer $t \geq 0$ the problem of the membership in $\mathcal{H}(\mathcal{P}, t) = \{G \mid \Gamma_{\mathcal{P}}(G) - \chi_{\mathcal{P}}(G) \leq t\}$ is coNP-complete. We conclude the paper with a new Brooks-type bound on the $\mathcal{P}$-Grundy number generalizing and strengthening the bounds given in [2,18].

2   Motivation, Critical Partitions and Minimal Graphs

Describing the structural properties of critical partitions we step towards a general technique that benefits from the knowledge of polynomial-time approximation algorithms solving diverse optimization problems for inputs in certain graph classes. We use these algorithms to develop new polynomial-time approximation algorithms that are applicable to all inputs in carefully constructed *superclasses* of the above-mentioned classes and preserve the order of original approximation ratios. More formally, let $\mathcal{C}(\mathcal{P}, k)$ denote a class of graphs $\mathcal{P}$-colorable with at most k colors. Consider a minimization problem Π for which there is a polynomial-time $\delta(n)$-approximation algorithm A_1 for inputs in $\mathcal{P}$, and assume that $(V_1, \ldots, V_k)$ is an arbitrary $\mathcal{P}$-coloring of $G \in \mathcal{C}(\mathcal{P}, k)$. Moreover, suppose that there exists a polynomial-time algorithm A_2 that, given the outputs of A_1 for each of the k instances $G[V_i]$, gives a solution of Π for G. A solution that admits such an algorithm A_2 is called *compositive* (note that the solutions of various domination and coloring problems are trivially compositive). We do not know, however, any simple realizations of this three-phase approach, since in general the $\mathcal{P}$-coloring problem is computationally hard [4] (with a sole exception of proper 2-COLORING) and our knowledge of $\boldsymbol{F}(\mathcal{C}(\mathcal{P}, k))$ is very limited. This directs our attention to a subclass $\mathcal{G}(\mathcal{P}, k)$ of $\mathcal{C}(\mathcal{P}, k)$, which consists of all graphs for which $\Gamma_{\mathcal{P}}(G) \leq k$. Since $\mathcal{P} \subset \mathcal{G}(\mathcal{P}, k) \subset \mathcal{C}(\mathcal{P}, k)$ for every $k \geq 2$, the algorithm PH that follows the above defined three phases and uses greedy $\mathcal{P}$-coloring in the first phase, works for all inputs in an *extension* $\mathcal{G}(\mathcal{P}, k)$ of an *arbitrary class* $\mathcal{P}$ and has an approximation ratio of the same order as A_1 on $\mathcal{P}$.

Proposition 1. *If Π admits a polynomial-time $\delta(n)$-approximation in $\mathcal{P}$ and its solutions are compositive, then PH is a polynomial-time $(k \cdot \delta(n))$-approximation algorithm for Π in $\mathcal{G}(\mathcal{P}, k)$.* $\square$

We say that a subset U of the vertices of a graph G is *strongly $\mathcal{P}$-dominating* in G if for every vertex $v \in V(G) \setminus U$ there exists a set $D \subseteq U$ such that $G[D \cup \{v\}] \in \boldsymbol{F}(\mathcal{P})$. For a given $\mathcal{P}$-coloring $(V_1, \ldots, V_k)$ of a graph G a vertex v is called a *Grundy vertex* if $v \in V_1$, or $v \in V_i$, $i \geq 2$ and every color class V_j with $j < i$ contains a set D_j such that $G[D_j \cup \{v\}] \in \boldsymbol{F}(\mathcal{P})$. Naturally, in every Grundy $\mathcal{P}$-coloring each vertex is a Grundy vertex.

Proposition 2. *Let $k \geq 2$ and let H be an induced subgraph of a graph G. If $\Gamma_{\mathcal{P}}(H) \geq k - 1$ and $V(G)$ contains a nonempty $\mathcal{P}$-independent set I that is disjoint from $V(H)$ and strongly $\mathcal{P}$-dominating in $G[I \cup V(H)]$, then $\Gamma_{\mathcal{P}}(G) \geq k$.*

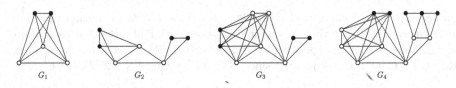

Fig. 1. The graphs $G_1, \ldots, G_4$ from Example 1.

Proof. Let $V(H) = \{x_1, \ldots, x_{n(H)}\}$, $I = \{y_1, \ldots, y_t\}$ and let $(x_1, \ldots, x_{n(H)})$ be an ordering of $V(H)$ that forces the greedy algorithm to produce $\mathcal{P}$-coloring φ of the graph H with $k-1$ colors. Now, consider the assignment of colors produced by the same algorithm for the following vertex ordering $(y_1, \ldots, y_t, x_1, \ldots, x_{n(H)})$. Naturally, since I is strongly $\mathcal{P}$-dominating in $G[I \cup V(H)]$, for each vertex $x \in V(H)$ there exists a subset D of I such that $G[D \cup \{x\}] \in \boldsymbol{F}(\mathcal{P})$. Moreover, since I is $\mathcal{P}$-independent and no vertex in I is preceded by a vertex in $V(H)$, all vertices in I will be colored 1. Consequently, for each vertex x of H the algorithm will use color $\varphi(x) + 1$. This yields $\Gamma_\mathcal{P}(G) \geq k$, since $G[I \cup V(H)] \leq G$.   □

The above proposition reveals the importance of specific bipartition of the vertex set in which both parts have certain coloring or domination properties. We focus on how they come into play in graphs that are minimal with respect to the $\mathcal{P}$-Grundy number under taking induced subgraphs. For $k \geq 1$ by $\mathcal{U}(\mathcal{P}, k)$ we denote the class of graphs G for which $\Gamma_\mathcal{P}(G) \geq k$. Naturally $\mathcal{U}(\mathcal{P}, k) = \mathcal{I} \setminus \mathcal{G}(\mathcal{P}, k-1)$, where $\mathcal{I}$ denotes the class of all graphs. In what follows $\boldsymbol{C}_k(\mathcal{P})$ stands for the family of *minimal graphs* in the class $\mathcal{U}(\mathcal{P}, k)$, that is, the graphs for which $\Gamma_\mathcal{P}(G) = k$ and $\Gamma_\mathcal{P}(G - v) < k$, where v is an arbitrary vertex of G. Note that $\boldsymbol{C}_1(\mathcal{P}) = \{K_1\}$, $\boldsymbol{C}_2(\mathcal{P}) = \boldsymbol{F}(\mathcal{P})$ and that forbidding all graphs in $\boldsymbol{C}_k(\mathcal{P})$ defines $\mathcal{G}(\mathcal{P}, k-1)$; more formally $\boldsymbol{C}_k(\mathcal{P}) = \boldsymbol{F}(\mathcal{G}(\mathcal{P}, k-1))$. Minimal graphs play a crucial role in the greedy coloring process; in fact they determine the number of colors used in the worst case, and they *characterize* classes $\mathcal{G}(\mathcal{P}, k)$. For proper coloring the classes $\mathcal{G}(\mathcal{P}, 2)$ and $\mathcal{G}(\mathcal{P}, 3)$ were characterized by Gyárfás et al. [11] who proved that $\boldsymbol{C}_3(\mathcal{P}) = \{K_3, P_4\}$ and listed all 22 graphs in $\boldsymbol{C}_4(\mathcal{P})$.

The key notion in our analysis of minimal graphs is a class $\mathcal{F}(\mathcal{P}, k)$ consisting of all graphs G for which there exists a partition (I, C) of $V(G)$ such that I is nonempty, $\mathcal{P}$-independent and strongly $\mathcal{P}$-dominating in G, and $G[C] \in \boldsymbol{C}_{k-1}(\mathcal{P})$, where $k \geq 2$. Such a partition is called a *critical partition* of graph G. Note that critical partition of G is not necessarily unique.

Example 1. Consider the graphs $G_1, \ldots, G_4$ in Fig. 1 and the class $\mathcal{P}$ for which $\boldsymbol{F}(\mathcal{P}) = \{K_3\}$. The graphs G_1 and G_2 are the elements of $\boldsymbol{C}_3(\mathcal{P})$; their Grundy $\mathcal{P}$-coloring with 3 colors can be easily obtained using Proposition 2 with black vertices forming the set I. Additionally, if we denote by C the set of white vertices, then we get a critical partition (I, C). It is not hard to verify that deleting any vertex results in the $\mathcal{P}$-Grundy number equal to 2. The graphs G_3 and G_4 belong to $\mathcal{F}(\mathcal{P}, 4)$, since they admit critical partitions with black and white vertices forming the sets I and C, respectively. Indeed, white vertices induce G_1 or G_2, while black ones do not induce K_3 which means that I is $\mathcal{P}$-independent. To see that I is strongly $\mathcal{P}$-dominating observe that each white vertex belongs to a triangle with two black vertices. Again, it is not hard to see that $G_3 \in C_4(\mathcal{P})$. Consequently, since $G_3 \leq G_4$, we get $G_4 \in \mathcal{F}(\mathcal{P}, 4) \setminus C_4(\mathcal{P})$.

Theorem 1. *If $\mathcal{P}$ is a class of graphs and $k \geq 2$ is an integer, then*

$$C_k(\mathcal{P}) = \min_{\leq} \mathcal{F}(\mathcal{P}, k).$$

Proof. Recall that $C_k(\mathcal{P})$ is defined as a family of minimal graphs in the class $\mathcal{U}(\mathcal{P}, k)$ in which for every graph G it holds $\Gamma_{\mathcal{P}}(G) \geq k$. If $G \in \mathcal{F}(\mathcal{P}, k)$, then by the definition of the class $\mathcal{F}(\mathcal{P}, k)$ and Proposition 2 we have $\Gamma_{\mathcal{P}}(G) \geq k$. Thus $\mathcal{F}(\mathcal{P}, k) \subseteq \mathcal{U}(\mathcal{P}, k)$. Consequently, for every graph $H' \in \min_{\leq} \mathcal{F}(\mathcal{P}, k)$ there exists a graph $H \in \min_{\leq} \mathcal{U}(\mathcal{P}, k)$ such that $H \leq H'$. To finish the proof it remains to show that $H \in \mathcal{F}(\mathcal{P}, k)$. Note that by minimality $\Gamma_{\mathcal{P}}(H) = k$. Now, if $(V_1, \ldots, V_k)$ is an arbitrary Grundy $\mathcal{P}$-coloring of H with k colors, we can simply construct a partition (I, C) of $V(H)$ by setting $I = V_1$ and $C = \bigcup_{i=2}^{k} V_i$. Naturally, the set I of the partition is nonempty, $\mathcal{P}$-independent, and by the definition of Grundy $\mathcal{P}$-coloring it is also strongly $\mathcal{P}$-dominating in H. Since $(V_2, \ldots, V_k)$ is a Grundy $\mathcal{P}$-coloring of $H[C]$, we have $\Gamma_{\mathcal{P}}(H[C]) \geq k - 1$. We shall show that $H[C] \in C_{k-1}(\mathcal{P})$. Suppose, contrary to our claim, that there exists a vertex $v \in C$ and an ordering of $C' = C \setminus \{v\}$ that forces the algorithm to color $H[C']$ with $k - 1$ colors. Consequently, from Proposition 2 applied to I and $H[C']$ it follows that $\Gamma_{\mathcal{P}}(H[I \cup C']) \geq k$. This contradicts minimality of H in the definition of $C_k(\mathcal{P})$. Hence $H \in \mathcal{F}(\mathcal{P}, k)$. □

3   The Complexity of Grundy $(\mathcal{P}, k)$-Coloring

In what follows, we have to carefully distinguish between the two important cases of finite and infinite family $F(\mathcal{P})$.

Proposition 3. *Let $\mathcal{P}$ be a class of graphs with finite $F(\mathcal{P})$. If α and β denote the minimum and maximum order of a graph in $F(\mathcal{P})$, respectively, then for every $k \geq 1$ and every graph $G \in C_k(\mathcal{P})$ it holds $(\alpha - 1)(k - 1) + 1 \leq n(G) \leq \beta^{k-1}$.*

Proof. We use induction on k. The above inequalities are evidently fulfilled for $k \in \{1, 2\}$. Assume that the statement is true for the parameters smaller than k, where $k \geq 3$. Let $G \in C_k(\mathcal{P})$. By Theorem 1 there exists a partition (I, C) of $V(G)$ such that I is nonempty, $\mathcal{P}$-independent and strongly $\mathcal{P}$-dominating in G. Moreover $G[C] \in C_{k-1}(\mathcal{P})$, which by the induction hypothesis implies

$$(\alpha - 1)(k - 2) + 1 \leq |C| \leq \beta^{k-2}. \tag{1}$$

Since $V(G) = I \cup C$ and $I \cap C = \emptyset$, the proof will be completed by showing $\alpha - 1 \leq |I| \leq \beta^{k-2}(\beta - 1)$. The inequality $\alpha - 1 \leq |I|$ is obvious because C is nonempty and I is strongly $\mathcal{P}$-dominating in G. Suppose to the contrary that $|I| \geq \beta^{k-2}(\beta - 1) + 1$. Let $\vartheta : C \to 2^I$ be a mapping that assigns to each $x \in C$ a subset $\vartheta(x)$ of I such that $G[\vartheta(x) \cup \{x\}] \in \boldsymbol{F}(\mathcal{P})$. Note that the existence of the set $\vartheta(x)$ follows immediately by the assumption that I is strongly $\mathcal{P}$-dominating in G. Naturally $|\vartheta(x)| \leq \beta - 1$ for each $x \in C$ and the sets $\vartheta(x_1), \vartheta(x_2)$ need not be distinct when $x_1 \neq x_2$. Let I' denote the following union $\bigcup_{x \in C} \vartheta(x)$. Thus

$$|I'| = |\bigcup_{x \in C} \vartheta(x)| \leq \sum_{x \in C} |\vartheta(x)| \leq (\beta - 1)|C| \overset{(1)}{\leq} (\beta - 1)\beta^{k-2}.$$

Consequently as a proper subset of I, the set I' is $\mathcal{P}$-independent in G and hence it is $\mathcal{P}$-independent in each induced subgraph of G. Therefore, if $G' = G[I' \cup C]$, then I' is strongly $\mathcal{P}$-dominating in G' by its definition, and it is $\mathcal{P}$-independent in G' by our earlier consideration. Finally, it follows that $G' \in \mathcal{F}(\mathcal{P}, k)$ and that G' is a proper induced subgraph of G. This contradicts the fact that by Theorem 1 the graph G is minimal in $\mathcal{F}(\mathcal{P}, k)$. □

In what follows we need the class $\mathcal{T}(\mathcal{P}, k)$ of *forcing* $(\mathcal{P}, k)$-*trees*. Namely, if $k = 2$, then $\mathcal{T}(\mathcal{P}, 2) = \boldsymbol{C}_2(\mathcal{P})$. For $k \geq 3$ let B be a graph in $\mathcal{T}(\mathcal{P}, k-1)$ and let $F_1, \ldots, F_{n(B)}$ be disjoint graphs, each being isomorphic to a graph in $\boldsymbol{F}(\mathcal{P})$. A graph G belongs to $\mathcal{T}(\mathcal{P}, k)$ if it can be obtained from $B, F_1, \ldots, F_{n(B)}$ by an identification of *each* vertex of B with an arbitrary vertex of some graph F_j, $j \in \{1, \ldots, n(B)\}$ in such a way that every F_j takes part in exactly one identification. As the *root* of a forcing $(\mathcal{P}, k)$-tree we take any vertex of $B \in \mathcal{T}(\mathcal{P}, 2)$. It is known that all graphs in $\mathcal{T}(\mathcal{P}, k)$ are minimal in $\mathcal{F}(\mathcal{P}, k)$.

Theorem 2. *For every class $\mathcal{P}$ and every integer $k \geq 2$, the set $\boldsymbol{C}_k(\mathcal{P})$ is finite if and only if $\boldsymbol{F}(\mathcal{P})$ is finite.*

Proof. If $\boldsymbol{F}(\mathcal{P})$ is finite, then simply observe that by Proposition 3 the set $\boldsymbol{C}_k(\mathcal{P})$ is finite, too. Now, suppose to the contrary that $\boldsymbol{F}(\mathcal{P})$ is infinite and that for some k the set $\boldsymbol{C}_k(\mathcal{P})$ is finite. Since $\boldsymbol{C}_2(\mathcal{P}) = \boldsymbol{F}(\mathcal{P})$, it remains to consider $k \geq 3$. Let n^* be the largest order of a graph in $\boldsymbol{C}_k(\mathcal{P})$ and let $F \in \boldsymbol{F}(\mathcal{P})$ be a graph such that $n(F) > \sqrt[k-1]{n^*}$. Note that by the finiteness of $\boldsymbol{C}_k(\mathcal{P})$ and the infiniteness of $\boldsymbol{F}(\mathcal{P})$ such a number and such a graph always exist. Next, consider $T_k \in \mathcal{T}(\mathcal{P}, k)$ constructed in $k - 1$ steps, starting with $T_2 = F$, and such that for each step $i \in \{2, \ldots, k-1\}$, in which we obtain T_{i+1}, it holds $F_1 = \cdots = F_{n(T_i)} = F$. Since $T_k \in \boldsymbol{C}_k(\mathcal{P})$, the construction of T_k and the assumption on $n(F)$ imply $n(T_k) = (n(F))^{k-1} > (\sqrt[k-1]{n^*})^{k-1} = n^*$. A contradiction with a choice of n^*. □

The above finite basis theorem allows for the proof of the following result on the computational complexity of GRUNDY $(\mathcal{P}, k)$-COLORING.

Theorem 3. *If $\mathcal{P}$ is a class with finite $\boldsymbol{F}(\mathcal{P})$ and $k > 0$ is a fixed integer, then* GRUNDY $(\mathcal{P}, k)$-COLORING *admits a polynomial-time certifying algorithm.*

Proof. For a fixed k by Theorem 2 the number of graphs in $\boldsymbol{C}_k(\mathcal{P})$ is finite. Moreover, checking whether a graph H of order p is an induced subgraph of a given graph G of order n can be done by brute force in $O(n^p)$ time. Since for finite $F(\mathcal{P})$ by Proposition 3 the order of any graph in $C_k(\mathcal{P})$ is bounded from above by β^{k-1}, we can check if H is contained in G in $O(n^{\beta^{k-1}})$ time. If G does not contain a graph in $C_k(\mathcal{P})$, then application of the greedy algorithm results in a Grundy $\mathcal{P}$-coloring with at most $k-1$ colors, while in the opposite case, i.e., when $\Gamma_{\mathcal{P}}(G) \geq k$ we get an induced subgraph (a YES certificate) that can be used to force a Grundy $\mathcal{P}$-coloring of G with at least k colors.   □

For the state-of-the-art survey on the computational complexity of coloring graphs with forbidden subgraphs, including certification, see Golovach et al. [9].

4   The Complexity of Grundy $\mathcal{P}$-Coloring

In this section we prove that GRUNDY $\mathcal{P}$-COLORING is NP-complete for every class $\mathcal{P}$ defined by $F(\mathcal{P}) = \{K_p\}$ with $p \geq 3$. In our proof we use a polynomial-time reduction from 3-COLORING of planar graphs with vertex degree at most 4 (for NP-completeness see Garey et al. [8]). In fact we need a slight strengthening of their result consisting in restricting the class to planar graphs with vertex degree at most 4, size $m \equiv 1(\bmod\ r)$ and every vertex belonging to at least one triangle; the class of such graphs we denote by $\mathcal{L}(r)$ (the proof of this result is omitted due to a limited space).

Theorem 4. *For every fixed $r \geq 2$, 3-COLORING is NP-complete in $\mathcal{L}(r)$.*   □

Theorem 5. *For every class $\mathcal{P}$ such that $F(\mathcal{P}) = \{K_p\}$ with $p \geq 3$, GRUNDY $\mathcal{P}$-COLORING is NP-complete.*

In order to prove the above theorem we give Construction 1 and prove several lemmas. Let $p \geq 3$, $F(\mathcal{P}) = \{K_p\}$, and let G be an instance of 3-COLORING in $\mathcal{L}(p-1)$. We construct a graph G' and calculate an integer k such that G' has a Grundy $\mathcal{P}$-coloring with k colors if and only if G has a proper 3-coloring.

Construction 1. First, we define the vertices of G'. For each vertex $v_i \in V(G)$, $i \in \{1, \ldots, n(G)\}$ create a set of vertices $U_i = \{u_1^i, \ldots, u_{p-1}^i\}$. Similarly, for each edge $e_i \in E(G)$, $i \in \{1, \ldots, m(G)\}$ create a set of vertices $W_i = \{w_1^i, \ldots, w_{p-1}^i\}$ and a single vertex x_i. Let Q denote the set $\{x_1, \ldots, x_{m(G)}\}$ and let $U = \bigcup_{1 \leq i \leq n(G)} U_i$, $W = \bigcup_{1 \leq i \leq m(G)} W_i$. Finally, set $V(G') = Q \cup W \cup U$. Now, we define the edges of G'. First, create all edges so that Q induces a complete graph. Then, for every edge $e_t \in E(G)$, $t \in \{1, \ldots, m(G)\}$ with endvertices v_i, v_j join the corresponding vertex x_t with all vertices in U_i, U_j and W_t, and create all edges so that $U_i \cup W_t$ and $U_j \cup W_t$ induce complete graphs. In what follows a graph induced in G' by $W_t \cup U_i \cup U_j \cup \{x_t\}$ is denoted by $X_{t,i,j}$ and called the *gadget* corresponding to the edge e_t. The construction is completed by setting $k = \lceil m(G)/(p-1) \rceil + 3$. For an example of the construction see Fig. 2.   □

Property 1. *Let $X_{t,i,j}$ be a gadget and let $H = K_p$ with $p \geq 3$. If $V(H) \cap W_t \neq \emptyset$, then (a) $V(H) \subseteq V(X_{t,i,j})$, (b) $V(H) \subseteq U_\tau \cup W_t \cup \{x_t\}$, when H contains a vertex $u \in U_\tau$, $\tau \in \{i, j\}$.*   □

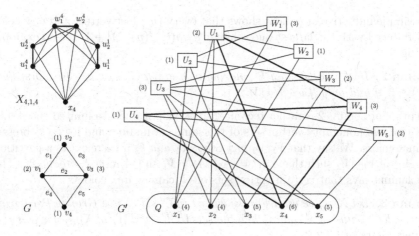

Fig. 2. An example of Construction 1 for $F(\mathcal{P}) = \{K_3\}$; the numbers in braces denote colors, while the lines in a drawing of G' represent joins.

We say that in a $\mathcal{P}$-colored graph H a set $U \subseteq V(H)$ *uses color* ℓ if every vertex in U has color ℓ.

Lemma 1. *If G has a proper coloring with 3 colors, then G' admits a Grundy $\mathcal{P}$-coloring with k colors.*

Proof. Suppose that φ is a proper 3-coloring of G. In order to define the corresponding Grundy $\mathcal{P}$-coloring φ' of G' we set $\varphi'(u) = \varphi(v_i)$ for all vertices u in the set U_i corresponding to the vertex v_i of G. Consequently, each set U_i uses one of the colors in $\{1, 2, 3\}$, which is feasible since U is $\mathcal{P}$-independent. Let $X_{t,i,j}$ be an arbitrary gadget in G'. Since $X_{t,i,j}$ corresponds to the edge $v_i v_j$ and φ is proper, the sets U_i and U_j use distinct colors, say a and b, respectively. Thus for every vertex $w \in W_t$ its colored neighbors in $U_i \cup U_j$ use colors in $\{a, b\}$ and hence the color $c \in \{1, 2, 3\} \setminus \{a, b\}$ is feasible for w. We set $\varphi'(w) = c$. Following Construction 1, for every vertex $w \in W_t$ it holds $G'[U_i \cup \{w\}] = K_p$ and $G'[U_j \cup \{w\}] = K_p$. It remains to observe that independently of color permutation, each color smaller than $\varphi'(w)$ is used by U_i or U_j. Hence, if $\varphi'(w) = c$, then w is a Grundy vertex. Now, without loss of generality consider a vertex u in U_i. Since by assumption every vertex of G belongs to a triangle, say induced by $\{v_i, v_j, v_\tau\}$, the colors used by the corresponding sets U_i, U_j and U_τ are distinct and uniquely determine distinct colors used by the sets W_t and $W_{t'}$ that correspond to the edges $v_i v_j$ and $v_i v_\tau$, respectively. Naturally, u is a Grundy vertex, which follows by similar argument as above (note $G'[W_t \cup \{u\}] = K_p$ and $G'[W_{t'} \cup \{u\}] = K_p$). Thus we have proved that $\varphi'|_{W \cup U}$ is a Grundy $\mathcal{P}$-coloring. It remains to extend $\varphi'|_{W \cup U}$ to $V(G')$ by processing vertices of Q in an arbitrary order and $\mathcal{P}$-coloring them greedily. Since for each $x_t \in Q$ the sets W_t, U_i, U_j of $X_{t,i,j}$ use three distinct colors in $\{1, 2, 3\}$ and $G'[A \cup \{x_t\}] = K_p$ for every $A \in \{U_i, U_j, W_t\}$, and $G'[D \cup \{x_t\}] = K_p$ for every $(p-1)$-element subset D of

Q, a simple inductive argument shows that every $(p-1)$st vertex of Q new color is introduced with the largest one achieving $\lceil m(G)/(p-1)\rceil + 3$. For an example see Fig. 2. □

Lemma 2. *If φ' is a Grundy $\mathcal{P}$-coloring of a graph G' with $k \geq 6$ colors, then $\varphi'(v) \leq 3$ if and only if $v \in U \cup W$.* □

The proof of Lemma 2 is rather technical so we skip it due to a limited space. Let V_1, V_2 be two nonempty vertex sets of the same cardinality, and let ℓ_1, ℓ_2 be two distinct colors. We say that V_1, V_2 *mix colors* ℓ_1 *and* ℓ_2 if there exists a partition (A_1, A_2) of $V_1 \cup V_2$ such that $|A_1| = |A_2|, A_1 \neq V_i$ and A_i uses $\ell_i, i \in \{1,2\}$. The next lemma says that certain vertex sets of G' cannot mix colors.

Lemma 3. *Let $X_{t,i,j}$ be an arbitrary gadget in G'. If φ' is a Grundy $\mathcal{P}$-coloring of a graph G' with $k \geq 6$ colors, then the sets U_i, U_j and W_t of $X_{t,i,j}$ use pairwise distinct colors in $\{1,2,3\}$.*

Proof. First we show that $V(X_{t,i,j}) \setminus \{x_t\}$ contains the three sets such that each of them uses a distinct color in $\{1,2,3\}$. By Lemma 2 we have $\varphi(x_t) \geq 4$. Hence, since x_t is a Grundy vertex, for each $\ell \in \{1,2,3\}$ there exists a set D_ℓ that uses color ℓ in coloring φ' and $G'[D_\ell \cup \{x_t\}] = K_p$. By the same claim it follows that each of these sets is contained in $N_{G'}(x_t) \cap (U \cup W)$ while from the construction of G' it is easy to see that $N_{G'}(x_t) \cap (U \cup W) = V(X_{t,i,j}) \setminus \{x_t\}$. Clearly, the sets D_ℓ are pairwise disjoint and hence by their cardinalities it follows that $D_1 \cup D_2 \cup D_3 = V(X_{t,i,j}) \setminus \{x_t\}$. It remains to prove that $\{D_1, D_2, D_3\} = \{U_i, U_j, W_t\}$. Using Property 1(b) it is not hard to argue that for every gadget $X_{t,i,j}$, considered *independently of other gadgets*, either (a) U_i, U_j, W_t use distinct colors in $\{1,2,3\}$, or (b) U_i, W_t (resp. U_j, W_t) mix colors $a, b \in \{1,2,3\}$ and U_j (resp. U_i) uses the color in $\{1,2,3\} \setminus \{a,b\}$. Now, we show that if one considers $X_{t,i,j}$ as a subgraph of G', then only the former condition is feasible. Let v_i be an arbitrary vertex of G. By assumption v_i belongs to a triangle, say induced by v_i, v_j and v_τ. Let $X_{t,i,j}, X_{t',i,\tau}$ and $X_{t'',j,\tau}$ be the gadgets corresponding to the edges of this triangle. Suppose to the contrary that U_i, W_t mix colors $a, b \in \{1,2,3\}$. Following (b) for $X_{t',i,\tau}$ we get that $U_i, W_{t'}$ have to mix a and b. Again by (b) applied to $X_{t,i,j}$ and $X_{t',i,\tau}$, it follows that U_j and U_τ have to use the third color, say c. This results in $2(p-1)$ vertices of $X_{t'',j,\tau}$ that use color c and hence contradicts the property that $V(X_{t'',j,\tau})$ contains the three sets of cardinality $p-1$ such that each of them uses a distinct color in $\{1,2,3\}$. □

Lemma 4. *If G' has a Grundy $\mathcal{P}$-coloring with k colors, then G admits a proper coloring with 3 colors.*

Proof. By the construction of G', each set U_i of G' corresponds to a unique vertex v_i of G, $i \in \{1, \ldots, n(G)\}$. Let φ' be a Grundy $\mathcal{P}$-coloring of G' with k colors and let ℓ_i denote the color used by the set U_i in φ'. In order to define an appropriate proper coloring φ of the graph G with 3 colors, for each vertex v_i of G we set $\varphi(v_i) = \ell_i$. From Lemma 3 it follows that $\ell_i \in \{1,2,3\}$ and that for every edge $e_t = v_i v_j$ of G the corresponding sets U_i, U_j of the gadget $X_{t,i,j}$ use distinct colors. Thus φ is a proper coloring of G with 3 colors. □

Proof of Theorem 5. If $F(\mathcal{P})$ is finite, we can use the arguments similar to those in the proof of Theorem 3 to show that a certificate function $\varphi : V(G) \to \mathbb{N}$ can be verified in polynomial time for being a Grundy $\mathcal{P}$-coloring with k colors. Hence GRUNDY $\mathcal{P}$-COLORING belongs to $\mathbb{NP}$. Now, Lemmas 1 and 4, and the fact that Construction 1 is polynomial in the size of G imply $\mathbb{NP}$-hardness of the problem, which finishes the proof. $\qquad\qquad\square$

5 coNP-Completeness of the Membership in $\mathcal{H}(\mathcal{P}, t)$

In this section we prove that the problem of deciding the membership in the class $\mathcal{H}(\mathcal{P}, t)$ is coNP-complete for every $t \geq 0$ and every class $\mathcal{P}$ for which $F(\mathcal{P}) = \{K_p\}$ with $p \geq 3$. We present a polynomial-time reduction from GRUNDY $\mathcal{P}$-COLORING of graphs in the class $\mathcal{Q}(p-1)$, where by $\mathcal{Q}(p-1)$ we mean a class of graphs obtained by the application of Construction 1 to all graphs G in $\mathcal{L}(p-1)$ for which $m(G) \geq 4p - 3$ (note that such a restriction on the size of graphs does not influence the hardness of GRUNDY $\mathcal{P}$-COLORING, i.e., it remains NP-complete even in $\mathcal{Q}(p-1)$). Before proving the hardness of the above membership problem we need some facts on the $\mathcal{P}$-Grundy number of graphs in $\mathcal{Q}(p-1)$.

Lemma 5. *Let $\mathcal{P}$ be a class of graphs such that $F(\mathcal{P}) = \{K_p\}$ with $p \geq 3$. If G' is a graph in $\mathcal{Q}(p-1)$, then $\chi_{\mathcal{P}}(G') = \lceil \omega(G')/(p-1) \rceil$.* $\qquad\square$

Remark 1. Note that if $m(G) \geq 4p - 3$, then $\omega(G') = m(G)$ and hence $\chi_{\mathcal{P}}(G') = k - 3$. Thus, the problem of determining the $\mathcal{P}$-Grundy number of graphs in $\mathcal{Q}(p-1)$ can be solved in polynomial time. Moreover, since by Lemma 1 we have $\Gamma_{\mathcal{P}}(G') \geq k$, for every graph $G' \in \mathcal{Q}(p-1)$ it holds $\chi_{\mathcal{P}}(G') < \Gamma_{\mathcal{P}}(G')$.

Theorem 6. *If $\mathcal{P}$ is a class of graphs such that $F(\mathcal{P}) = \{K_p\}$ with $p \geq 3$, then for every $t \geq 0$ the problem of the membership in $\mathcal{H}(\mathcal{P}, t)$ is coNP-complete.*

Proof. In order to show that our problem belongs to coNP it is enough to observe that a graph does not belong to $\mathcal{H}(\mathcal{P}, t)$ if and only if it admits two Grundy $\mathcal{P}$-colorings with k_1 and k_2 colors respectively, and $k_2 > k_1 + t$. Equivalently, two orderings of the vertex set leading to the above-mentioned colorings could also serve as an appropriate NO certificate. Clearly, due to the general assumption that the membership in $\mathcal{P}$ can be checked in polynomial time, the problem of the verification of our certificate is also polynomial.

Now, we present a polynomial-time reduction from GRUNDY $\mathcal{P}$-COLORING of graphs in $\mathcal{Q}(p-1)$ to the problem of the membership in $\mathcal{I} \setminus \mathcal{H}(\mathcal{P}, t)$. Given an arbitrary instance (G, κ) of the former problem, for every $t \geq 0$ we construct a graph G' such that $G' \notin \mathcal{H}(\mathcal{P}, t)$ if and only if $\Gamma_{\mathcal{P}}(G) \geq \kappa$.

Construction 2. First, for a given graph G with the vertex set $\{v_1, \ldots, v_{n(G)}\}$ we construct a graph H_t. Let $T_1, \ldots, T_{n(G)}$ be disjoint graphs such that each of them is isomorphic to the forcing $(\mathcal{P}, t+1)$-tree T (note that since $F(\mathcal{P}) = \{K_p\}$, such a graph T is unique) and let x_i denote the root of T_i, $i \in \{1, \ldots, n(G)\}$. The graph H_t we construct from the graphs $G, T_1, \ldots, T_{n(G)}$ by the identification of each vertex v_i of G with the root x_i of the corresponding T_i, $i \in \{1, \ldots, n(G)\}$. Next, following Lemma 5 we calculate $\chi_{\mathcal{P}}(G)$. If $\chi_{\mathcal{P}}(G) \geq \kappa$, then we set $G' = H_t$. On the other hand, if $\chi_{\mathcal{P}}(G) < \kappa$, then to construct G' we take a complete graph K of order $\kappa(p-1)$ and a set $S \subseteq V(K)$ such that $|S| = p-1$ and join each vertex of H_t with all vertices in S. Since the order of G' is at most $n(G) \cdot p^t + \kappa = O(n(G))$ and by Lemma 5 determining $\chi_{\mathcal{P}}(G)$ admits a polynomial-time algorithm, the graph G' can be constructed in polynomial time. $\qquad \square$

In what follows we use the following properties of H_t. Namely, for every $t \geq 0$: (i) $\chi_{\mathcal{P}}(H_t) = \chi_{\mathcal{P}}(G)$, (ii) $\Gamma_{\mathcal{P}}(H_t) = \Gamma_{\mathcal{P}}(G) + t$. The former follows easily by the fact that every forcing $(\mathcal{P}, t+1)$-tree admits a $\mathcal{P}$-coloring with 2 colors. For a lower bound in the latter one it is enough to consider one of the "from the leaves up" orderings, that forces the greedy algorithm to use $t + 1$ colors on each of the forcing $(\mathcal{P}, t+1)$-trees $T_1, \ldots, T_{n(G)}$, while an upper bound follows by the analysis of graphs F in $C_k(\mathcal{P})$ contained in H_t and such that $k > \chi_{\mathcal{P}}(F)$.

Suppose that $\chi_{\mathcal{P}}(G) \geq \kappa$. Naturally, $\chi_{\mathcal{P}}(G) \geq \kappa$ implies $\Gamma_{\mathcal{P}}(G) \geq \kappa$ and hence (G, κ) is a YES instance of GRUNDY $\mathcal{P}$-COLORING. Since by Remark 1 for every graph $G \in \mathcal{Q}(p-1)$ it holds $\chi_{\mathcal{P}}(G) < \Gamma_{\mathcal{P}}(G)$, by (i) and (ii) (recall $G' = H_t$) we get $\chi_{\mathcal{P}}(G') = \chi_{\mathcal{P}}(G) < \Gamma_{\mathcal{P}}(G) = \Gamma_{\mathcal{P}}(G') - t$. Thus $G' \notin \mathcal{H}(\mathcal{P}, t)$; in other words G' is a YES instance of the problem of the membership in $\mathcal{I} \setminus \mathcal{H}(\mathcal{P}, t)$.

Now, assume that $\chi_{\mathcal{P}}(G) < \kappa$. Let us consider graph $H' = H_t + G'[S]$. It is not hard to see that $\chi_{\mathcal{P}}(H') \leq \chi_{\mathcal{P}}(H_t) + 1$ and hence by (i) we easily get $\chi_{\mathcal{P}}(H') \leq \chi_{\mathcal{P}}(G) + 1$. On the other hand, observe that $\chi_{\mathcal{P}}(H') \geq \lceil \omega(H')/(p-1) \rceil$. Since $\omega(H') = \omega(H_t) + p - 1$ and $\omega(H_t) = \omega(G)$, we have $\chi_{\mathcal{P}}(H') \geq \lceil \omega(G)/(p-1) \rceil + 1$, which, by Lemma 5, results in $\chi_{\mathcal{P}}(H') \geq \chi_{\mathcal{P}}(G) + 1$. Finally, $\chi_{\mathcal{P}}(H') = \chi_{\mathcal{P}}(G) + 1$. Now, consider G' and observe that $\chi_{\mathcal{P}}(G') = \max\{\chi_{\mathcal{P}}(H'), \chi_{\mathcal{P}}(K)\}$. Hence, since $\chi_{\mathcal{P}}(K) = \kappa$ and by assumption $\chi_{\mathcal{P}}(G) < \kappa$, we get $\chi_{\mathcal{P}}(G') = \max\{\chi_{\mathcal{P}}(G) + 1, \kappa\} = \kappa$. Moreover, by Proposition 2 and simple analysis of maximum degrees of appropriate subgraphs of G' we obtain $\Gamma_{\mathcal{P}}(H') = \Gamma_{\mathcal{P}}(H_t) + 1$. Similar argument results in $\Gamma_{\mathcal{P}}(G') = \max\{\Gamma_{\mathcal{P}}(H'), \Gamma_{\mathcal{P}}(K)\}$, and since $\Gamma_{\mathcal{P}}(K) = \kappa$ and from (ii) it follows that $\Gamma_{\mathcal{P}}(H_t) = \Gamma_{\mathcal{P}}(G) + t$, we obtain $\Gamma_{\mathcal{P}}(G') = \max\{\Gamma_{\mathcal{P}}(G) + t + 1, \kappa\}$. Now, let us continue by considering the two cases. If $\Gamma_{\mathcal{P}}(G) \geq \kappa$, then $\Gamma_{\mathcal{P}}(G') \geq \max\{\kappa + t + 1, \kappa\} = \kappa + t + 1 > \chi_{\mathcal{P}}(G') + t$, which implies $G' \notin \mathcal{H}(\mathcal{P}, t)$. On the other hand, if $\Gamma_{\mathcal{P}}(G) < \kappa$, then $\Gamma_{\mathcal{P}}(G') \leq \max\{\kappa - 1 + t + 1, \kappa\} = \kappa + t = \chi_{\mathcal{P}}(G') + t$ and hence $G' \in \mathcal{H}(\mathcal{P}, t)$. $\qquad \square$

6   An Upper Bound on the $\mathcal{P}$-Grundy Number

We conclude our paper with a new upper bound on the $\mathcal{P}$-Grundy number. For a vertex v of a graph G we use $\mathcal{D}(G, v)$ to denote the family of all subsets of $V(G) \setminus \{v\}$ such that for every D in $\mathcal{D}(G, v)$ it holds $G[D \cup \{v\}] \in \mathbf{F}(\mathcal{P})$. The $\mathcal{P}$-degree of a vertex v of G, denoted by $d_G(\mathcal{P}, v)$, we mean the cardinality of a largest subfamily of $\mathcal{D}(G, v)$ consisting of pairwise disjoint sets. A natural upper bound $\Gamma_{\mathcal{P}}(G) \leq \max_{v \in V(G)} d_G(\mathcal{P}, v) + 1$ follows directly from the greedy rule. This can be significantly improved by careful analysis of a specific function on the vertex set of a graph. Let $\phi_G^0(\mathcal{P}, v) = d_G(\mathcal{P}, v)$. For every integer $r \geq 1$, let $\phi_G^r(\mathcal{P}, v)$ be the largest k for which there exist k pairwise disjoint sets $D_1, \ldots, D_k$ in $\mathcal{D}(G, v)$ such that for each of them $\lambda_G^{r-1}(D_i) \geq i$, where $\lambda_G^{r-1}(D_i)$ denotes the $(r-1)$st *intensity* of D_i defined as follows $\lambda_G^{r-1}(D_i) = \min\{\phi_G^{r-1}(\mathcal{P}, u) \mid u \in D_i\}$. A simple inductive argument shows that $\phi_G^r(\mathcal{P}, v) \leq \phi_G^{r-1}(\mathcal{P}, v)$ for every vertex v and every $r \geq 1$. This implies the existence of an integer $t \geq 0$ such that $\phi_G^t(\mathcal{P}, v) = \phi_G^r(\mathcal{P}, v)$ for every vertex v and every $r \geq t$. Thus $\phi_G^t(\mathcal{P}, v)$ is uniquely determined. The *potential of a vertex* v, denoted by $\phi_G(\mathcal{P}, v)$, is given by $\phi_G(\mathcal{P}, v) = \phi_G^t(\mathcal{P}, v)$, while $\Phi_\mathcal{P}(G)$ stands for the *potential of a graph* G, defined by $\Phi_\mathcal{P}(G) = \max\{\phi_G(\mathcal{P}, v) \mid v \in V(G)\}$. Using the properties of minimal graphs (see Sect. 2) it is possible to prove the following upper bound on $\Gamma_\mathcal{P}(G)$.

Theorem 7. *For every class $\mathcal{P}$ and every graph G it holds*

$$\Gamma_\mathcal{P}(G) \leq \Phi_\mathcal{P}(G) + 1.$$

It is also not hard to prove that for every $\eta > 0$ there exists an infinite number of graphs G for which $\max\{d_G(\mathcal{P}, v) \mid v \in V(G)\} - \Phi_\mathcal{P}(G) > \eta$.

Acknowledgements. Special thanks to D. Dereniowski, E. Drgas-Burchardt and anonymous referees for their remarks on preliminary version of this paper, and to S. Vishwanathan for sending the manuscript [10].

References

1. Araujo, J., Linhares-Sales, C.: On the Grundy number of graphs with few P_4's. Discret. Appl. Math. **160**(18), 2514–2522 (2012). doi:10.1016/j.dam.2011.08.016
2. Borowiecki, P., Rautenbach, D.: New potential functions for greedy independence and coloring. Discret. Appl. Math. **182**, 61–72 (2015). doi:10.1016/j.dam.2013.12.011
3. Borowiecki, P., Sidorowicz, E.: Dynamic coloring of graphs. Fund. Inform. **114**(2), 105–128 (2012). doi:10.3233/FI-2012-620
4. Brown, J.I.: The complexity of generalized graph colorings. Discret. Appl. Math. **69**(3), 257–270 (1996). doi:10.1016/0166-218X(96)00096-0
5. Choudum, S.A., Karthick, T.: First-Fit coloring of $\{P_5, K_4 - e\}$-free graphs. Discret. Appl. Math. **158**(6), 620–626 (2010). doi:10.1016/j.dam.2009.12.009
6. Christen, C.A., Selkow, S.M.: Some perfect coloring properties of graphs. J. Comb. Theory Ser. B **27**(1), 49–59 (1979). doi:10.1016/0095-8956(79)90067-4
7. Cockayne, E.J., Miller, G.G., Prins, G.: An interpolation theorem for partitions which are complete with respect to hereditary properties. J. Comb. Theory Ser. B **13**, 290–297 (1972)
8. Garey, M.R., Johnson, D.S., Stockmeyer, L.J.: Some simplified NP-complete graph problems. Theoret. Comput. Sci. **1**(3), 237–267 (1976)
9. Golovach, P.A., Johnson, M., Paulusma, D., Song, J.: A survey on the complexity of coloring graphs with forbidden subgraphs. J. Graph Theory **84**, 331–363 (2017)
10. Goyal, N., Vishwanathan, S.: NP-completeness of undirected Grundy numberings and related problems (1998, unpublished manuscript)

11. Gyárfás, A., Király, Z., Lehel, J.: On-line 3-chromatic graphs II. Critical graphs. Discret. Math. **177**(1–3), 99–122 (1997). doi:10.1016/S0012-365X(96)00359-7
12. Gyárfás, A., Lehel, J.: First-Fit and on-line chromatic number of families of graphs. Ars Comb. **29C**, 168–176 (1990)
13. Hedetniemi, S.M., Hedetniemi, S.T., Beyer, T.: A linear algorithm for the Grundy (coloring) number of a tree. Congr. Numer. **36**, 351–363 (1982)
14. Kierstead, H.A.: Coloring graphs on-line. In: Fiat, A., Woeginger, G.J. (eds.) Online Algorithms. LNCS, vol. 1442, pp. 281–305. Springer, Heidelberg (1998). doi:10.1007/BFb0029574
15. Kortsarz, G.: A lower bound for approximating Grundy numbering. Discret. Math. Theor. Comput. Sci. **9**, 7–22 (2007)
16. Narayanaswamy, N.S., Babu, R.S.: A note on first-fit coloring of interval graphs. Order **25**(1), 49–53 (2008). doi:10.1007/s11083-008-9076-6
17. Zaker, M.: Results on the Grundy chromatic number of graphs. Discret. Math. **306**(23), 3166–3173 (2006). doi:10.1016/j.disc.2005.06.044
18. Zaker, M.: New bounds for the chromatic number of graphs. J. Graph Theory **58**(2), 110–122 (2008). doi:10.1002/jgt.20298

Maximum Edge Bicliques in Tree Convex Bipartite Graphs

Hao Chen and Tian Liu[✉]

Key Laboratory of High Confidence Software Technologies (MOE),
Institute of Software, School of Electronic Engineering and Computer Science,
Peking University, Beijing 100871, China
lt@pku.edu.cn

Abstract. We show that the computational complexity of the maximum edge biclique (MEB) problem in tree convex bipartite graphs depends on the associated trees. That is, MEB is $\mathcal{NP}$-complete for star convex bipartite graphs, but polynomial time solvable for tree convex bipartite graphs whose associated trees have a constant number of leaves. In particular, MEB is polynomial time solvable for triad convex bipartite graphs. Moreover, we show that the same algorithm strategy may not work for circular convex bipartite graphs, and triad convex bipartite graphs are incomparable with respect to chordal bipartite graphs.

Keywords: Maximum edge biclique · Tree convex bipartite graphs · Star convex bipartite graphs · Triad convex bipartite graphs · $\mathcal{NP}$-completeness · Polynomial-time

1 Introduction

In the maximum edge biclique (MEB) problem, a bipartite graph $G = (A, B, E)$ and a positive integer k are given, the question is to decide whether there exist two subsets $R \subseteq A$ and $S \subseteq B$ such that, $|R| * |S| \geq k$ and (R, S) induces a complete bipartite subgraph (biclique) in G. MEB was firstly introduced in [6] and shown $\mathcal{NP}$-complete for bipartite graphs in [27]. MEB has many applications in molecular biology, web community discovery, manufacturing optimization, text mining, and conjunctive clustering, see e.g. [1,11,26]. Polynomial time algorithms for MEB in convex bipartite graphs and chordal bipartite graphs were developed in [2,7,8,10,26]. An algorithm of MEB for random graphs was developed in [11]. The (in)approximability of MEB was also investigated in [3,9,12].

In a tree convex bipartite graph $G = (A, B, E)$, there is an associated tree $T = (A, F)$ such that, for each vertex b in B, its neighborhood $N_G(b)$ induces a subtree on T. When the associated tree T is a *star* (a number of edges with a common end) or a *triad* (three paths with a common end), the tree convex bipartite graph is called *star convex* or *triad convex*, respectively. Tree convex

---

T. Liu—Partially supported by Natural Science Foundation of China (Grant Nos. 61370052 and 61370156).

M. Xiao and F. Rosamond (Eds.): FAW 2017, LNCS 10336, pp. 47–55, 2017.
DOI: 10.1007/978-3-319-59605-1_5

bipartite graphs were introduced in [14] as a generalization of convex bipartite graphs. In a convex bipartite graph, the associated tree is a path [13].

Many graph problems are still $\mathcal{NP}$-complete for bipartite graphs, but tractable for convex bipartite graphs, such as the minimum feedback vertex set (FVS), the minimum dominating set, treewidth, hamiltonicity, etc. For tree convex bipartite graphs, it turns out that the computational complexity of these problems depends on the associated trees. For example, star convex bipartite graphs were introduced in [14], triad convex bipartite graphs were introduced in [16], and it was shown that FVS is still $\mathcal{NP}$-complete for star convex bipartite graphs [14,31], but tractable for triad convex bipartite graphs [16]. Similar results for the minimum dominating set and its variants such as independent dominating set or connected dominating sets, as well as for the treewidth and hamiltonicity, were obtained in [5,20,22,24,28–30]. (In)approximability of the minimum dominating set for star convex bipartite graphs was also obtained in [28].

Besides the above mentioned graph problems, when taken tree convex bipartite graphs as hypertrees or tree convex set systems, similar results for the minimum set cover, the minimum hitting set and the maximum set packing were obtained in [23]. The union-closed sets conjecture was shown to hold for tree convex sets [21].

In this paper, we show that the computational complexity of the maximum edge biclique (MEB) problem in tree convex bipartite graphs depends on the associated trees. That is, MEB is $\mathcal{NP}$-complete for star convex bipartite graphs, but polynomial time solvable for tree convex bipartite graphs whose associated trees have a constant number of leaves. In particular, MEB is polynomial time solvable for triad convex bipartite graphs. Moreover, we show that the same algorithm strategy may not work for circular convex bipartite graphs, and triad convex bipartite graphs are incomparable with respect to chordal bipartite graphs.

This paper is structured as follows. After introducing necessary definitions and facts in Sect. 2, the $\mathcal{NP}$-completeness of MEB for star convex bipartite graph classes is shown in Sect. 3, the tractability of MEB for tree convex bipartite graphs whose associated trees have a constant number of leaves and for triad convex bipartite graphs is shown in Sect. 4, the comparison between triad convex bipartite graphs and chordal bipartite graphs is shown in Sect. 5 and finally are concluding remarks in Sect. 6.

2 Preliminaries

A graph $G = (V, E)$ has a vertex set V and an edge set E. We use $V(G)$ to denote the vertex set of G, and $E(G)$ the edge set of G. Each edge $e = (u, v)$ has its two ends u and v in V, and these two ends u and v are *adjacent*. The neighborhood of a vertex x, denoted by $N_G(x)$, is the set of all adjacent vertices to x. For a graph $G = (V, E)$ and a subset $U \subseteq V$, we use $G[U]$ to denote the induced subgraph $(U, \{(u, v) \in E | u \in U, v \in U\})$. In a *complete* graph, every two vertices are adjacent. A *clique* in a graph is an induced complete subgraph.

A *bipartite* graph $G = (A, B, E)$ has a bipartition $A \cup B = V$ with no adjacent vertices in A (B, respectively). For a bipartite graph $G = (A, B, E)$

and two subset $R \subseteq A$ and $S \subseteq B$, we use $G[R, S]$ to denote the induced bipartite subgraph $(R, S, \{(u, v) \in E | u \in R, v \in S\})$. In a *complete* bipartite graph $G = (A, B, E)$, every vertex in A is adjacent to every vertex in B. A *biclique* in a bipartite graph is an induced complete bipartite subgraph.

A *path* is a vertex sequence with every two consecutive vertices adjacent. A *cycle* is a path where the first vertex is equal to the last vertex in the path. A graph is *connected* if every two vertices are connected by a path. A *tree* is a connected cycle-less graph. Two special kinds of trees are stars and triads. A *star* is a set of edges with a common end called its center. A *triad* is three paths with a common end also called its center.

Given a cycle, a *chord* is an edge whose two endpoints are not consecutive on the cycle. In a *chordal bipartite* graph, every cycle of length at least six has a chord. It is known that all convex bipartite graphs are chordal bipartite graphs, and all chordal bipartite graphs are tree convex bipartite graphs [4,15]. In a *circular convex* bipartite graph $G = (A, B, E)$, there is an associated circular ordering on A such that, for each vertex b in B, its neighborhood $N_G(b)$ induces an arc or interval on A under the circular ordering [17,19,25].

3   Hardness

In this section, MEB is shown $\mathcal{NP}$-complete for star convex bipartite graphs.

Theorem 1. *MEB is $\mathcal{NP}$-complete for star convex bipartite graphs.*

Proof. The proof in [27], showing $\mathcal{NP}$-completeness of MEB in bipartite graphs, made a reduction from CLIQUE to MEB and produced a bipartite graph which is already a star convex bipartite graph. For completeness, we repeat the reduction here and check the star convexity of the produced bipartite graph.

Given an instance (G, k) of CLIQUE, we can assume that $G = (V, E)$ and, without loss of generality, $k = \frac{1}{2}|V|$. An instance (G', k') of MEB is constructed by $G' = (A, B, E')$, $A = V$, $B = E \cup W$, $|W| = \frac{1}{2}k^2 - k$, $k' = k^3 - \frac{3}{2}k^2$, and

$$E' = \{(v, e) | v \in V, e \in E, v \notin e\} \cup \{(v, w) | v \in V, w \in W\}.$$

The correctness of this reduction was shown in [27]. To see that G' is a star convex bipartite graph with an associated star T on B, note that each vertex w in W is adjacent to every vertex in A, thus we can take any w in W as the center of T and make all vertices in $B \setminus \{w\}$ the leaves of T. The proof is finished. $\square$

Since star convex bipartite graphs is a subclass of tree convex bipartite graphs, MEB is also $\mathcal{NP}$-complete for tree convex bipartite graphs. Another subclass of bipartite graphs is the so-called perfect elimination bipartite graphs, which is incomparable with respect to tree convex bipartite graphs by the results in [18]. We note that $\mathcal{NP}$-completeness of MEB for perfect elimination bipartite graphs or for the so-called comb convex bipartite graphs [29] is still unknown.

4   Tractability

In this section, MEB is shown polynomial time solvable for tree convex bipartite graphs whose associated trees are given and have a bounded number of leaves. In particular, MEB is polynomial time solvable for triad convex bipartite graphs.

To this end, we first show a structural property of optimal solutions of MEB for tree convex bipartite graphs. Recall that a solution of an instance (G, k) of MEB is a pair of sets (R, S), such that $R \subseteq A$, $S \subseteq B$, $|R| * |S| \geq k$, and $G[R, S]$ is a biclique, where $G[R, S]$ is the subgraph of G induced by (R, S). We call a solution (R, S) to be *optimal*, if $|R| * |S|$ is maximized among all the solutions.

Lemma 1. *If (R, S) is an optimal solution to an instance (G, k) of MEB and $G = (A, B, E)$ is a tree convex bipartite graph, with an associated tree $T = (A, F)$, then R is a vertex set of a subtree in T, that is, $V(T[R]) = R$.*

Proof. Recall that $R \subseteq A$ and $S \subseteq B$. For any two vertices x, y in R, there is a unique path P_{xy} in T connecting x and y. For every w in S, $N_G(w)$ induces a subtree $T[N_G(w)]$ in T, which contains both x and y. So any vertex z in P_{xy} is also in $N_G(w)$ for all w in S, and $(R \cup \{z\}, S)$ is also a solution to (G, k). If z is not in R, then $|R \cup \{z\}| * |S| > |R| * |S|$, a contradiction to the optimality of (R, S). Thus, R is already a vertex set of a subtree in T.

The proof is finished.   □

Lemma 2. *In a tree with bounded number of leaves, the number of subtrees is bounded by a polynomial.*

Proof. The number of leaves in a subtree will never exceeds the number of leaves in the tree. A subtree is uniquely determined by listing all its leaves. Indeed, a subtree is just the union of all pairs shortest paths between its leaves. If a tree with n vertices has $L = O(1)$ leaves, then the number of its subtrees is at most $O(n^L) = n^{O(1)}$. The proof is finished.   □

Theorem 2. *MEB is polynomial time solvable in tree convex bipartite graphs whose associated trees are given and have a constant number of leaves.*

Proof. A polynomial time algorithm based on enumeration is as follows.

Input: (G, k), where $G = (A, B, E)$ is a tree convex bipartite graph with an associated tree T on A. The number of leaves of T is a constant L.

Output: Yes, if there is (R, S) such that $G[R, S]$ is a biclique and $|R| * |S| \geq k$; No, otherwise.

Algorithm:

1. Enumerate all subtrees of T;
2. For each subtree with vertex set R, let $S_R = \{w \in B | R \subseteq N_G(w)\}$.
3. Record (R, S_R) with the maximum $|R| * |S_R|$.
4. Return Yes, if $|R| * |S_R| \geq k$; No, otherwise.

To enumerate all subtrees of T, we first enumerate and record all pairs shortest paths of T in $O(n^3)$ time, where n is the input size, in any standard way. Note that for a tree T, the shortest path P_{xy} between two vertices x and y in T is also the unique path connecting x and y in T. Then we enumerate all subsets A' of size at most L of A in $O(n^L)$ time, where L is the number of leaves of T. For each subset A' of A, we compute the union of all shortest paths P_{xy} in T for all pairs x and y in A' in $O(nL^2)$ time, since we have $O(L^2)$ paths and each path has length $O(n)$. This union is a subtree T' of T, and any subtree T' of T is obtained in this way by putting all leaves of T' into A'.

For each subtree T' with vertex set R, we can find the set S_R and compute $|R| * |S_R|$ in $O(n^2)$ time. So the total running time is $O(n^{L+2})$.

The correctness of this algorithm is guaranteed by Lemmas 1 and 2. The proof is finished. □

Theorem 3. *MEB is $O(n^5)$ time solvable in triad convex bipartite graphs.*

Proof. By Lemma 2 and the fact that all triads have three leaves. □

The number of subtrees in a star with n vertices is $O(2^n)$, and this seems to be a reason for the hardness of MEB for star convex bipartite graphs. We guess that MEB is also hard for comb convex bipartite graphs by the same reason, that is, an exponential number of subtrees.

We note that Lemma 1 does not hold for circular convex bipartite graphs. We construct a circular convex bipartite graph G by $G = (A, B, E)$, where $A = \{a_1, a_2, a_3, a_4\}$, $B = \{b_1, b_2, b_3\}$, $E = \{(a_i, b_j) | 1 \le i \le 4, 1 \le j \le 3\} \setminus \{(a_1, b_3), (a_3, b_1)\}$, and the associated circular ordering on A is T, as shown in Fig. 1. The maximum biclique in G is induced by $R = \{a_2, a_4\}$ and $S = B$, but R does not induce an interval in T. Due to this break down of the connectedness of the optimal solutions of MEB for circular convex bipartite graphs, the above enumeration algorithm does not work for circular convex bipartite graphs. We note that it is still unknown whether MEB is polynomial time solvable for circular convex bipartite graphs.

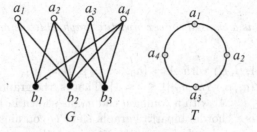

Fig. 1. A circular convex bipartite graph G whose optimal solution does not induce an interval in T.

5 Comparison

In this section, we show that triad convex bipartite graphs are incomparable with respect to chordal bipartite graphs.

Lemma 3. *There is a triad convex bipartite graph which is not a chordal convex bipartite graph.*

Proof. Let $G = (A, B, E)$ with $A = \{a_0, a_1, a_2, a_3\}$, $B = \{b_1, b_2, b_3\}$ and $E = \{(b_i, a_0), (a_i, b_i), (a_i, b_{i-1 \bmod 3}) | 1 \le i \le 3\}$. That is, the graph G is a cycle $C = a_1 b_1 a_2 b_2 a_3 b_3 a_1$ plus a star with center a_0 and leaves b_1, b_2, b_3, as shown in Fig. 2.

Apparently, G is not a chordal bipartite graph, since G has a cycle C of length 6 but without a chordal. We can easily check that G is a triad convex bipartite graph, with the associated triad T, where $V(T) = \{a_0, a_1, a_2, a_3\}$ and $E(T) = \{(a_0, a_1), (a_0, a_2), (a_0, a_3)\}$, respectively. The proof is finished. □

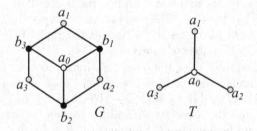

Fig. 2. A triad convex bipartite graph G which is not a chordal bipartite graph.

We note that this graph G is used to separate tree convex bipartite graphs from chordal bipartite graphs in [18]. Actually, G separates triad convex bipartite graphs and star convex bipartite graphs from chordal bipartite graphs, as shown above.

Lemma 4. *There is a chordal convex bipartite graph which is not a triad convex bipartite graph.*

Proof. Let $G = (A, B, E)$ with $A = \{a_0, \ldots, a_4\}$, $B = \{u_1, \ldots, u_4, w_1, \ldots, w_4\}$ and $E = \{(w_i, a_i), (a_i, u_i), (u_i, a_0) | 1 \le i \le 4\}$. That is, the graph G is four paths $P_i = w_i a_i u_i a_0$ $(1 \le i \le 4)$ with a common end a_0, as shown in Fig. 3.

Apparently, G is a chordal bipartite graph, since G contains no cycle at all. To show that G is not a triad convex bipartite graph, by symmetry of G, it is enough to show that none of w_1, a_1, u_1, a_0 is the center of the associated triad T. Indeed, if w_1 is the center of T, since $N_G(a_0) = \{u_1, u_2, u_3, u_4\}$, then u_1, u_2, u_3, u_4 must be grouped into a subpath P in T. Since $N_G(a_i) = \{w_i, u_i\}$ for $2 \le i \le 4$, each of w_2, w_3, w_4 must be consecutive to the path P, but this is impossible,

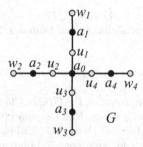

Fig. 3. A chordal bipartite graph G which is not a triad convex bipartite graph.

since P only has two ends. If a_1 is the center of T, since $N_G(u_i) = \{a_i, a_0\}$ for $2 \le i \le 4$, then each of a_2, a_3, a_4 must be consecutive to a_0 in T, but this is impossible, since a_0 has at most two neighbors in T. If u_1 is the center of T, then by Pigeonhole Principle, two of w_1, u_2, u_3, u_4 must be on a subpath of T, but this is impossible, since $N_G(a_1) = \{w_1, u_1\}$ and $N_G(a_0) = \{u_1, u_2, u_3, u_4\}$, any two of w_1, u_2, u_3, u_4 must be consecutive to u_1 simultaneously and thus can not be on a subpath of T. If a_0 is the center of T, then by Pigeonhole Principle again, two of a_1, a_2, a_3, a_4 must be on a subpath of T, but this is impossible, since $N_G(u_i) = \{a_i, a_0\}$ for $1 \le i \le 4$, each of a_1, a_2, a_3, a_4 must be consecutive to a_0 in T. The proof is finished. $\qquad\square$

6   Conclusions

We have shown that MEB is $\mathcal{NP}$-complete for star convex bipartite graphs, but polynomial time solvable for tree convex bipartite graphs whose associated trees have a constant number of leaves. In particular, MEB is $O(n^3)$ polynomial time solvable for triad convex bipartite graphs. We also have show that the enumeration algorithm may not work for circular convex bipartite graphs, and triad convex bipartite graphs are incomparable with respect to chordal bipartite graphs.

We list some open problems. First, it is unknown that whether MEB is $\mathcal{NP}$-complete for comb convex bipartite graphs. Second, it is unknown that whether MEB is tractable for circular convex bipartite graphs. Third, it is unknown that whether the $O(n^5)$ time bound of MEB for triad convex bipartite graphs can be lowered to $O(n(\log n)^k)$ or $O(n^2(\log n)^k)$.

Acknowledgments. We thank the unknown reviewers whose comments are helpful to improve our presentations.

References

1. Acuña, V., Ferreira, C.E., Freire, A.S., Moreno, E.: Solving the maximum edge biclique packing problem on unbalanced bipartite graphs. Discret. Appl. Math. **164**, 2–12 (2014)

2. Alexe, G., Alexe, S., Crama, Y., Foldes, S., Hammer, P.L., Simeone, B.: Consensus algorithms for the generation of all maximal bicliques. Discret. Appl. Math. **145**(1), 11–21 (2004)
3. Ambühl, C., Mastrolilli, M., Svensson, O.: Inapproximability resullts for maximum edge biclique, minimum linear arrangement, and sparse cut. SIAM J. Comput. **40**(2), 567–596 (2011)
4. Brandstad, A., Le, V.B., Spinrad, J.P.: Graph Classes - A Survey. Society for Industrial and Applied Mathematics, Philadelphia (1999)
5. Chen, H., Lei, Z., Liu, T., Tang, Z., Wang, C., Xu, K.: Complexity of domination, hamiltonicity and treewidth for tree convex bipartite graphs. J. Comb. Optim. **32**(1), 95–110 (2016)
6. Dawande, M., Keskinocak, P., Swaminathan, J.M., Tayur, S.: On bipartite and multipartite clique problems. J. Algorithms **41**, 388–403 (2001)
7. Dias, V.M., de Figueiredo, C.M., Szwarcfiter, J.L.: Generating bicliques of a graph in lexicographic order. Theoret. Comput. Sci. **337**(1C3), 240–248 (2005)
8. Dias, V.M., de Figueiredo, C.M., Szwarcfiter, J.L.: On the generation of bicliques of a graph. Discret. Appl. Math. **155**(14), 1826–1832 (2007)
9. Feige, U.: Relations between average case complexity and approximation complexity. In: Proceedings of STOC 2002, pp. 534–543 (2002)
10. Gély, A., Nourine, L., Sadi, B.: Enumeration aspects of maximal cliques and bicliques. Discret. Appl. Math. **157**(7), 1447–1459 (2009)
11. Gillis, N., Glineur, F.: A continuous characterization of the maximum edge biclique problem. J. Glob. Optim. **58**, 439–464 (2014)
12. Goerdt, A., Lanka, A.: An approximation hardness result for bipartite clique. Electronic Colloquium on Computation Complexity TR2004-048 (2004)
13. Grover, F.: Maximum matching in a convex bipartite graph. Nav. Res. Logist. Q. **14**, 313–316 (1967)
14. Jiang, W., Liu, T., Ren, T., Xu, K.: Two hardness results on feedback vertex sets. In: Atallah, M., Li, X.-Y., Zhu, B. (eds.) FAW-AAIM 2011. LNCS, vol. 6681, pp. 233–243. Springer, Heidelberg (2011). doi:10.1007/978-3-642-21204-8_26
15. Jiang, W., Liu, T., Wang, C., Xu, K.: Feedback vertex sets on restricted bipartite graphs. Theoret. Comput. Sci. **507**, 41–51 (2013)
16. Jiang, W., Liu, T., Xu, K.: Tractable feedback vertex sets in restricted bipartite graphs. In: Wang, W., Zhu, X., Du, D.-Z. (eds.) COCOA 2011. LNCS, vol. 6831, pp. 424–434. Springer, Heidelberg (2011). doi:10.1007/978-3-642-22616-8_33
17. Liang, Y.D., Blum, N.: Circular convex bipartite graphs: maximum matching and Hamiltonian circuits. Inf. Process. Lett. **56**, 215–219 (1995)
18. Liu, T.: Restricted bipartite graphs: comparison and hardness results. In: Gu, Q., Hell, P., Yang, B. (eds.) AAIM 2014. LNCS, vol. 8546, pp. 241–252. Springer, Cham (2014). doi:10.1007/978-3-319-07956-1_22
19. Liu, T., Lu, M., Lu, Z., Xu, K.: Circular convex bipartitegraphs: feedback vertex sets. Theoret. Comput. Sci. **556**, 55–62 (2014)
20. Liu, T., Lu, Z., Xu, K.: Tractable connected domination for restricted bipartite graphs. J. Comb. Optim. **29**(1), 247–256 (2015)
21. Liu, T., Xu, K.: Union closed tree convex sets. In: Wang, J., Yap, C. (eds.) FAW 2015. LNCS, vol. 9130, pp. 198–203. Springer, Cham (2015). doi:10.1007/978-3-319-19647-3_19
22. Lu, M., Liu, T., Xu, K.: Independent domination: reductions from circular- and triad-convex bipartite graphs to convex bipartite graphs. In: Fellows, M., Tan, X., Zhu, B. (eds.) AAIM/FAW -2013. LNCS, vol. 7924, pp. 142–152. Springer, Heidelberg (2013). doi:10.1007/978-3-642-38756-2_16

23. Lu, M., Liu, T., Tong, W., Lin, G., Xu, K.: Set cover, set packing and hitting set for tree convex and tree-like set systems. In: Gopal, T.V., Agrawal, M., Li, A., Cooper, S.B. (eds.) TAMC 2014. LNCS, vol. 8402, pp. 248–258. Springer, Cham (2014). doi:10.1007/978-3-319-06089-7_17
24. Lu, Z., Liu, T., Xu, K.: Tractable connected domination for restricted bipartite graphs (extended abstract). In: Du, D.-Z., Zhang, G. (eds.) COCOON 2013. LNCS, vol. 7936, pp. 721–728. Springer, Heidelberg (2013). doi:10.1007/978-3-642-38768-5_65
25. Lu, Z., Lu, M., Liu, T., Xu, K.: Circular convex bipartite graphs: feedback vertex set. In: Widmayer, P., Xu, Y., Zhu, B. (eds.) COCOA 2013. LNCS, vol. 8287, pp. 272–283. Springer, Cham (2013). doi:10.1007/978-3-319-03780-6_24
26. Nussbaum, D., Pu, S., Sack, J., Uno, T., Zarrabi-Zadeh, H.: Finding maximum edge bicliques in convex bipartite graphs. Algorithmica **64**, 311–325 (2012)
27. Peeters, R.: The maximum edge biclique problem is NP-complete. Discret. Appl. Math. **131**, 651–654 (2003)
28. Pandey, A., Panda, B.S.: Domination in some subclasses of bipartite graphs. In: Ganguly, S., Krishnamurti, R. (eds.) CALDAM 2015. LNCS, vol. 8959, pp. 169–180. Springer, Cham (2015). doi:10.1007/978-3-319-14974-5_17
29. Song, Y., Liu, T., Xu, K.: Independent domination on tree convex bipartite graphs. In: Snoeyink, J., Lu, P., Su, K., Wang, L. (eds.) AAIM/FAW -2012. LNCS, vol. 7285, pp. 129–138. Springer, Heidelberg (2012). doi:10.1007/978-3-642-29700-7_12
30. Wang, C., Chen, H., Lei, Z., Tang, Z., Liu, T., Xu, K.: Tree convex bipartite graphs: $\mathcal{NP}$-complete domination, hamiltonicity and treewidth. In: Chen, J., Hopcroft, J.E., Wang, J. (eds.) FAW 2014. LNCS, vol. 8497, pp. 252–263. Springer, Cham (2014). doi:10.1007/978-3-319-08016-1_23
31. Wang, C., Liu, T., Jiang, W., Xu, K.: Feedback Vertex Sets on Tree Convex Bipartite Graphs. In: Lin, G. (ed.) COCOA 2012. LNCS, vol. 7402, pp. 95–102. Springer, Heidelberg (2012). doi:10.1007/978-3-642-31770-5_9

Complete Submodularity Characterization in the Comparative Independent Cascade Model

Wei Chen[1] and Hanrui Zhang[2(✉)]

[1] Microsoft Research, Beijing, China
weic@microsoft.com
[2] Tsinghua University, Beijing, China
zhang-hr13@mails.tsinghua.edu.cn, segtree@gmail.com

Abstract. We study the propagation of comparative ideas in social network. A full characterization for submodularity in the comparative independent cascade (Com-IC) model of two-idea cascade is given, for competing ideas and complementary ideas respectively. We further introduce One-Shot model where agents show less patience toward ideas, and show that in One-Shot model, only the stronger idea spreads with submodularity.

1 Introduction

Propagation of information in social networks has been extensively studied over the past decades, along with its most prominent algorithmic aspect - influence maximization. The cascade procedure of ideas in a network is usually modeled by a stochastic process, and influence maximization seeks to maximize the expected influence of a certain idea by choosing k agents (the seed set) in the network to be early adopters of the idea. The seed set then initiates the propagation through the network structure.

Influence maximization is proven to be NP-hard [7] in almost any non-trivial setting. Most research therefore focuses on approximation algorithms, some particularly successful ones out of which are based on the celebrated $(1 - \frac{1}{e})$-approximate submodular maximization [11]. Submodularity of influence in the seed set therefore plays a central role in such optimization.[1]

Nevertheless, submodularity appears harder to tract when there are multiple ideas interacting with each other. Most prior work focuses on single-idea cascade, or completely competing propagation of ideas. These models somewhat fails in modeling real world behavior of agents. Lu et al. [9] introduce a general model called *comparative independent cascade (Com-IC) model*, which covers the entire spectrum of two item cascades from full competition to full complementarity. This full spectrum is crucially characterized by four probability parameters called *global adoption probabilities (GAP)*, and their space is called the GAP space. However, they only provide submodularity analysis in a few marginal cases of

---

[1] We say a function $f : 2^U \to \mathbb{R}$ is submodular, if for any $S \subseteq U, a, b \in U, f(S) + f(S \cup \{a, b\}) \leq f(S \cup \{a\}) + f(S \cup \{b\})$.

© Springer International Publishing AG 2017
M. Xiao and F. Rosamond (Eds.): FAW 2017, LNCS 10336, pp. 56–67, 2017.
DOI: 10.1007/978-3-319-59605-1_6

the entire GAP space, and a full submodularity characterization for the entire
GAP space is left as an open problem discussed in their conclusion section.

Our Contribution. In this paper, we provide a full characterization of the submodularity of the Com-IC model in both the mutually competing case and the
mutually complementary case (Theorems 1, 2, and 3). Our results show that in
the entire continuous GAP space, the parameters satisfying submodularity only
has measure zero. Next, we introduce a slightly modified One-Shot model for
the mutual competing case where agents are less patient: they would reject the
second item if they get influenced but failed to adopt the first item. We provide
the full submodularity characterization of the parameter space for this model
(Theorem 4), which contains a nontrivial half space satisfying submodularity,
contrasting the result for the Com-IC model. Our techniques for establishing
these characterization results may draw separate interests from the technical
aspect for the study of submodularity for various influence propagation models.

Related Work. Single-idea models, where there is only one propagating entity for
social network users to adopt, has been thoroughly studied. Some examples are
the classic Independent Cascade (IC) and Linear Thresholds (LT) models [7].
Some other work studies pure competition between ideas. See, e.g. [1–4,6,8].
Beside competing settings, Datta et al. [5] study influence maximization of independently propagating ideas, and Narayanam and Nanavati [10] discuss a per-
fectly complementary setting, which is extended in [9].

2 The Model

We first recapitulate the independent cascade model for comparative ideas
(Com-IC).

First recall that in the classic Independent Cascade (IC) model, the social
network is described by a directed graph $G = (V, E, p)$ with probabilities
$p : E \rightarrow [0, 1]$ on each edge. Each vertex in V stands for an agent, an edge
for a connection, whose strength is characterized by the associated probability.
Cascading proceeds at each time step $0, 1, \ldots$. At time 0, only the seed set is
active. At time t, each vertex u activated at time $t - 1$ tries to activate its neighbor v, and succeeds with probability $p(u, v)$. The procedure ends when no new
vertices are activated at some time step.

In comparative IC (Com-IC henceforth) model, there are two ideas, A and B,
spreading simultaneously in the network, and therefore 9 states of each vertex:

$$\{A\text{-idle}, A\text{-adopted}, A\text{-rejected}\} \times \{B\text{-idle}, B\text{-adopted}, B\text{-rejected}\}.$$

When an A-proposal reaches an A-idle vertex u, if u is previously B-adopted,
it adopts A w.p. $q_{A|B}$. Otherwise, it adopts A w.p. $q_{A|\emptyset}$. The rules for idea B
is totally symmetric. The four probabilities, $q_{A|\emptyset}, q_{B|\emptyset}, q_{A|B}, q_{B|A}$, therefore fully
characterize strengths of the two ideas and the relationship between them: when
A and B are mutually competing ideas, $q_{A|\emptyset} \geq q_{A|B}$ and $q_{B|\emptyset} \geq q_{B|A}$; when they
are mutually complementary ideas, $q_{A|\emptyset} \leq q_{A|B}$ and $q_{B|\emptyset} \leq q_{B|A}$. These four

probability parameters are referred as global adoption probabilities (GAP), and their space as the GAP space.

For tie-breaking, we generate a random ordering of all in-going edges for each vertex, and let proposals which reach at the same time try according to that order. If a vertex adopts two ideas at a same time step, it proposes the two ideas to its neighbors in the order adopted. We refer interested readers to [9] for more details of Com-IC model.

3   Notations

Let the set of possible worlds (the complete state of the network and vertices after fixing all randomness) be $\mathcal{W}$. For a possible world $W \in \mathcal{W}$, A-seed set S_A and B-seed set S_B (unless otherwise specified), let $\sigma_A(S_A, S_B, W)$ (resp. $\sigma_B(S_A, S_B, W)$) be the number of vertices which adopt A (resp. B) at the end of cascading in possible world W. $\sigma_A(S_A, S_B) = \mathbb{E}[\sigma_A(S_A, S_B, W)]$ (resp. $\sigma_B(S_A, S_B) = \mathbb{E}[\sigma_B(S_A, S_B, W)]$) then stands for the expected influence of A (resp. B) after cascading. Similarly, let $\sigma_A^u(S_A, S_B, W)$ be 1 if A affects u in W, and 0 if not, and $\sigma_A^u(S_A, S_B) = \mathbb{E}[\sigma_A^u(S_A, S_B, W)]$ the probability that A affects u. Parameters are ignored when in clear context.

4   Submodularity in the Mutually Competing Case

Recall that when the two ideas are competing, we have $q_{A|\emptyset} \geq q_{A|B}, q_{B|\emptyset} \geq q_{B|A}$. We are naturally interested in submodularity of $\sigma_A(S_A, S_B)$ in S_A fixing S_B. It turns out that this kind of submodularity is guaranteed only in a 0-measure subset of the parameter space. Formally, we have the following theorem:

Theorem 1 (Submodularity Characterization for the Mutually Competing Case). *When the two ideas are mutually competing, for a fixed S_B, σ_A is submodular in S_A whenever one of the following holds:*

- *$q_{A|\emptyset} \in \{0,1\}$,*
- *$q_{B|\emptyset} = 0$,*
- *$q_{A|\emptyset} = q_{A|B}$,*
- *$q_{B|\emptyset} = q_{B|A}$.*

And when none of these conditions hold, submodularity is violated, i.e., there exists (G, S_A, S_B, u, v) such that for each group of $(q_{A|\emptyset}, q_{B|\emptyset}, q_{A|B}, q_{B|A})$,

$$\sigma_A(S_A, S_B) + \sigma_A(S_A \cup \{u, v\}, S_B) > \sigma_A(S_A \cup \{u\}, S_B) + \sigma_A(S_A \cup \{v\}, S_B).$$

Proof. First we prove the negative (non-submodular) half of the theorem by given an counterexample, illustrated in Fig. 1. The basic seed sets for A and B are $S_A = \{a\}$ and $S_B = \{b\}$ respectively. In order to show non-submodularity, we consider the marginals of u at t when v is an A-seed and when v is not.

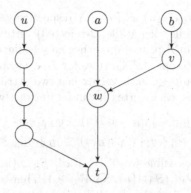

Fig. 1. Counterexample used in the proofs of Theorem 1 and Theorem 2

Note that considering submodularity at a single vertex suffices for establishing a global proof, since we could duplicate the vertex such that it dominates the expected influence. Also, we assume $p(u, v) = 1$ for each $(u, v) \in E$, since all positive (submodularity) proofs can be partially derandomized and done in each partial possible world, and for counterexamples, we simply set the probabilities to be 1.

Formally, define

$$M_1 = \sigma_A^t(S_A \cup \{u\}, S_B) - \sigma_A^t(S_A, S_B),$$
$$M_2 = \sigma_A^t(S_A \cup \{u, v\}, S_B) - \sigma_A^t(S_A \cup \{v\}, S_B).$$

Submodularity is violated if we show $M_1 < M_2$. We now calculate M_1 and M_2 separately. When v is not a seed, u has a marginal at t iff a fails to activate w and idea A succeeds in affecting t from u. That is,

$$M_1 = (1 - q_{A|\emptyset})[(1 - q_{B|\emptyset}^3)q_{A|\emptyset}^4 + q_{B|\emptyset}^3 q_{A|\emptyset}^3 q_{A|B}].$$

Similarly, when v is an A-seed, we have

$$M_2 = (1 - q_{A|\emptyset})[(1 - q_{B|\emptyset}q_{B|A}q_{B|\emptyset})q_{A|\emptyset}^4 + q_{B|\emptyset}q_{B|A}q_{B|\emptyset}q_{A|\emptyset}^3 q_{A|B}].$$

Taking the difference, we get

$$M_2 - M_1 = q_{A|\emptyset}^3 q_{B|\emptyset}^2 (1 - q_{A|\emptyset})(q_{A|B} - q_{A|\emptyset})(q_{B|A} - q_{B|\emptyset}).$$

It is easy to see, when none of the conditions listed in Theorem 1 hold, $M_2 - M_1 > 0$, and σ_A is not submodular in the seed set of A.

We now show case by case, that whenever one of the conditions holds, σ_A is submodular in the seed set of A.

- $q_{A|\emptyset} \in \{0, 1\}$. When $q_{A|\emptyset} = 0, \sigma_A$ is always the size of S_A, so submodularity is obvious. Now suppose $q_{A|\emptyset} = 1$. Consider an equivalent formulation of the model: each vertex u draws two independent numbers uniformly at random from $[0, 1]$, denoted by $\alpha_A(u)$ and $\alpha_B(u)$ respectively. When an A-proposal reaches an (A-idle, B-idle) or (A-idle, B-rejected) vertex u, if $\alpha_A(u) \leq q_{A|\emptyset}, u$ will accept A. When an A-proposal reaches an (A-idle, B-adopte) vertex u, if $\alpha_A(u) \leq q_{A|B}, u$ will accept A. The rules for B are symmetric.

After fixing all randomness, each vertex has two attributes for ideas A and B respectively. That is, each vertex u can be in exactly one state out of

$$\{\alpha_A(u) \leq q_{A|B}, q_{A|B} < \alpha_A(u) \leq q_{A|\emptyset}, q_{A|\emptyset} < \alpha_A(u)\} \times$$
$$\{\alpha_B(u) \leq q_{B|A}, q_{B|A} < \alpha_B(u) \leq q_{B|\emptyset}, q_{B|\emptyset} < \alpha_B(u)\}.$$

We show that in any possible world W, if $\sigma_A^t(S_A \cup \{u, v\}, S_B, W) = 1$, then $\sigma_A^t(S_A \cup \{u\}, S_B, W) + \sigma_A^t(S_A \cup \{v\}, S_B, W) \geq 1$. That is, if t is reachable by A when u and v are both A-seeds, then it is reachable by A when u or v alone is an A-seed. Submodularity then follows from monotonicity of $\sigma_A^t(S_A, S_B, W)$ in S_A and convex combination of possible worlds.

Let $p = (w_1, \ldots, w_k)$ be the A-path which reaches t when u and v are both A-seeds, where w_1 is an A-seed, and $w_k = t$. W.l.o.g. $v \notin p$. We argue that for each $w \in p$, if w is not B-adopted at the time A reaches it when u and v are A-seeds, then w is not B-adopted at the time A reaches it when only u is an A-seed, so p remains A-affected even if v is not an A-seed. Suppose not. Let w be the vertex closest to w_1 on p, which becomes affected by B when v is not a seed, p' be the B-path through which w is affected by B. Let x be the closest vertex to the B-seed on p', which is affected by A at the time the B-proposal reaches when v is an A-seed, and is affected by B when v is not a seed (such a vertex must exist). Then because $q_{A|\emptyset} = 1$, the subpath $[x, t] \subseteq p'$ must be completely A-affected when v is an A-seed, and reaches t earlier than p does, a contradiction.

Now since each vertex $w \in p$ which is not affected by B when v is an A-seed remains not affected when v is not, idea A can pass through the entire path p from some seed vertex to t just like when v is an A-seed, so t is still A-affected. In other words, w.l.o.g. $\sigma_A^t(S_A \cup \{u\}, S_B, W) = 1$.

- $q_{B|\emptyset} = 0$. B does not propagate at all. We simply remove S_B from the graph and consider the equivalent IC procedure of A alone. Submodularity is then easy.
- $q_{A|\emptyset} = q_{A|B}$. B does not affect the propagation of A. Again the propagation of A is equivalent as an IC procedure, and submodularity follows directly.
- $q_{B|\emptyset} = q_{B|A}$. We use the possible world model discussed in the first case, where $q_{A|\emptyset} \in \{0, 1\}$. Still, let $p = \{w_1, \ldots, w_k\}$ be the path through which t is affected by A when both u and v are A-seeds, and w.l.o.g. $v \notin p$. We apply induction on i to prove that A reaches w_i still at the $(i-1)$-th time slot when v is not an A-seed.

When $i = 1$, the statement holds evidently as w_1 is an A-seed. Assume at time $i - 1, w_i$ has just been reached by A and become A-adopted. Since the propagation of B is not affected by the A seed set, w_{i+1} is in the same state w.r.t. B as when v is a seed, so the A-proposal to w_{i+1} from w_i ends up just in the same way, and w_{i+1} becomes A-adopted at time i. And as a result, t is eventually A-adopted, i.e. $\sigma_A^t(S_A \cup \{u\}, S_B, W) = 1$.

5 Submodularity in the Mutually Complimentary Case

When the two ideas are complementary, i.e. when $q_{A|\emptyset} \leq q_{A|B}$ and $q_{B|\emptyset} \leq q_{B|A}$, enlarging the seed set of one idea helps the propagation of both the idea itself and that of the other idea. We discuss in this section the self and cross effect of the seed set of an idea.

5.1 Self Submodularity

Fixing S_B, we are interested in submodularity of σ_A in S_A, i.e., submodularity of the influence of some idea w.r.t. its own seed set, fixing the seed set of the other idea.

Theorem 2 (Self-Submodularity Characterization for the Mutually Complementary Case). *When the two ideas are complementary, for a fixed* S_B, σ_A *is submodular in* S_A *whenever one of the following holds:*

- $q_{A|\emptyset} \in \{0,1\}$,
- $q_{B|\emptyset} = 0$,
- $q_{A|\emptyset} = q_{A|B}$,
- $q_{B|\emptyset} = q_{B|A}$.

And when none of these conditions hold, submodularity is violated, i.e., there exists (G, S_A, S_B, u, v) *such that for each group of* $(q_{A|\emptyset}, q_{B|\emptyset}, q_{A|B}, q_{B|A})$,

$$\sigma_A(S_A, S_B) + \sigma_A(S_A \cup \{u,v\}, S_B) > \sigma_A(S_A \cup \{u\}, S_B) + \sigma_A(S_A \cup \{v\}, S_B).$$

Proof. We first show the negative part. Recall that in the proof of Theorem 1, we calculate that for the graph in Fig. 1,

$$M_2 - M_1 = q_{A|\emptyset}^3 q_{B|\emptyset}^2 (1 - q_{A|\emptyset})(q_{A|B} - q_{A|\emptyset})(q_{B|A} - q_{B|\emptyset}),$$

which remains exactly the same no matter whether A and B are competing or complementary. If none of the conditions in Theorem 2 hold, then $M_2 - M_1 > 0$, and σ_A^t is not submodular in the seed set of A.

Now we prove case by case the positive cases.

- $q_{A|\emptyset} \in \{0,1\}$. When $q_{A|\emptyset} = 1, \sigma_A$ is simply all vertices reachable from the A seed set, so submodularity is trivial. Now consider $q_{A|\emptyset} = 0$, and take again the possible world view. The fact that $q_{A|\emptyset} = 0$ gives us two messages: that A spreads only by following B, and consequently that A does not affect the propagation of B. We use the same notations as in the proof of Theorem 1. Assume that in possible world W, when both u and v are A-seeds, t is affected by A (or $\sigma_A^t(S_A \cup \{u,v\}, S_B, W) = 1$), and let $p = \{w_1, \ldots, w_k\}$ be the path through which A reaches t, where w.l.o.g. $v \notin p$. We prove by induction that when v is not an A-seed, A affects w_i exactly at time slot $i - 1$ for $i \in [k]$. Then statement is trivial when $i = 1$. Assume that w_i becomes A-adopted at time $i - 1$. Since the propagation of B is not interfered by the A seed set, w_{i+1} is in the same B-state as when v is an A-seed at time $i - 1$, and the A-proposal from w_i gets the same reaction at w_{i+1}, i.e. acceptance. So t is eventually A-adopted, or $\sigma_A^t(S_A \cup \{u\}, S_B, W) = 1$.

- $q_{B|\emptyset} = 0$. That is, B spreads only through A-adopted vertices, and thus does not affect the propagation of A. The equivalent IC cascade procedure gives submodularity directly.
- $q_{A|\emptyset} = q_{A|B}$. Again, B does not affect A, and submodularity is trivial.
- $q_{B|\emptyset} = q_{B|A}$. The proof is totally similar to the case where $q_{A|\emptyset} = 0$.

Note 1. The conuterexample used in the proof of Theorem 2 is exactly the same as that used in the proof of Theorem 1. This versatility of the counterexample comes from the factor $(q_{A|\emptyset} - q_{A|B})(q_{B|\emptyset} - q_{B|A})$. In each case, $q_{A|\emptyset} - q_{A|B}$ and $q_{B|\emptyset} - q_{B|A}$ are of the same sign.

5.2   Cross Submodularity

Fixing S_A, because of the complementary nature of the two ideas, we are also curious about submodularity of σ_A in S_B, i.e., submodularity of the influence of some idea w.r.t. the seed set of the other idea, fixing its own seed set.

Theorem 3 (Cross-Submodularity Characterization for the Mutually Complementarity Case). *When the two ideas are complementary, for a fixed S_A, σ_A is submodular in S_B whenever one of the following holds:*

- $q_{A|\emptyset} \in \{0, 1\}$,
- $q_{B|\emptyset} = 1$,
- $q_{A|\emptyset} = q_{A|B}$.

And when none of these conditions hold, submodularity is violated, i.e., there exists (G, S_A, S_B, u, v) such that for each group of $(q_{A|\emptyset}, q_{B|\emptyset}, q_{A|B}, q_{B|A})$,

$$\sigma_A(S_A, S_B) + \sigma_A(S_A, S_B \cup \{u, v\}) > \sigma_A(S_A, S_B \cup \{u\}) + \sigma_A(S_A, S_B \cup \{v\}).$$

Proof. We prove the negative part first. Consider the counterexample presented in Fig. 2, and let the basic seed sets of A and B be $S_A = \{a\}, S_B = \{b\}$.

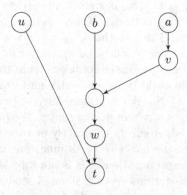

Fig. 2. Counterexample used in the proof of Theorem 3

We consider the marginals of u as a B-seed when v is a B-seed and when v is not. Let

$$M_1 = \sigma_A^t(S_A, S_B \cup \{u\}) - \sigma_A^t(S_A, S_B),$$
$$M_2 = \sigma_A^t(S_A, S_B \cup \{u, v\}) - \sigma_A^t(S_A, S_B \cup \{v\}).$$

When v is not a B-seed, u has a non-zero marginal iff it helps the propagation of A at t (while b does not). That is, a reaches w, b does not reach w, and u reaches t. Formally,

$$M_1 = q_{B|\emptyset}(q_{A|B} - q_{A|\emptyset})[(1 - q_{B|\emptyset})q_{A|\emptyset}^3 + q_{B|\emptyset}(1 - q_{B|\emptyset})q_{A|B}q_{A|\emptyset}^2].$$

And when v is a B-seed, everything is the same except that it becomes easier for A to activate v. And therefore

$$M_2 = q_{B|\emptyset}(q_{A|B} - q_{A|\emptyset})[(1 - q_{B|\emptyset})q_{A|B}q_{A|\emptyset}^2 + q_{B|\emptyset}(1 - q_{B|\emptyset})q_{A|B}^2 q_{A|\emptyset}].$$

Taking the difference,

$$M_2 - M_1 = (q_{A|B} - q_{A|\emptyset})^2 q_{A|\emptyset}(1 - q_{A|\emptyset})(q_{A|\emptyset} + q_{B|\emptyset}q_{A|B}).$$

It is clear that when no conditions stated in Theorem 3 hold, $M_2 - M_1 > 0$ and submodularity fails.

Now we look at the positive cases.

- $q_{A|\emptyset} \in \{0, 1\}$. When $q_{A|\emptyset} = 1$, submodularity is trivial. Now suppose $q_{A|\emptyset} = 0$, i.e., A spreads only by following B and does not affect the propagation of B. We prove that for any possible world W, where $\sigma_A^t(S_A, S_B \cup \{u, v\}, W) = 1$, we have $\sigma_A^t(S_A, S_B \cup \{u\}, W) + \sigma_A^t(S_A, S_B \cup \{v\}, W) \geq 1$. That is, when t is A-adopted when both u and v are B-seeds, t will still be activated either when u alone is a B-seed or v alone is.
  When both u and v are seeds, let $p = \{w_1, \ldots, w_k\}$ be the path through which A reaches t, where w.l.o.g. $v \notin p$. Clearly in our possible world view, for any $w \in p, \alpha_B(w) \leq q_{B|\emptyset}$. When u alone is a B-seed, w_1 remains a B-seed, and can activate every vertex w on the path just as when v is also a B-seed. It follows that A still can reach t through p.
- $q_{B|\emptyset} = 1$. We follow the same manner as we do in the first bullet point. The same argument works in a sense that we still have $\alpha_B(w) \leq q_{B|\emptyset} = 1$ here for $w \in p$. Sobmodularity follows.
- $q_{A|\emptyset} = q_{A|B}$. That means b has nothing to do with the propagation of A. Submodularity of IC model then carries over directly.

6   The One-Shot Model

In foregoing sections, properties of a model with somewhat rational agents are discussed. The agents are rational, in a sense that when a first proposal of some idea fails, they still allow the other idea a chance to propose; and when a first proposal succeeds, they do not accept/reject the possible proposal from the other idea instantly. In this section, we look at a model where agents act more extremely.

6.1   The Model

As in the Com-IC model, there is a backbone network $G = (V, E, p)$. The model also has four parameters as the GAP parameters in Com-IC. We only consider the mutually competing case for the One-Shot model. The key difference here is that an idle vertex considers only the first proposal that reaches it. Each vertex has 4 possible states: idle, exhausted, A-adopted, B-adopted.

Cascading proceeds in the following fashion: when an A (resp. B) proposal reaches an idle vertex, the vertex adopts A (resp. B) w.p. $q_{A|\emptyset}$ (resp. $q_{B|\emptyset}$), and becomes exhausted w.p. $1 - q_{A|\emptyset}$ (resp. $1 - q_{B|\emptyset}$). Once a vertex becomes exhausted, it no longer considers any further proposals. Since A and B are competing ideas, an A-adopted (resp. B-adopted) vertex no longer considers proposals of B (resp. A). $(q_{A|\emptyset}, q_{B|\emptyset})$ therefore completely characterizes the strengths of the ideas.

6.2   Submodularity in One-Shot Model

The characterization of sumodularity in One-Shot model appears to be more interesting. It demonstrates a dichotomy over the GAP space of One-Shot model, i.e., only the stronger idea propagates with submodularity.

Theorem 4. *In One-Shot model, when $q_{A|\emptyset} \geq q_{B|\emptyset}$ or $q_{A|\emptyset} = 0, \sigma_A$ is submodular in S_A; when $0 < q_{A|\emptyset} < q_{B|\emptyset}$, submodularity is violated. To be specific, when $0 < q_{A|\emptyset} < q_{B|\emptyset}$, there exists (G, S_A, S_B, u, v) such that*

$$\sigma_A(S_A, S_B) + \sigma_A(S_A \cup \{u, v\}, S_B) > \sigma_A(S_A \cup \{u\}, S_B) + \sigma_A(S_A \cup \{v\}, S_B).$$

Proof. We prove the negative part first. Consider the network shown in Fig. 3, where the basic seed sets are $S_A = \emptyset$ and $S_B = \{b\}$. We calculate the marginals of u at t when v is an A-seed and when v is not. Formally, let

$$M_1 = \sigma_A^t(S_A \cup \{u\}, S_B) - \sigma_A^t(S_A, S_B),$$
$$M_2 = \sigma_A^t(S_A \cup \{u, v\}, S_B) - \sigma_A^t(S_A \cup \{v\}, S_B).$$

When v is not a seed, u has a positive marginal iff b fails to reach t and u successfully reaches t. That is,

$$M_1 = q_{A|\emptyset}^{k+3}(1 - q_{B|\emptyset}^{k+2}).$$

And when v is an A-seed, t has a positive marginal iff v fails to reach t and u succeeds. So,

$$M_2 = q_{A|\emptyset}^{k+3}(1 - q_{A|\emptyset}^{k+1}).$$

Taking the difference,

$$M_2 - M_1 = q_{A|\emptyset}^{k+3}(q_{B|\emptyset}^{k+2} - q_{A|\emptyset}^{k+1}).$$

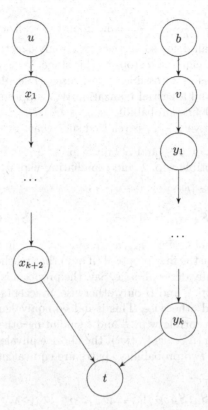

Fig. 3. Counterexample used in the proof of Theorem 4

As $q_{A|\emptyset} < q_{B|\emptyset}$,

$$\lim_{k \to \infty} \frac{q_{B|\emptyset}^{k+2}}{q_{A|\emptyset}^{k+1}} = \infty,$$

so when $q_{A|\emptyset} > 0$, there is some k such that $M_2 - M_1 > 0$, and submodularity is violated.

We prove the positive part now. When $q_{A|\emptyset} = 0, \sigma_A = |S_A|$ is clearly submodular in S_A. Now we consider the case where $q_{A|\emptyset} \geq q_{B|\emptyset}$. We take a different possible world view here, i.e., each vertex flips two independent coins and decide whether it accepts A-proposals and B-proposals. Each vertex has 4 possible realizations: A-only, B-only, susceptible and repudiating, indicating that the vertex accepts A-proposals only, B-proposals only, all proposals, and none respectively.

First we consider a partial realization of the world. We realize all susceptible and repudiating vertices first. To do so, for each vertex v, we flip a coin and determined with probability $q_{A|\emptyset}q_{B|\emptyset} + (1 - q_{A|\emptyset})(1 - q_{B|\emptyset})$ that v is eventually realized to be either susceptible or repudiating. If so, we then flip another coin to determine whether it is susceptible (w.p. $\frac{q_{A|\emptyset}q_{B|\emptyset}}{q_{A|\emptyset}q_{B|\emptyset}+(1-q_{A|\emptyset})(1-q_{B|\emptyset})}$) or repudiating (otherwise). For vertices remaining not realized, we flip a coin and decide it to be

A-only w.p. $\frac{q_{A|\emptyset}(1-q_{B|\emptyset})-q_{B|\emptyset}(1-q_{A|\emptyset})}{q_{A|\emptyset}(1-q_{B|\emptyset})+q_{B|\emptyset}(1-q_{A|\emptyset})}$. Now upon full realization, each of the rest of vertices (which we call deferred vertices) is A-only exactly w.p. $\frac{1}{2}$ and B-only otherwise. The partial realization stops at this stage. We remove all repudiating vertices, leaving vertices in 3 possible states: susceptible, A-only, and deferred.

Now for S_A, S_B, t and a partial realization W_p, suppose there are k deferred vertices, $w_1, \ldots, w_k$. Define probability spaces $\Omega_0, \Omega_1, \ldots, \Omega_k$ in the following fashion: in Ω_i, deferred vertex w_j is realized such that:

- If $j > i, w_i$ is A-only w.p. $\frac{1}{2}$ and B-only w.p. $\frac{1}{2}$.
- If $j \leq i, w_i$ is susceptible w.p. $\frac{1}{2}$ and repudiating w.p. $\frac{1}{2}$.

We show that for all $i \in [k]$,

$$\mathbb{E}_{W_i \leftarrow \Omega_i}[\sigma_A^t(S_A, S_B, W_i)] = \mathbb{E}_{W_{i-1} \leftarrow \Omega_{i-1}}[\sigma_A^t(S_A, S_B, W_{i-1})].$$

Consider fixing randomness of $w_1, \ldots, w_{i-1}, w_{i+1}, \ldots, w_k$ in Ω_{i-1}. After doing so, we are able to determine the first proposal (if any) that reaches w_i, since that part of the propagation is fully deterministic. Say, the proposal is an A-proposal, then because w_i is A-only w.p. $\frac{1}{2}$ and B-only otherwise, it accepts the proposal w.p. $\frac{1}{2}$, and becomes exhausted otherwise. This is indeed equivalent w.r.t. propagation of A to making w_i susceptible w.p. $\frac{1}{2}$ and repudiating otherwise. Since for any partial realization of $\{w_1, \ldots, w_k\} \setminus \{w_i\}$, the above equivalence always holds, we may conclude that the two probability spaces are equivalent w.r.t. the influence of A. Formally,

$$\mathbb{E}_{W_i \leftarrow \Omega_i}[\sigma_A^t(S_A, S_B, W_i)] = \mathbb{E}_{W_{i-1} \leftarrow \Omega_{i-1}}[\sigma_A^t(S_A, S_B, W_{i-1})].$$

Now we only need to show submodularity in Ω_k. We fix all randomness, remove repudiating vertices, and establish submodularity in each possible world. In each possible world W_k drawn from Ω_k, there are possibly 2 types of vertices: susceptible ones and A-only ones. We show that for S_A, S_B, u, v, t,

$$\sigma_A^t(S_A \cup \{u, v\}, S_B, W_k) = 1 \Rightarrow \sigma_A^t(S_A \cup \{u\}, S_B, W_k) + \sigma_A^t(S_A \cup \{v\}, S_B, W_k) \geq 1.$$

Let $p = \{w_1, \ldots, w_k\}$ be the A-path through which A reaches t when both u and v are A-seeds. W.l.o.g. $v \notin p$. In the competing case, let w be the vertex closest to w_1 on p, which becomes not A-adopted (and in fact, B-adopted) when v is not a seed. w must be reachable from v. Let p' be the shortest path from v to w, and x the closest vertex to v on p' which becomes B-adopted when v is not a seed. Since v blocks B from affecting x through path $[v, x] \subseteq p'$, and when v is not a seed, x blocks w from being affected by A through path $[x, w] \subseteq p'$, clearly p' is a shorter A-path (recall that A can pass through every vertex in the world) from the A seed set to w than $[w_1, w] \subseteq p$ when v is an A-seed, a contradiction.

Note 2. Unlike in Theorems 1, 2 or 3, the counterexample needed for Theorem 4 has to be constructed after fixing $q_{A|\emptyset}$ and $q_{B|\emptyset}$.

Acknowledgment. We would like to thank Yingru Li for some early discussions on the subject.

References

1. Bharathi, S., Kempe, D., Salek, M.: Competitive influence maximization in social networks. In: Deng, X., Graham, F.C. (eds.) WINE 2007. LNCS, vol. 4858, pp. 306–311. Springer, Heidelberg (2007). doi:10.1007/978-3-540-77105-0_31
2. Borodin, A., Filmus, Y., Oren, J.: Threshold models for competitive influence in social networks. In: Saberi, A. (ed.) WINE 2010. LNCS, vol. 6484, pp. 539–550. Springer, Heidelberg (2010). doi:10.1007/978-3-642-17572-5_48
3. Budak, C., Agrawal, D., El Abbadi, A.: Limiting the spread of misinformation in social networks. In: Proceedings of the 20th International Conference on World Wide Web, pp. 665–674. ACM (2011)
4. Chen, W., Collins, A., Cummings, R., Ke, T., Liu, Z., Rincon, D., Sun, X., Wang, Y., Wei, W., Yuan, Y.: Influence maximization in social networks when negative opinions may emerge and propagate. In: SDM, vol. 11, pp. 379–390. SIAM (2011)
5. Datta, S., Majumder, A., Shrivastava, N.: Viral marketing for multiple products. In: 2010 IEEE International Conference on Data Mining, pp. 118–127. IEEE (2010)
6. He, X., Song, G., Chen, W., Jiang, Q.: Influence blocking maximization in social networks under the competitive linear threshold model. In: SDM, pp. 463–474. SIAM (2012)
7. Kempe, D., Kleinberg, J., Tardos, É.: Maximizing the spread of influence through a social network. In: Proceedings of the Ninth ACM SIGKDD International Conference on Knowledge Discovery and Data Mining, pp. 137–146. ACM (2003)
8. Lu, W., Bonchi, F., Goyal, A., Lakshmanan, L.V.: The bang for the buck: fair competitive viral marketing from the host perspective. In: Proceedings of the 19th ACM SIGKDD International Conference on Knowledge Discovery and Data Mining, pp. 928–936. ACM (2013)
9. Lu, W., Chen, W., Lakshmanan, L.V.: From competition to complementarity: comparative influence diffusion and maximization. Proc. VLDB Endow. $9(2)$, 60–71 (2015)
10. Narayanam, R., Nanavati, A.A.: Viral marketing for product cross-sell through social networks. In: Flach, P.A., Bie, T., Cristianini, N. (eds.) ECML PKDD 2012. LNCS, vol. 7524, pp. 581–596. Springer, Heidelberg (2012). doi:10.1007/978-3-642-33486-3_37
11. Nemhauser, G.L., Wolsey, L.A., Fisher, M.L.: An analysis of approximations for maximizing submodular set functions–I. Math. Program. $14(1)$, 265–294 (1978)

A Risk–Reward Model for On-line Financial Leasing Problem with an Interest Rate

Xiaoli Chen and Weijun Xu[(⊠)]

School of Business Administration, South China University of Technology,
Guangzhou 510640, Guangdong, People's Republic of China
chenxiaoli871122@163.com, xuweijun75@163.com

Abstract. As an important financing tool, the financial lease can help the lessee obtain the ownership of equipment after paying the rent till the expiration of the lease. Because the lessee does not know the exact length of using the equipment, the financial leasing problem can be viewed as an on-line problem. In this paper, we consider the on-line financial leasing problem with an interest rate where there are two lease options: financial lease and operating lease. We first discuss the traditional deterministic optimal competitive strategy by competitive analysis method. Next, we introduce the risk tolerance and forecast of the decision maker(the lessee) into this problem and acquire the optimal risk–reward strategy. Finally, we give numerical examples and the results show that the introduction of the interest rate and risk tolerance has a significant influence on the deterministic optimal strategy and risk–reward strategy.

Keywords: Financial leasing · Interest rate · Risk tolerance · Risk–reward model · Competitive analysis

1 Introduction

The leasing industry as a sunrise industry has demonstrated its resilience since the global economic crisis and the outlook is cautiously optimistic [1]. A company or an individual without enough money to buy certain equipment can own the right to use the equipment by leasing. To decide whether leasing is a beneficial way to use the equipment or not, we should determine the duration the equipment will be used. However, in practice, it is hard to know the exact duration. This shows the on-line feature of leasing. Fortunately, researchers explore the competitive analysis and on-line algorithm [2,3] to study the on-line problems and evaluate their strategies. In this framework, we can find the appropriate strategy through competitive ratio, which is the ratio of the cost paid by our on-line strategy to the cost paid by the optimal off-line strategy (which is obtained when everything is known in advance).

The classical on-line leasing problem is "ski rental" problem: to lease a pair of skis needs 1 per day and to buy a pair of skis needs s, then which strategy is optimal? By the competitive analysis method Karp [4] gave the optimal on-line

© Springer International Publishing AG 2017
M. Xiao and F. Rosamond (Eds.): FAW 2017, LNCS 10336, pp. 68–78, 2017.
DOI: 10.1007/978-3-319-59605-1_7

strategy which is to lease for the first $s - 1$ days then to buy the skis if the skier continues to ski in the s-th day. And the optimal competitive ratio is $2 - 1/s$, which means the on-line strategy never pays more than $2 - 1/s$ times of the optimum. Then a lot of researchers discussed the on-line leasing problem based on Karp's model. For example, they introduced the interest rate [5], tax rate [6], price fluctuation [7] and so on. Moreover, Fujiwara et al. [8] and Lotker et al. [9] analyzed the multi-slope ski rental problem and gave the randomized algorithms and the best possible competitive ratio, respectively.

Because the risk preference of the decision maker can not be ignored, Al-Binali [10] introduced the decision maker's risk tolerance and forecast for the future into the "ski rental" problem and defined the risk and reward of a competitive algorithm. In this model, if the input σ is an instance of the problem Σ and the cost ratio of the on-line algorithm A and the optimal algorithm OPT is denoted by $R_A(\sigma)$, then the competitive ratio of A on the problem Σ is $R_A = \sup_{\sigma \in \Sigma} R_A(\sigma)$ and the optimal competitive ratio for the problem Σ is $R^* = \inf_A R_A$. Then the risk of A is defined as R_A/R^*. If the decision maker's risk tolerance is $\lambda(\lambda \geq 1)$, then the set of risk tolerable strategies is $I_\lambda = \{A | R_A \leq \lambda R^*\}$. And if the decision maker has a forecast $F \subset \Sigma$, then the restricted ratio of A is $\overline{R}_A = \sup_{\sigma \in F} R_A(\sigma)$ and the reward of A is $R^*/\overline{R}_A$ when the forecast is correct. Then the risk-reward model is to maximize $R^*/\overline{R}_A$ subject to $A \in I_\lambda$. From then on, many researchers discussed the on-line leasing problem based on Al-Binali's risk–reward framework. For example, Zhang et al. [11] introduced the depreciation of equipment and obtained the optimal risk–reward strategies with and without an interest rate.

In the above studies, the lease option is an operating lease which can not transfer the ownership of the equipment to the lessee at the end of leasing period. In the leasing market, there is another important leasing mode: the financial lease. In a financial lease, the lessee can obtain the ownership of the equipment finally, which is preferred by the lessee. Considering the necessity to study the on-line financial leasing problem, Dai et al. [12] analyzed the financial leasing problem that contains operating lease and financial lease. They gave the optimal deterministic on-line strategy for this problem. In this paper, we analyze the on-line financial leasing problem with an interest rate and give the optimal risk–reward strategy based on Al-Binali's framework.

The remainder of this paper is organized as follows. We first give the optimal deterministic competitive strategy for the on-line financial leasing problem with an interest rate in Sect. 2. Then we discuss the risk–reward strategy according to the decision maker's forecast in Sect. 3. Numerical examples are given in Sect. 4. Finally, a summary of this paper is presented in Sect. 5.

2   Optimal Deterministic Competitive Strategy

In this section, we compute the optimal deterministic competitive strategy for this problem with the interest rate. First of all, we give the precise definition of the on-line financial leasing problem. In this problem there are two ways for the decision maker to obtain the use of an equipment. One is operating lease, and the other is financial lease. The operating lease is to lease the equipment at a cost of c per period. The financial lease is to pay $r(r > c)$ for each period and the decision maker can obtain the ownership of the equipment after M periods, and then no longer needs to pay for the use of this equipment. Then which strategy is optimal for the decision maker? Here we assume that when the decision maker ends the financial lease before obtaining the ownership of the equipment, he does not need to pay any fees and the lessor still owns the equipment.

As the decision maker has no knowledge of the actual length n of using the equipment, it is smart to choose the operating lease first. Let $A(k)$ be the strategy that the decision maker uses the equipment by operating lease for the first k periods and by financial lease thereafter. We define OPT as the optimal off-line strategy. Let i be the interest rate and $\beta = 1/(i+1)$ is the discount factor. For strategies $A(k)$ and OPT, let $Cost_{A(k)}(n)$ and $Cost_{OPT}(n)$ be the present value of the total cost spent by $A(k)$ and OPT from the start to time n, respectively. Then according to competitive analysis we can know that $A(k)$ is ρ-competitive if for all possible n there is a constant α such that

$$Cost_{A(k)}(n) \leq \rho Cost_{OPT}(n) + \alpha.$$

And the smallest ρ such that $A(k)$ is ρ-competitive is called the competitive ratio of $A(k)$. All over this paper the competitive ratios are the strict competitive ratios which is obtained under $\alpha = 0$.

The off-line adversary knows exactly how long the equipment will be used. He can obtain the least cost according to n. Let $n^* = \ln[1 - r(1 - \beta^M)/c]/\ln\beta$, then the cost of the optimal off-line strategy OPT is

$$Cost_{OPT}(n) = \begin{cases} \frac{c(1-\beta^n)}{1-\beta}, & n \leq n^*; \\ \frac{r(1-\beta^M)}{1-\beta}, & n > n^*. \end{cases} \qquad (1)$$

We assume that $1 - r(1 - \beta^M)/c > 0$, that is $c - r + r\beta^M > 0$. Otherwise, operating lease is always better than financial lease, then no one will choose financial lease. For the on-line strategy $A(k)$, the cost is

$$Cost_{A(k)}(n) = \begin{cases} \frac{c(1-\beta^n)}{1-\beta}, & n \leq k; \\ \frac{c(1-\beta^k)}{1-\beta} + \frac{r\beta^k(1-\beta^{n-k})}{1-\beta}, & k < n \leq k+M; \\ \frac{c(1-\beta^k)}{1-\beta} + \frac{r\beta^k(1-\beta^M)}{1-\beta}, & n > k+M. \end{cases} \qquad (2)$$

According to competitive analysis and the cost of on-line and off-line strategies, we can obtain the following theorem:

Theorem 1. *The optimal competitive strategy for the on-line financial leasing problem with an interest rate i is $A(n^* - M)$ and the competitive ratio is $R^* = r/c - (r-c)^2/(cr\beta^M)$, where $\beta = 1/(1+i)$ and $n^* = \ln[1 - r(1 - \beta^M)/c]/\ln\beta$.*

Proof. Let $R(k,n) = Cost_{A(k)}(n)/Cost_{OPT}(n)$, $R(k) = \sup_n R(k,n)$, then $R(k)$ is the competitive ratio of $A(k)$. Next we divide the value of k into three cases to compute the optimal competitive ratio.

Case 1. If $k < n^* - M$, then we have

$$R(k,n) = \begin{cases} 1, & n \leq k; \\ \frac{c(1-\beta^k)+r\beta^k(1-\beta^{n-k})}{c(1-\beta^n)}, & k < n \leq k+M; \\ \frac{c(1-\beta^k)+r\beta^k(1-\beta^M)}{c(1-\beta^n)}, & k+M < n \leq n^*; \\ \frac{c(1-\beta^k)+r\beta^k(1-\beta^M)}{r(1-\beta^M)}, & n > n^*. \end{cases}$$

Through simple derivation we find that $R(k,n)$ increases in the second interval and decreases in the third interval. Besides $R(k,n)$ is a piecewise continuous function. So $R(k,n)$ reaches its greatest value with regard to n at $n = k+M$. Then the competitive ratio of strategy $A(k)$ in this case is

$$R_1(k) = R(k,k+M) = \frac{c(1-\beta^k)+r\beta^k(1-\beta^M)}{c(1-\beta^{k+M})}.$$

Take the derivation of $R_1(k)$ we have

$$\begin{aligned} R_1'(k) &= \frac{\left[\beta^k(r-r\beta^M-c)(1-\beta^{k+M})+\beta^{k+M}[c+\beta^k(r-r\beta^M-c)]\right]\ln\beta}{c(1-\beta^{k+M})^2} \\ &= \frac{\left[(r-r\beta^M-c)(1-\beta^{k+M})+c\beta^M+\beta^{k+M}(r-r\beta^M-c)\right]\beta^k\ln\beta}{c(1-\beta^{k+M})^2} \\ &= \frac{(r-c)(1-\beta^M)\beta^k\ln\beta}{c(1-\beta^{k+M})^2} < 0 \end{aligned}$$

So $R_1(k)$ decreases with respect to k. Then when $k \to n^* - M$, $R_1(k)$ attains its minimum R_1, where

$$\begin{aligned} R_1 &= \lim_{k \to n^*-M} R_1(k) \\ &= \frac{(c-r)(1-\beta^{n^*-M})+r(1-\beta^{n^*})}{c(1-\beta^{n^*})} \\ &= \frac{r}{c} - \frac{r-c-(r-c)\left[1-r(1-\beta^M)/c\right]/\beta^M}{r(1-\beta^M)} \\ &= \frac{r}{c} - \frac{(r-c)\left[\beta^M-1+r(1-\beta^M)/c\right]}{r\beta^M(1-\beta^M)} \\ &= \frac{r}{c} - \frac{(r-c)^2}{cr\beta^M}. \end{aligned}$$

Case 2. If $n^* - M \leq k < n^*$, we have

$$R(k,n) = \begin{cases} 1, & n \leq k; \\ \frac{c(1-\beta^k)+r\beta^k(1-\beta^{n-k})}{c(1-\beta^n)}, & k < n \leq n^*; \\ \frac{c(1-\beta^k)+r\beta^k(1-\beta^{n-k})}{r(1-\beta^M)}, & n^* < n \leq k+M; \\ \frac{c(1-\beta^k)+r\beta^k(1-\beta^M)}{r(1-\beta^M)}, & n > k+M. \end{cases}$$

It is easy to find that $R(k,n)$ is also a piecewise continuous function and it increases in both the second and third intervals. Then the competitive ratio $R_2(k)$ of on-line strategy $A(k)$ can be reached at $n = n^*$ or $k + M$ in this case, namely

$$
R_2(k) = \max\left\{ \frac{c(1-\beta^k)+r\beta^k(1-\beta^{n^*-k})}{c(1-\beta^{n^*})}, \frac{c(1-\beta^k)+r\beta^k(1-\beta^M)}{r(1-\beta^M)} \right\}
$$

$$
= \max\left\{ \frac{c(1-\beta^k)+r\beta^k(1-\beta^{n^*-k})}{r(1-\beta^M)}, \frac{c(1-\beta^k)+r\beta^k(1-\beta^M)}{r(1-\beta^M)} \right\}
$$

$$
= \frac{c(1-\beta^k)+r\beta^k(1-\beta^M)}{r(1-\beta^M)}.
$$

Because $R_2(k)$ is an increasing function, it attains its smallest value R_2 at $k = n^* - M$, namely

$$
R_2 = R_2(n^* - M) = \frac{c + \beta^{n^*-M}\left[r(1-\beta^M)-c\right]}{r(1-\beta^M)}
$$

$$
= \frac{c\beta^M + \left[r(1-\beta^M)-c\right]\left[1-r(1-\beta^M)/c\right]}{r\beta^M(1-\beta^M)}
$$

$$
= \frac{r}{c} - \frac{(r-c)^2}{cr\beta^M} = R_1
$$

Case 3. If $k \geq n^*$, we have

$$
R(k,n) = \begin{cases}
1, & n \leq n^*; \\
\frac{c(1-\beta^n)}{r(1-\beta^M)}, & n^* < n \leq k; \\
\frac{c(1-\beta^k)+r\beta^k(1-\beta^{n-k})}{r(1-\beta^M)}, & k < n \leq k+M; \\
\frac{c(1-\beta^k)+r\beta^k(1-\beta^M)}{r(1-\beta^M)}, & n > k+M.
\end{cases}
$$

According to the monotonicity of each section of $R(k,n)$ we can get the competitive ratio $R_3(k)$ of on-line strategy $A(k)$ in this case. It is

$$
R_3(k) = \max\left\{ \frac{c(1-\beta^k)}{r(1-\beta^M)}, \frac{c(1-\beta^k)+r\beta^k(1-\beta^M)}{r(1-\beta^M)} \right\} = \frac{c(1-\beta^k)}{r(1-\beta^M)} + \beta^k.
$$

Through simple derivation of $R_3(k)$ we find that it increases with respect to k. Then $R_3(k)$ attains its smallest value R_3 at $k = n^*$, namely

$$
R_3 = R_3(n^*) = \frac{c(1-\beta^{n^*})}{r(1-\beta^M)} + \beta^{n^*} = 1 + \beta^{n^*} = 2 - \frac{r(1-\beta^M)}{c}.
$$

Because

$$
R_3 - R_1 = 2 - \frac{r(1-\beta^M)}{c} - \frac{r}{c} - \frac{(r-c)^2}{cr\beta^M} = \frac{(r-c-r\beta^M)^2}{cr\beta^M} > 0,
$$

that is $R_3 > R_1$. So the optimal competitive ratio of the financial leasing problem is $R^* = R_1$. And the optimal on-line strategy is $A(n^* - M)$. Then the theorem is proved.

According to Theorem 1 we can obtain that the competitive ratio of the financial leasing problem without the interest rate is $2 - c/r$ and if we let $n^{*\prime} = (r/c-1)M$, then the corresponding optimal strategy is $A(n^{*\prime})$, which agrees with the results in Dai et al. [12]. In addition we find that the introduction of the interest rate can reduce the competitive ratio of the financial leasing problem.

3   Risk–Reward Strategy with an Interest Rate

In this section, we introduce the decision maker's forecast for the length of using the equipment and his risk tolerance into the on-line financial leasing problem. Based on the risk–reward framework of Al-Binali [10] we obtained the optimal risk–reward strategy with the interest rate for this on-line problem.

Let $\lambda(\lambda > 1)$ be the risk tolerance of the decision maker. Then the set of risk tolerable strategies is $I_\lambda = \{A(k)|R(k) \leq \lambda R^*\}$ according to Al-Binali's risk–reward framework. As n^* and $n^* - M$ are both critical points, we assume that there are three forecasts for n: $F_1 = \{n : n \leq n^* - M\}$, $F_2 = \{n : n^* - M < n \leq n^*\}$ and $F_3 = \{n : n > n^*\}$. Let $\overline{R}_j(k) = \sup\limits_{n \in F_j} \dfrac{Cost_{A(k)}(n)}{Cost_{OPT}(n)}$ for $j = 1, 2, 3$. For these three forecasts we have the following theorem.

Theorem 2. *If there are three forecasts $F_1 = \{n : n \leq n^* - M\}$, $F_2 = \{n : n^* - M < n \leq n^*\}$ and $F_3 = \{n : n > n^*\}$ for the decision maker to choose and the risk tolerance is $\lambda(\lambda > 1)$, then the risk–reward strategy for the on-line financial leasing problem is as follows:*

(1) When the decision maker chooses forecast F_1, the optimal risk–reward strategy is $A(n^* - M)$ and the restricted ratio is $\overline{R}^* = 1$ if F_1 comes true.

(2) When the decision maker chooses F_2, if $1 < \lambda \leq \Delta$, then the optimal risk–reward strategy is $A(k_2)$ and the restricted ratio is $\overline{R}^* = \overline{R}_2(k_2)$ if F_2 comes true. Else the optimal strategy is $A(n^*)$ and the restricted ratio is $\overline{R}^* = 1$ if forecast F_2 comes true.

(3) When the decision maker chooses F_3, the corresponding optimal risk–reward strategy is $A(k_1)$ and the restricted ratio is $\overline{R}^* = \overline{R}_3(k_1)$ when forecast F_3 comes true.

Wherein k_1, k_2, Δ and functions $\overline{R}_2(\cdot)$, $\overline{R}_3(\cdot)$ refer to the proof.

Proof. First of all, we compute the set of risk tolerable strategies I_λ. According to on-line algorithm and the analysis in Sect. 2, we have

$$R(k) = \begin{cases} \dfrac{c + \beta^k (r - r\beta^M - c)}{c(1 - \beta^{k+M})}, & k < n^* - M; \\[2ex] \dfrac{c + \beta^k (r - r\beta^M - c)}{r(1 - \beta^M)}, & k \geq n^* - M. \end{cases}$$

Then when $k < n^* - M$ we have

$$R(k) \leq \lambda R^* \Leftrightarrow \frac{c + \beta^k(r - r\beta^M - c)}{c(1 - \beta^{k+M})} \leq \lambda \left[\frac{r}{c} - \frac{(r-c)^2}{cr\beta^M} \right]$$

$$\Leftrightarrow cr\beta^M + r\beta^{k+M}(r - r\beta^M - c) \leq \lambda[r^2\beta^M - (r-c)^2](1 - \beta^{k+M})$$

$$\Leftrightarrow (k + M)\ln\beta \leq \ln \frac{\lambda r(r\beta^M - r + c) + \lambda c(r - c) - cr\beta^M}{(\lambda - 1)r(r\beta^M - r + c) + \lambda c(r - c)}$$

$$\Leftrightarrow k \geq \frac{1}{\ln\beta} \ln \frac{\lambda r(r\beta^M - r + c) + \lambda c(r - c) - cr\beta^M}{(\lambda - 1)r(r\beta^M - r + c) + \lambda c(r - c)} - M \triangleq k_1$$

That is to say, the set of risk tolerable strategies is $I_\lambda^1 = \{A(k)|k_1 \leq k < n^* - M\}$ when $k < n^* - M$.

Similarly, when $k \geq n^* - M$ we have

$$R(k) \leq \lambda R^* \Leftrightarrow \frac{c + \beta^k(r - r\beta^M - c)}{r(1 - \beta^M)} \leq \lambda \left[\frac{r}{c} - \frac{(r-c)^2}{cr\beta^M} \right]$$

$$\Leftrightarrow \beta^k(r\beta^M - r + c) \geq c - \frac{\lambda(1 - \beta^M)[r^2\beta^M - (r-c)^2]}{c\beta^M}$$

$$\Leftrightarrow \beta^k \geq \frac{\lambda(r\beta^M - r + c)^2 - (\lambda - 1)c^2\beta^M}{c\beta^M(r\beta^M - r + c)}$$

$$\Leftrightarrow k \leq \frac{1}{\ln\beta} \ln \frac{\lambda(r\beta^M - r + c)^2 - (\lambda - 1)c^2\beta^M}{c\beta^M(r\beta^M - r + c)} \triangleq k_2$$

That is to say, the set of risk tolerable strategies is $I_\lambda^2 = \{A(k)|n^* - M \leq k \leq k_2\}$ when $k \geq n^* - M$.

To sum up, we know that for arbitrary k the set of risk tolerable strategies is $I_\lambda = I_\lambda^1 \bigcup I_\lambda^2 = \{A(k)|k_1 \leq k \leq k_2\}$.

Next, we discuss the optimal risk–reward strategies for the three forecasts respectively.

For the forecast F_1, we can obtain the restricted ratio $\overline{R}_1(k)$ as

$$\overline{R}_1(k) = \sup_{\sigma \in F_1} \frac{Cost_{A(k)}(\sigma)}{Cost_{OPT}(\sigma)}$$

$$= \begin{cases} 1, & k \geq n^* - M; \\ \sup\limits_{n \leq n^* - M} \left\{ \frac{c(1-\beta^k)+r\beta^k(1-\beta^{n-k})}{c(1-\beta^n)} \right\}, & n^* - 2M \leq k < n^* - M; \\ \sup\limits_{n \leq n^* - M} \left\{ \frac{c(1-\beta^k)+r\beta^k(1-\beta^{n-k})}{c(1-\beta^n)}, \frac{c(1-\beta^k)+r\beta^k(1-\beta^M)}{c(1-\beta^n)} \right\}, & k < n^* - 2M. \end{cases}$$

$$= \begin{cases} 1, & k \geq n^* - M; \\ \frac{(c-r)(1-\beta^k)+r(1-\beta^{n^*-M})}{c(1-\beta^{n^*-M})}, & n^* - 2M \leq k < n^* - M; \\ \frac{(c-r)(1-\beta^k)+r(1-\beta^{k+M})}{c(1-\beta^{k+M})}, & k < n^* - 2M. \end{cases} \quad (3)$$

It is easy to find that $\overline{R}_1(k)$ is a piecewise continuous function and non-increasing with respect to k. So when the forecast F_1 comes true the optimal restricted ratio is $\overline{R}^* = \min \overline{R}_1(k) = 1$. And the strategies $\{A(k)|n^* - M \leq k \leq k_2\}$ can obtain the restricted ratio 1. But bigger k can bring larger risk, so the optimal strategy is $A(k)$ with $k = n^* - M$.

For the forecast F_2, the corresponding restricted ratio $\overline{R}_2(k)$ is

$$
\overline{R}_2(k) = \sup_{\sigma \in F_2} \frac{Cost_{A(k)}(\sigma)}{Cost_{OPT}(\sigma)}
$$

$$
= \begin{cases} \frac{c(1-\beta^k)+r\beta^k(1-\beta^M)}{c(1-\beta^{k+M})}, & k < n^* - M; \\ \frac{c(1-\beta^k)+r\beta^k(1-\beta^{n^*-k})}{r(1-\beta^M)}, & n^* - M \leq k < n^*; \\ 1, & k \geq n^*. \end{cases} \tag{4}
$$

Through simple derivation we can find that $\overline{R}_2(k)$ is a non-increasing continuous function. So when the forecast F_2 is true the optimal restricted ratio is

$$
\overline{R}^* = \begin{cases} \overline{R}_2(k_2), & n^* \geq k_2; \\ 1, & n^* < k_2. \end{cases}
$$

By simplification we have

$$
n^* \geq k_2 \Leftrightarrow \ln \frac{r\beta^M - r + c}{c} \leq \ln \frac{\lambda(r\beta^M - r + c)^2 - (\lambda - 1)c^2\beta^M}{c\beta^M(r\beta^M - r + c)}
$$

$$
\Leftrightarrow \frac{r\beta^M - r + c}{c} \leq \frac{\lambda(r\beta^M - r + c)^2 - (\lambda - 1)c^2\beta^M}{c\beta^M(r\beta^M - r + c)}
$$

$$
\Leftrightarrow \lambda[c^2\beta^M - (r\beta^M - r + c)^2] \leq r\beta^M(1 - \beta^M)(r\beta^M - r + 2c)
$$

Because $c\beta^{M/2} - (r\beta^M - r + c) > (c - r)(\beta^{M/2} - 1) > 0$ and $r\beta^M - r + 2c > 0$, so if $1 < \lambda \leq \frac{c^2\beta^M - \beta^M(r\beta^M - r + c)^2}{c^2\beta^M - (r\beta^M - r + c)^2} \triangleq \Delta$, we have $n^* \geq k_2$ and $\overline{R}^* = \overline{R}_2(k_2)$; if $\lambda > \Delta$, we have $n^* < k_2$ and $\overline{R}^* = 1$.

For the forecast F_3, the corresponding restricted ratio $\overline{R}_3(k)$ is

$$
\overline{R}_3(k) = \sup_{\sigma \in F_3} \frac{Cost_{A(k)}(\sigma)}{Cost_{OPT}(\sigma)} = \frac{c(1 - \beta^k) + r\beta^k(1 - \beta^M)}{r(1 - \beta^M)} \tag{5}
$$

As $\overline{R}_3(k)$ is an increasing function, the optimal restricted ratio is $\overline{R}^* = \overline{R}_3(k_1)$ when the forecast F_3 comes true.

Above all, we can obtain the optimal strategies with different forecasts. Then the theorem is proved.

According to Theorem 2 we can obtain the following on-line algorithm.

Algorithm 1. The optimal risk–reward strategy for on-line financial leasing problem

Input: the parameters (c, r, M) in the on-line financial leasing problem, the interest rate i, risk tolerance λ and the forecast $F_j (j = 1, 2, 3)$ of the decision maker.

Output: the optimal risk–reward strategy $A(k^*)$ and the restricted ratio $\overline{R}^*$.

Compute $\beta = \frac{1}{1+i}, n^* = \frac{1}{\ln\beta}\ln[1 - r(1 - \beta^M)/c], \Delta = \frac{c^2\beta^M - \beta^M(r\beta^M - r + c)^2}{c^2\beta^M - (r\beta^M - r + c)^2}$,

$k_1 = \frac{1}{\ln\beta}\ln\frac{\lambda r(r\beta^M - r + c) + \lambda c(r - c) - cr\beta^M}{(\lambda - 1)r(r\beta^M - r + c) + \lambda c(r - c)} - M, k_2 = \frac{1}{\ln\beta}\ln\frac{\lambda(r\beta^M - r + c)^2 - (\lambda - 1)c^2\beta^M}{c\beta^M(r\beta^M - r + c)}$,

and functions $\overline{R}_2(\cdot), \overline{R}_3(\cdot)$ in Equation (4) and Equation (5), respectively.

1. If the forecast of the decision maker is F_1,
   then $k^* := n^* - M$, $\overline{R}^* := 1$.
2. If the forecast of the decision maker is F_2, then
   (a) If $1 < \lambda \leq \Delta$, then $k^* := k_2$, $\overline{R}^* := \overline{R}_2(k_2)$.
   (b) Else $k^* := n^*$, $\overline{R}^* := 1$.
3. If the forecast of the decision maker is F_3,
   then $k^* := k_1$, $\overline{R}^* := \overline{R}_3(k_1)$.

Then the decision maker can make decisions flexibly according to Algorithm 1.

4   Numerical Examples

In this section, we give numerical examples to show the influence of the interest rate i and risk tolerance λ on the optimal risk–reward strategy. Because the largest improvement is R^*, we refer to the study of Zhang et al. [11] and take $imp = \frac{R^* - \overline{R}^*}{R^* - 1}$ as the improvement of the risk-reward strategy over the deterministic competitive ratio R^*. Here, we only consider the on-line financial leasing problem with $c = 1$, $r = 3$ and $M = 10$. The results are shown in Table 1. Since the optimal restricted ratio is 1 when the forecast F_1 is true, the data $\overline{R}^*$ and imp are not included in Table 1. For the on-line financial leasing problem with different c, r and M, the results can be obtained similarly according to Algorithm 1.

Table 1. Numerical examples when $c = 1, r = 3, M = 10$

i	R^*	λ	Under F_1	Under F_2			Under F_3		
			k^*	k^*	$\overline{R}^*$	imp	k^*	$\overline{R}^*$	imp
0.003	1.6261	1.05	20.9462	23.7652	1.4477	0.2849	17.5319	1.5267	0.1588
		1.1		26.6082	1.2693	0.5699	14.7965	1.4464	0.2871
		1.5		30.9462	1	1	3.8171	1.1170	0.8131
0.005	1.5985	1.05	21.6334	24.7075	1.4113	0.3127	18.1557	1.5066	0.1536
		1.1		27.8294	1.2241	0.6255	15.3679	1.4317	0.2786
		1.5		31.6334	1	1	4.1599	1.1202	0.7992
0.007	1.5703	1.05	22.3724	25.7451	1.3735	0.3451	18.8223	1.4857	0.1485
		1.1		29.1900	1.1767	0.6902	15.9758	1.4163	0.2702
		1.5		32.3724	1	1	4.5206	1.1225	0.7852

According to Table 1 we can find that larger interest rate can bring smaller optimal deterministic competitive ratio. No matter which forecast the decision maker chooses, the optimal switching time delays and the optimal restricted ratio decreases when the interest rate increases. Besides, for a given interest rate bigger risk tolerance can lead to later switching time but smaller restricted ratio under the same forecast. In addition, we can find that the introduction of the forecast and risk tolerance can improve the optimal competitive strategy, sometimes the improvement can reach to about 80%. Above all, the interest rate and risk tolerance of the decision maker have significant influence on the optimal on-line financial leasing problem.

5 Conclusion

Financial lease is more and more popular with the public. However, most of previous studies consider only operating lease. In this paper, we consider the operating lease and financial lease at the same time. Since the decision maker always can not know the exact length of using the equipment, this financial leasing problem is on-line. Considering the value of money and decision maker's risk preference, we introduce the interest rate and risk tolerance of the decision maker into the on-line financial leasing problem. And we analysis the traditional competitive strategy and the risk–reward strategy for this problem. We also give numerical examples to show the influence of the interest rate i and risk tolerance λ on the risk–reward strategy. In the future research, more financial leasing options can be considered and more general models can be introduced.

Acknowledgments. This paper was financially supported by the National Natural Science Foundation of China under Grant no. 71471065.

References

1. White, E.: White Clarke Global Leasing Report. White Clarke Group, Buckinghamshire (2016)
2. Borodin, A., El-Yaniv, R.: Online Algorithms and Competitive Analysis. Cambridge University Press, Cambridge (1998)
3. Alvers, S.: Online algorithms: a survey. Math. Program. **97**, 3–26 (2003)
4. Karp, R.: Online algorithms versus offline algorithms: how much is it worth to know the future? In: IFIP 12th World Computer Congress, pp. 416–429. North-Holland Publishing Co., Amsterdam (1992)
5. El-Yaniv, R., Kaniel, R., Linial, N.: Competitive optimal online leasing. Algorithmica **25**, 116–140 (1999)
6. Xu, Y.F., Xu, W.J., Li, H.Y.: On the on-line rent-or-buy problem in probabilistic environments. J. Glob. Optim. **38**(1), 1–20 (2007)
7. Epstein, L., Zebedat-Haider, H.: Rent or buy problems with a fixed time horizon. Theory Comput. Syst. **56**(2), 309–329 (2015)
8. Lotker, Z., Patt-Shamir, B., Rawitz, D.: Rent, lease, or buy: randomized algorithms for multislope ski rental. SIAM J. Discret. Math. **26**(2), 718–736 (2012)

9. Fujiwara, H., Kitano, T., Fujito, T.: On the best possible competitive ratio for the multislope ski-rental problem. J. Comb. Optim. **31**(2), 463–490 (2016)

10. Al-Binali, S.: A risk-reward framework for the competitive analysis of financial games. Algorithmica **25**, 99–115 (1999)

11. Zhang, Y., Zhang, W.G., Xu, W.J., Li, H.Y.: A risk-reward model for the on-line leasing of depreciable equipment. Inf. Process. Lett. **111**(6), 256–261 (2011)

12. Dai, W.Q., Dong, Y.C., Zhang, X.T.: Competitive analysis of the online financial lease problem. Eur. J. Oper. Res. **250**, 865–873 (2016)

Designing and Implementing Algorithms
for the Closest String Problem

Shota Yuasa[1], Zhi-Zhong Chen[1(✉)], Bin Ma[2], and Lusheng Wang[3]

[1] Division of Information System Design, Tokyo Denki University,
Hatoyama, Saitama 350-0394, Japan
zzchen@mail.dendai.ac.jp
[2] School of Computer Science, University of Waterloo,
200 University Ave. W, Waterloo, ON N2L3G1, Canada
binma@uwaterloo.ca
[3] Department of Computer Science, City University of Hong Kong,
Tat Chee Avenue, Kowloon, Hong Kong SAR
cswangl@cityu.edu.hk

Abstract. Given a set of n strings of length L and a radius d, the closest string problem (CSP for short) asks for a string t_{sol} that is within a Hamming distance of d to each of the given strings. It is known that the problem is NP-hard and its optimization version admits a polynomial time approximation scheme (PTAS). A number of parameterized algorithms have been then developed to solve the problem when d is small. Among them, the relatively new ones have not been implemented before and their performance in practice was unknown. In this study, we implement all of them by careful engineering. For those that have been implemented before, our implementation is much faster. For some of those that have not been implemented before, our experimental results show that there exist huge gaps between their theoretical and practical performances. We also design a new parameterized algorithm for the binary case of CSP. The algorithm is deterministic and runs in $O\left(nL + n^2 d \cdot 6.16^d\right)$ time, while the previously best deterministic algorithm runs in $O\left(nL + nd^3 \cdot 6.731^d\right)$ time.

1 Introduction

An instance of the closest string problem (CSP for short) is a pair (S, d), where S is a set of strings of the same length L over an alphabet Σ and d is a nonnegative integer. The objective is to find a string t_{sol} of length L such that $d(t_{sol}, s) \leq d$ for every $s \in S$. The problem is fundamental and has been extensively studied in a variety of applications in bioinformatics, such as finding signals in DNA or protein family, and modif finding. Unfortunately, it is NP-hard [4,7].

Although CSP is NP-hard in general, we can still solve CSP exactly in reasonable amount of time via parameterized algorithms when d is small. Indeed, Gramm, Niedermeier and Rossmanith [5] designed the first parameterized algorithm that runs in $O(nL + nd \cdot d^d)$ time. Ma and Sun [8] present an algorithm whose time complexity is $O(nL + nd \cdot (16|\Sigma|)^d)$. Chen and Wang [2]

© Springer International Publishing AG 2017
M. Xiao and F. Rosamond (Eds.): FAW 2017, LNCS 10336, pp. 79–90, 2017.
DOI: 10.1007/978-3-319-59605-1_8

improve the time complexity to $O(nL + nd \cdot 8^d)$ for binary strings and to $O(nL + nd \cdot (\sqrt{2}|\Sigma| + \sqrt[4]{8}(\sqrt{2} + 1)(1 + \sqrt{|\Sigma| - 1}) - 2\sqrt{2})^d)$ for arbitrary alphabets Σ. Chen, Ma, and Wang [1] further improved the time complexity to $O(nL+nd^3 \cdot 6.731^d)$ for binary strings and to $O(nL+nd \cdot (1.612(|\Sigma|+\beta^2+\beta-2))^d)$ for arbitrary alphabets, where $\beta = \alpha^2+1-2\alpha^{-1}+\alpha^{-2}$ with $\alpha = \sqrt[3]{\sqrt{|\Sigma| - 1} + 1}$. In theory, this algorithm has the best time complexity among all known deterministic parameterized algorithms for CSP when $|\Sigma|$ is small (such as binary strings and DNA strings). Chen, Ma, and Wang [3] designed a randomized algorithm that runs in $O^*(9.81^d)$ and $O^*(40.1^d)$ expected time for DNA and protein strings, respectively. In particular, for binary strings, their randomized algorithm runs in $O(nL + n\sqrt{d} \cdot 5^d)$ expected time. Hence, in theory, the randomized algorithms in [3] look faster than the deterministic algorithms in [1,2].

All the aforementioned algorithms for CSP have been rigorously analyzed and hence their theoretical performance is known. On the other hand, there is another type of algorithms for CSP which solve the problem exactly but their theoretical performance has never been rigorously analyzed. For convenience, we refer to algorithms of this type as *heuristic* algorithms. Among heuristic algorithms, *TraverString* [11] and *qPMS9* [9] are known to have the best performance in practice. Indeed, they are much faster than *Provable*, which is obtained by implementing the algorithm for CSP in [2].

Ideally, we want to find an algorithm for CSP which not only has a good theoretical time-bound but also can be implemented into a program which runs faster than the other known algorithms for CSP (including the best-known heuristic algorithms, namely, *TraverString* [11] and *qPMS9* [9]). To find such an algorithm, one way is to implement the known algorithms (by careful engineering) which have good theoretical time-bounds. Unfortunately, the recent algorithms in [1,3] for CSP had not been implemented previously and hence their performance in practice was previously unknown. Moreover, although the algorithm for CSP in [2] has the best theoretical time-bound for large alphabets (such as protein strings), its simple implementation done in [2] only yields a program that is much slower than the best-known heuristic algorithms for CSP.

So, in this paper, we re-implement the algorithm for CSP in [2] by careful engineering. For convenience, we refer to this algorithm as the *2-string* algorithm as in [1]. We also carefully implement the *3-string* algorithm in [1] and the two best randomized algorithms (namely, *NonRedundantGuess* and *LargeAlphabet*) in [3], because they have good theoretical time-bounds. Our experimental results show that *NonRedundantGuess* and *LargeAlphabet* are actually much slower than the deterministic algorithms, although their theoretical time-bounds look better. Of special interest is that for large alphabets (such as protein strings), our careful implementation of the *2-string* algorithm is much faster than all the other algorithms including *TraverString* and *qPMS9*. Hence, for large alphabets, the *2-string* algorithm is an ideal algorithm because it not only has the best-known theoretical time-bound but also can be implemented into a program that outperforms all the other known programs for CSP in practice.

We also design and implement a new algorithm for the binary case of CSP. The new algorithm runs in $O(nL + n^2d \cdot 6.16^d)$ time and is hence faster than the previously best deterministic algorithm (namely, the *3-string* algorithm). It is worth pointing out that although the best-known randomized algorithm (namely, *NonRedundantGuess*) for the binary case runs in $O(nL + n\sqrt{d} \cdot 5^d)$ expected time, its running time is random and it may fail to find a solution even if one exists. Hence, one cannot say that *NonRedundantGuess* is better than our new deterministic algorithm. Indeed, as aforementioned, it turns out that *NonRedundantGuess* is actually very slow in practice. Another drawback of randomized algorithms is that they cannot be used to enumerate all solutions. In real applications of CSP, we actually need to enumerate all solutions rather than finding a single one. We can claim that all the deterministic algorithms can be used for this purpose with their time bounds remaining intact.

Nishimura and Simjour [10] designed an algorithm for (implicitly) enumerating all solutions of a given instance (S, d). For binary strings, their algorithm runs in time $O\left(nL + nd \cdot ((n+1)(d+1))^{\lceil \log_{(1-\delta/2)} \epsilon \rceil} 5^{d(1+\epsilon+\delta)}\right)$ for any $0 < \delta \leq 0.75$ and $0 \leq \epsilon \leq 1$, and they claim that their time-bound is asymptotically better than $O(nL + nd^3 \cdot 6.731^d)$ which is achieved by the 3-string algorithm. In order for their claim to hold, $\epsilon + \delta < \log_5 6.731 - 1$ and $5^d \geq (n+1)^\gamma(d+1)^\rho$, where γ and ρ are the minimum values of the functions $\frac{\log_{(1-\delta/2)} \epsilon}{\log_5 6.731 - 1 - \epsilon - \delta}$ and $\frac{(\log_{(1-\delta/2)} \epsilon) - 2}{\log_5 6.731 - 1 - \epsilon - \delta}$ under the condition $\epsilon + \delta < \log_5 6.731 - 1$, respectively. Since $\gamma \geq 1795$ and $\rho \geq 1765$, their claim holds only when $5^d \geq (n+1)^{1795}(d+1)^{1765}$ (i.e., the parameter d is very large). In other words, their claim is false when $5^d < (n+1)^{1795}(d+1)^{1765}$ (i.e., the parameter d is not very large, which is often the case for a fixed-parameter algorithm to be meaningful).

2   Notations

Throughout this paper, Σ denotes a fixed alphabet and a string always means one over Σ. For a string s, $|s|$ denotes the length of s. For each $i \in \{1, 2, \ldots, |s|\}$, $s[i]$ denotes the letter of s at its i-th position. A *position set* of a string s is a subset of $\{1, 2, \ldots, |s|\}$. For two strings s and t of the same length, $d(s,t)$ denotes their Hamming distance.

Two strings s and t of the same length L *agree* (respectively, *differ*) *at a position* $i \in \{1, 2, \ldots, L\}$ if $s[i] = t[i]$ (respectively, $s[i] \neq t[i]$). The *position set where s and t agree* (respectively, *differ*) is the set of all positions $i \in \{1, 2, \ldots, L\}$ where s and t agree (respectively, differ). The proofs in the paper frequently use the position sets where a few strings are the same as or different from each other. The following special notations will be very useful. For two or more strings $s_1, \ldots, s_h$ of the same length, $\{s_1 \equiv s_2 \equiv \cdots \equiv s_h\}$ denotes the position set where s_i and s_j agree for all pairs (i, j) with $1 \leq i < j \leq h$, while $\{s_1 \not\equiv s_2 \not\equiv \cdots \not\equiv s_h\}$ denotes the position set where s_i and s_j differ for *all* pairs (i, j) with $1 \leq i < j \leq h$. Moreover, for a sequence $s_1, \ldots, s_h, t_1, \ldots, t_k$ of strings of the same length with $h \geq 2$ and $k \geq 1$, $\{s_1 \equiv s_2 \equiv \cdots \equiv s_h \not\equiv t_1 \not\equiv t_2 \not\equiv \cdots \not\equiv t_k\}$ denotes $\{s_1 \equiv s_2 \equiv \cdots \equiv s_h\} \cap \{s_h \not\equiv t_1 \not\equiv t_2 \not\equiv \cdots \not\equiv t_k\}$.

Another useful concept is that of a *partial string*, which is a string whose letters are only known at its certain positions. If s is a string of length L and P is a position set of s, then $s|_P$ denotes the partial string of length L such that $s|_P[i] = s[i]$ for each position $i \in P$ but $s|_P[j]$ is unknown for each position $j \in \{1, 2, \ldots, L\} \setminus P$. Let t be another string of length L. For a subset P of $\{1, 2, \ldots, L\}$, the *distance* between $s|_P$ and $t|_P$ is $|\{i \in P \mid s[i] \neq t[i]\}|$ and is denoted by $d(s|_P, t|_P)$. For two disjoint position sets P and Q of s, $s|_P + t|_Q$ denotes the partial string $r|_{P \cup Q}$ such that $r|_{P \cup Q}[i] = \begin{cases} s[i], \text{ if } i \in P; \\ t[i], \text{ if } i \in Q. \end{cases}$

At last, when an algorithm exhaustively tries all possibilities to find the right choice, we say that the algorithm *guesses* the right choice.

3   A New Algorithm for the Binary Case

A sequence $(x_1, \ldots, x_k)$ of nonnegative integers is *superdecreasing* if for all $1 \leq i \leq k - 1$, $x_i \geq \sum_{j=i+1}^{k} x_j$. For a nonnegative integer x and a positive integer k, let $\mathcal{S}_k(x)$ denote the set of all superdecreasing sequences $(x_1, \ldots, x_k)$ of k nonnegative integers with $\sum_{i=1}^{k} x_i = x$.

Lemma 1. *Let k be a positive integer, and x and X be two nonnegative integers. Consider the function f_X that maps each $(x_1, \ldots, x_k) \in \mathcal{S}_k(x)$ to $\prod_{i=1}^{k} \binom{X + x_i}{x_i}$. Then, f_X reaches its maximum value when $x_1 = \lceil \frac{x}{2} \rceil$.*

Lemma 2. *Let k, x, X, and f_X be as in Lemma 1. Then,*

$$\sum_{(x_1, \ldots, x_k) \in \mathcal{S}_k(x)} f_X(x_1, \ldots, x_k) \leq \binom{X + \lceil \frac{x}{2} \rceil}{\lceil \frac{x}{2} \rceil} \binom{X + \lfloor \frac{x}{2} \rfloor}{\lfloor \frac{x}{2} \rfloor} \cdot 2^x \cdot \left(\frac{4}{3}\right)^k.$$

Theorem 1. *The algorithm in Fig. 1 is correct and runs in $O^*(6.16^d)$ time.*

Proof. The algorithm is a simple modification of the *2-string* algorithm in [2] (see Fig. 2). The only difference between the two is the way of guessing $t_{sol}|_{A_1}$. In more details, the *2-string* algorithm guesses $t_{sol}|_{A_1}$ by guessing $\{t_{sol} \not\equiv s_0\} \cap A_1$ and further obtaining $t_{sol}[p]$ by flipping $s_0[p]$ for each $p \in \{t_{sol} \not\equiv s_0\} \cap A_1$. So, the algorithm is clearly correct.

We next analyze the time complexity. Fix an i with $2 \leq i \leq \log d$. For each $1 \leq j \leq i - 1$, let $\lambda_j = d(s_i|_{A_i}, t_{sol}|_{A_i})$. Since $d(s_i, t_{sol}) \leq d$ and $\lambda \leq \lambda_1$, we have

$$\lambda + \sum_{j=2}^{i-1} \lambda_j + |A_i| - \delta_i + \sum_{j=i+1}^{\log d} \delta_j \leq d. \tag{1}$$

Moreover, since $d(s_0, s_1)$ is maximized over all pairs of strings in S, $d(s_i, s_0) + d(s_i, s_1) \leq 2d(s_0, s_1) = 2|A_1|$. Summing up the contribution of the positions in

---

Input: An instance $\langle S, d \rangle$ of the binary case of CSP.
Output: A solution to $\langle S, d \rangle$ if one exists, or NULL otherwise.

1. Select a pair of strings from S whose Hamming distance is maximized over all pairs of strings in S. Without loss of generality, assume that (s_0, s_1) is such a pair.
2. Let $P = A_1 = \{s_0 \not\equiv s_1\}$.
3. Guess $s_h \in S$ (by trying every string in S) such that $d(s_h|_P, t_{sol}|_P)$ is minimized over all strings in S, where t_{sol} is a fixed (unknown) solution to $\langle S, d \rangle$. Let $\lambda = d(s_h|_P, t_{sol}|_P)$.
4. Guess $t_{sol}|_P$ (by trying all choices of λ positions in P and flipping the letters of s_h in the λ chosen positions). Let $t = t_{sol}|_P + s_0|_{\{1,2,\ldots,|s_0|\}\setminus P}$ and $b = d - d(s_0|_P, t|_P)$.
5. For $i = 2, 3, \ldots, \log d$ (in this order), perform the following steps:
   5.1. If every string in S is within a Hamming distance at most d from t, then output t and halt.
   5.2. Find a string in S whose Hamming distance from t is at least $d + 1$. Without loss of generality, assume that s_i is such a string. Let $A_i = \{s_0 \equiv s_1 \equiv \cdots \equiv s_{i-1} \not\equiv s_i\}$.
   5.3. Let $\ell = d(t, s_i) - d$ and $R = \{s_i \not\equiv t\} \setminus P$.
   5.4. If $\ell > \min\{b, |R|\}$, then return NULL.
   5.5. Guess $\delta_i = d(t|_{A_i}, t_{sol}|_{A_i})$ (by trying all integers in $\{\ell, \ell+1, \ldots, b\}$).
   5.6. Guess $t_{sol}|_{A_i}$ and modify t accordingly (by trying all choices of δ_i positions in R and flipping the letters of t in the chosen δ_i positions).
   5.7. Update $b = \min\{b - \delta_i, \delta_i - \ell\}$ and $P = P \cup R$.
6. Return NULL.

Fig. 1. The new algorithm for the binary case

Input: An instance $\langle \mathcal{S}, d, t, P, b \rangle$ of ECSP.
Output: A solution to $\langle \mathcal{S}, d, t, P, b \rangle$ if one exists, or NULL otherwise.

1. If there is no $s \in \mathcal{S}$ with $d(t, s) > d$, then output t and halt.
2. If $d = b$, then find a string $s \in \mathcal{S}$ such that $d(t, s)$ is maximized over all strings in $\mathcal{S}$; otherwise, find an arbitrary string $s \in \mathcal{S}$ such that $d(t, s) > d$.
3. Let $\ell = d(t, s) - d$ and $R = \{s \not\equiv t\} \setminus P$.
4. If $\ell > \min\{b, |R|\}$, then return NULL.
5. Guess $t_{sol}|_R$ by performing the following steps, where t_{sol} is a fixed (unknown) solution to $\langle \mathcal{S}, d, t, P, b \rangle$.
   5.1 Guess two sets X and Y such that $Y \subseteq X \subseteq R$, $\ell \leq |X| \leq b$, and $|Y| \leq |X| - \ell$.
   5.2 For each $i \in Y$, guess a letter z_i different from both $s[i]$ and $t[i]$. Let the partial string $\hat{s}|_Y$ be such that $\hat{s}|_Y[i] = z_i$ for all $i \in Y$.
   5.3 Let $t_{sol}|_R = \hat{s}|_Y + s|_{X \setminus Y} + t|_{R \setminus X}$.
6. Let $t' = t_{sol}|_R + t|_{\{1,2,\ldots,|t|\}\setminus R}$ and $b' = \min\{b - |X|, |X| - \ell - |Y|\}$. (*Comment:* $d(t, t') = |X|$.)
7. Solve $\langle \mathcal{S} \setminus \{s\}, d, t', P \cup R, b' \rangle$ recursively.
8. Return NULL.

Fig. 2. The 2-string algorithm given in [2]

each A_j $(1 \leq j \leq i)$ towards $d(s_i, s_0) + d(s_i, s_1)$, we have $|A_1| + \sum_{j=2}^{i-1} 2|\delta_j - \lambda_j| + 2|A_i| \leq d(s_i, s_0) + d(s_i, s_1) \leq 2|A_1|$. Thus,

$$\sum_{j=2}^{i-1}(\delta_j - \lambda_j) + |A_i| \leq \sum_{j=2}^{i-1}|\delta_j - \lambda_j| + |A_i| \leq \frac{|A_1|}{2}. \tag{2}$$

Adding up Eqs. 1 and 2, we have $\lambda + \sum_{j=2}^{i-1}\delta_j + 2|A_i| - \delta_i + \sum_{j=i+1}^{\log d}\delta_j \leq \frac{|A_1|}{2} + d$. Let $d' = \sum_{j=2}^{\log d}\delta_j$. Then, $\lambda + 2|A_i| - 2\delta_i + d' \leq \frac{|A_1|}{2} + d$. So, $|A_i| \leq \frac{|A_1|}{4} + \frac{d}{2} - \frac{d'}{2} - \frac{\lambda}{2} + \delta_i$. Since $|A_1| + 2d' = d(s_0, t_{sol}) + d(s_1, t_{sol}) \leq 2d$, we now have $|A_1| \leq 2(d - d')$ and hence $|A_i| \leq d - d' - \frac{\lambda}{2} + \delta_i$.

Let $X = d - d' - \frac{\lambda}{2}$ and $k = \log d$. By Lemma 2, the exponential factor in the time complexity of the algorithm is bounded from above by

$$\binom{|A_1|}{\lambda} \sum_{(\delta_2, \ldots, \delta_k) \in \mathcal{S}_{k-1}(d')} \prod_{i=2}^{k} \binom{X + \delta_i}{\delta_i}$$

$$\leq \binom{|A_1|}{\lambda}\binom{X + \lceil \frac{d'}{2} \rceil}{\lceil \frac{d'}{2} \rceil}\binom{X + \lfloor \frac{d'}{2} \rfloor}{\lfloor \frac{d'}{2} \rfloor} \cdot 2^{d'} \cdot \left(\frac{4}{3}\right)^{k-1}$$

$$\leq d \cdot \binom{2(d - d')}{\lambda}\binom{X + \lceil \frac{d'}{2} \rceil}{\lceil \frac{d'}{2} \rceil}\binom{X + \lfloor \frac{d'}{2} \rfloor}{\lfloor \frac{d'}{2} \rfloor} \cdot 2^{d'}$$

$$\leq d \cdot \binom{2(d - d')}{\lambda}\binom{d - \frac{\lambda}{2} - \lfloor \frac{d'}{2} \rfloor}{\lceil \frac{d'}{2} \rceil}\binom{d - \frac{\lambda}{2} - \lceil \frac{d'}{2} \rceil}{\lfloor \frac{d'}{2} \rfloor} \cdot 2^{d'}.$$

Now, using Stirling's formula, the exponential factor in the time complexity of the algorithm is bounded from above by

$$\frac{(2(d - d'))^{2(d-d')}}{\lambda^{\lambda} \cdot (2(d - d') - \lambda)^{2(d-d')-\lambda}} \cdot \left(\frac{\left(d - \frac{\lambda}{2} - \frac{d'}{2}\right)^{d - \frac{\lambda}{2} - \frac{d'}{2}}}{\left(\frac{d'}{2}\right)^{\frac{d'}{2}} \cdot \left(d - \frac{\lambda}{2} - d'\right)^{d - \frac{\lambda}{2} - d'}}\right)^2 \cdot 2^{d'}.$$

Let $a = \frac{\lambda}{d}$ and $b = \frac{d'}{d}$. Then, the above bound becomes c^d, where

$$c = \frac{(2(1 - b))^{2(1-b)}}{a^a \cdot (2(1 - b) - a)^{2(1-b)-a}} \cdot \left(\frac{\left(1 - \frac{a}{2} - \frac{b}{2}\right)^{1 - \frac{a}{2} - \frac{b}{2}}}{\left(\frac{b}{2}\right)^{\frac{b}{2}} \cdot \left(1 - \frac{a}{2} - b\right)^{1 - \frac{a}{2} - b}}\right)^2 \cdot 2^b$$

$$= \frac{2^{2-b}(1 - b)^{2(1-b)}(2 - a - b)^{2-a-b}}{a^a b^b (2 - a - 2b)^{2(2-a-2b)}}$$

Note that $a + b \leq 1$ for $\lambda + d' \leq d(s_0, t_{sol}) \leq d$. So, by numerical calculation, one can verify that $c \leq 6.16$ no matter what a and b are (as long as $a + b \leq 1$). Thus, the time complexity of the algorithm is $O^*(6.16^d)$.

4   Previous Algorithms

A number of parameterized algorithms whose time complexity has been rigorously analyzed have not been implemented. One objective of this paper is to implement the recent algorithms and see their performance in practice. We only sketch the *2-string* algorithm for CSP below.

The 2-String Algorithm has actually been implemented in [2]. However, as demonstrated in [11], the implementation in [2] yields a program (called *Provable*) which runs much slower than *TraverString*. In this paper, we give a different implementation of the *2-string* algorithm.

The algorithm is actually designed for a more general problem, called the *extended closest string problem* (ECSP for short). An instance of ECSP is a quintuple $(\mathcal{S}, d, t, P, b)$, where $\mathcal{S}$ is a set of strings of the same length L, t is a string of length L, d is a positive integer, P is a subset of $\{1, 2, \ldots, L\}$, and b is a nonnegative integer. The objective is to find a string t_{sol} of length L such that $t_{sol}|_P = t|_P$, $d(t_{sol}, t) \leq b$, and $\forall s \in \mathcal{S}$, $d(t_{sol}, s) \leq d$. Intuitively speaking, we want to transform t into a solution t_{sol} by modifying $\leq b$ positions of t outside P.

To solve a given instance (S, d) of CSP, it suffices to solve the instance $(S \setminus \{t\}, d, t, \emptyset, d)$ of ECSP, where t is an arbitrary string in S. The algorithm for ECSP is detailed in Fig. 2.

5   Implementing the Algorithms

Because each step of *NonRedundantGuess* and *LargeAlphabet* is very simple, their implementation is rather straightforward. So, we only describe the main ideas used in our implementation of the deterministic algorithms below.

Basically, each of the deterministic algorithms maintains a string t, a set P of fixed positions of t, and a bound b, and tries to transform t into a solution by selecting and modifying at most b *unfixed* positions (i.e., positions outside P) of t. It is possible that there is no way to transform t into a solution by modifying at most b unfixed positions of t. We want to efficiently decide if this is really the case. The next lemma can be used for this purpose.

Lemma 3 [11]. *Let u, v, and w be three strings of the same length K. Then, there is a string t_{sol} of length K such that $d(t_{sol}, u) \leq d_u$, $d(t_{sol}, v) \leq d_v$, and $d(t_{sol}, w) \leq d_w$ if and only if the following conditions hold:*

1. $d_u \geq 0$, $d_v \geq 0$, and $d_w \geq 0$.
2. $d(u, v) \leq d_u + d_v$, $d(u, w) \leq d_u + d_w$, and $d(v, w) \leq d_v + d_w$.
3. $d_u + d_v + d_w \geq |\{u \equiv v \not\equiv w\}| + |\{u \equiv w \not\equiv v\}| + |\{v \equiv w \not\equiv u\}| + 2|\{u \not\equiv v \not\equiv w\}|$.

As an example, we explain how to use Lemma 3 to prune a search tree for the *2-string algorithm*. Consider a call of the algorithm on input $\langle \mathcal{S}, d, t, P, b \rangle$. For convenience, let $Q = \{1, \ldots, L\} \setminus P$ and $K = L - |P|$. For each $u \in \mathcal{S}$, let $d_u = d - d(t|_P, u|_P)$. Recall that t has been obtained by modifying the fixed positions of some $\tilde{u} \in \mathcal{S}$. Hence, $b = d_{\tilde{u}}$. What the algorithm needs to do is to transform $t|_Q$ into a string t_{sol} of length K such that $d(t_{sol}, u|_Q) \leq d_u$ for all $u \in S$. To decide if such a transformation exists, we want to check if Conditions 1, through 3 in Lemma 3 hold for every triple $\{u, v, w\}$ of strings in S. However, there are $\Omega(|S|^3)$ such triples and hence it is time-consuming and wasteful to do the checking for all of them. So, in our implementation, we only check those triples (u, v, w) such that $u = \tilde{u}$, v is the string s selected in Step 2 of the algorithm, and $w \in S \setminus \{u, v\}$. If the checking fails for at least one such triple, Lemma 3 ensures that t cannot be transformed into a solution by selecting and modifying at most b unfixed positions of t.

5.1   Enumerating Subsets of Unfixed Positions

In Step 5 of the *2-string* algorithm, we need to decide which unfixed positions of t should be selected and further how to modify them. The other deterministic algorithms have the same issue. We only explain how to deal with this issue for the *2-string* algorithm below; the same can be done for the other algorithms.

In Step 5, we need to enumerate all subsets X of R with $\ell \leq |X| \leq b$. Then, for each enumerated X, we need to enumerate all subsets Y of X with $|Y| \leq |X| - \ell$. Furthermore, for each enumerated Y, we need to enumerate all valid ways of modifying the positions of t in Y. Roughly speaking, in the implementation of the *2-string* algorithm done in [2], only after enumerating X and Y, we start to enumerate all valid ways of modifying the positions of t in Y. So, in the implementation in [2], every possible combination of X and Y will be enumerated because only after modifying one or more positions of t in X, we can decide if it is unnecessary to make a certain recursive call on the modified t (i.e., if it is possible to prune a certain branch of the search tree). This seems to be the main reason whey the implementation in [2] yields a slow program for CSP.

To get over the above-mentioned drawback of the implementation in [2], our idea is to enumerate the elements of X and Y one by one and at the same time enumerate all possible ways of modifying each position in Y. In more details, we scan the positions in R one by one (in any order). When scanning a $p \in R$, we need to make two choices depending on whether to include p in X or not. If we decide to exclude p from X, then we proceed to the next position in R. Otherwise, we need to make two choices depending on whether to include p in Y or not. If we decide to exclude p from Y, then we modify t by changing $t[p]$ to $s[p]$ and then use Lemma 3 to check if the modified t can be further transformed into a solution by modifying at most $b - 1$ positions outside $P \cup \{p\}$. If the checking yields a "no" answer, then we can quit scanning the remaining positions in R and hence prune a certain branch of the search tree at an early stage. Similarly, if we decide to include p in Y, then we need to make $|\Sigma| - 2$ choices depending on to which letter we should change $t[p]$. For each of the choices, after modifying position p of t, we use Lemma 3 to check if the modified t can be further transformed into a solution by modifying at most $b - 1$ positions outside $P \cup \{p\}$. If the checking yields a "no" answer, then we can quit scanning the remaining positions in R and hence prune a certain branch of the search tree at an early stage.

---

5. Let p_1, p_2, ..., p_q be the positions in R. Initialize $X = \emptyset$ and $Y = \emptyset$. For $i = 1, 2, \ldots, q$ (in this order), guess $t_{sol}[p_i]$ by performing Steps 5.1 through 5.3:

   5.1 If $|X| + q - i + 1 = \ell$, then add p_i to X, set $t_{sol}[p_i] = s[p_i]$, and proceed to the next i (without performing Steps 5.2 and 5.3).

   5.2 If $|X| + q - i + 1 > \ell$, perform the following steps:

      5.2.1 Guess whether $p_i \in X$ or not.

      5.2.2 If the guess is $p_i \in X$, then add p_i to X and perform Steps 5.2.2.1 through 5.2.2.4:

         5.2.2.1 If $|X| + q - i - |Y| = \ell$, then set $t_{sol}[p_i] = s[p_i]$ and proceed to the next i (without performing Steps 5.2.2.2 through 5.2.2.4 and 5.3).

         5.2.2.2 Guess whether $p_i \in Y$ or not.

         5.2.2.3 If the guess is $p_i \notin Y$, then set $t_{sol}[p_i] = s[p_i]$; otherwise, add p_i to Y and guess $t_{sol}[p_i]$ from $\Sigma \setminus \{t[p_i], s[p_i]\}$.

         5.2.2.4 Proceed to the next i (without performing Step 5.3).

   5.3 Set $t_{sol}[p_i] = t[p_i]$.

Fig. 3. Modifying Step 5 of the 2-string algorithm in Figure 2

More formally, we modify Step 5 in the *2-string* algorithm as shown in Fig. 3. A crucial but missing detail in the modified Step 5 is that before we decide to set $t_{sol}[p_i]$ to be a certain letter $a \neq t[p_i]$ in Step 5.1, 5.2.2.1, or 5.2.2.3, we actually use Lemma 3 to check if setting $t_{sol}[p_i] = a$ can lead to a solution as follows. First, we obtain a string u from t by changing $t[p_i]$ to a and changing $t[p_j]$ to $t_{sol}[p_j]$ for all $j \in \{1, 2, \ldots, i-1\}$. We then compute $d_u = b - d(t, u)$, set $v = s$, and compute $d_v = d - d(v|_{P \cup \{p_1, \ldots, p_i\}}, u|_{P \cup \{p_1, \ldots, p_i\}})$. For all $w \in \mathcal{S}$, we further compute $d_w = d - d(w|_{P \cup \{p_1, \ldots, p_i\}}, u|_{P \cup \{p_1, \ldots, p_i\}})$ and now check if u, v, w, d_u, d_v, and d_w altogether satisfy Conditions 1 through 3 in Lemma 3. If this checking fails for at least one w, then we can conclude that setting $t_{sol}[p_i] = a$ cannot lead to a solution, and hence we can quit setting $t_{sol}[p_i] = a$ (i.e., can prune a certain branch of the search tree).

5.2 Sorting the Input Strings

Again, we explain the idea by using the *2-string* algorithm as an example. The idea also applies to the other deterministic algorithms. As mentioned in the above (immediately before Sect. 5.1), when we apply Lemma 3, we only check those triples (u, v, w) such that u is the input string from which the current t has been obtained, v is the string selected in Step 2 of the algorithm, and $w \in \mathcal{S} \setminus \{u, v\}$. So, there are $|\mathcal{S}| - 2$ choices for w and we can try the choices in any order. As one can expect, different orders lead to different speeds. In our implementation, we try the choices for w in descending order of the Hamming distance of w from t. Intuitively speaking, this order seems to enable us to find out that t cannot be transformed into a solution at an earlier stage than other orders.

5.3   On Implementing the Algorithm in Sect. 3

In Step 3 of our new algorithm, we need to guess an input string s_h such that $d(s_h|_P, t_{sol}|_P)$ is minimized over all input strings, where t_{sol} is a fixed solution. To guess s_h, a simple way is to try all input strings (in any order). When trying a particular s_h, we use Lemma 4 to cut unnecessary branches of the search tree.

Lemma 4. *Let P be as in Step 2 of the algorithm. Further let t_{sol} and s_h be as in Step 3 of the algorithm. Then, for every $s_j \in S$, $d(s_h|_D, t_{sol}|_D) \leq \frac{|D|}{2}$, where $D = \{s_h|_P \not\equiv s_j|_P\}$.*

To use Lemma 4, we first compute $m_j = \frac{d(s_h|_P, s_j|_P)}{2}$ for each $s_j \in S$ in Step 3. Later in Step 4, we scan the positions in P one by one (in any order). When scanning a position $p \in P$, we need to make two recursive calls – one of them corresponds to flipping the letter of s_h at position p while the other corresponds to keeping the letter of s_h at position p intact. Before making each of the calls, we decrease m_j by 1 for all $s_j \in S$ such that the letter of s_h at position p has become different from $s_j[p]$ after scanning p; if m_j becomes negative for some $s_j \in S$, then we know that it is unnecessary to make the recursive call. In this way, we are able to cut certain unnecessary branches of the search tree.

6   Results and Discussion

We have implemented the new algorithm in Sect. 3 and the previously known algorithms reviewed in Sect. 4. As the result, we have obtained a program (written in C) for each of the algorithms. We not only compare these programs against each other but also include *TraverString* and *qPMS9* in the comparison. We do not compare with the algorithm in [6], because its code is not available. The machine on which we ran the programs is an Intel Core i7-975 (3.33GHz, 6MB Cache, 6GB RAM in 64bit mode) Linux PC.

As in previous studies, we generate random instances of CSP as input to the programs. In the generation of instances, we fix $L = 600$ and $n = 20$ but varies d and $|\Sigma|$. As usual, the choices for $|\Sigma|$ we employ in our test are 2 (binary strings), 4 (DNA strings), and 20 (protein strings). Choosing d is less obvious. Clearly, the larger d is, the longer the programs run. To distinguish the programs from each other in terms of speed, we consider the following five ranges of d: (1) $10 \leq d \leq 15$, (2) $28 \leq d \leq 33$, (3) $80 \leq d \leq 85$, (4) $82 \leq d \leq 87$, and (5) $d \in \{89, 92, 95, 98, 101\}$. The choice of these ranges is based on the expectation that some of the programs can be slow even if d is small, while the others can show their significant difference in running time only if d is modest or even large.

Since some of the programs can be very slow for certain d, we set a time limit of 5 hours on each run of each program in our test. For each setting of parameters (e.g., $(L, n, |\Sigma|, d) = (600, 20, 4, 15)$), we generate five instances, pass them to the programs, and require each program to find *all* solutions for each instance. We will summarize our experimental results in several tables. If a program solves all

Table 1. Comparison of the programs for $|\Sigma| = 4,\ 20,\ 2$

	Program	$d=10$	$d=11$	$d=12$	$d=13$	$d=14$	$d=15$
	NonRedundantGuess	5.6m	48m	TL	TL	TL	TL
	LargeGuess	< 0.05s	< 0.05s	5.2s	23s	2.1m	8.7m
	2-string	< 0.05s	< 0.05s	< 0.05s	< 0.05s	< 0.05s	< 0.05s
D	3-string	< 0.05s	< 0.05s	< 0.05s	< 0.05s	< 0.05s	< 0.05s
	TraverString	< 0.05s	< 0.05s	< 0.05s	< 0.05s	< 0.05s	< 0.05s
N	qPMS9	5.9s	5.9s	6.0s	5.9s	5.9s	5.9s
	Program	$d=80$	$d=81$	$d=82$	$d=83$	$d=84$	$d=85$
A	2-string	10s	1.6m	37s	8m	4.8m	6.3m
	3-string	11s	1.7m	41s	8.4m	5.4m	7m
	TraverString	5.7m	TL	11m	TL	TL	TL
	qPMS9	5.9s	5.9s	5.9s	5.9s	5.9s	5.9s
	Program	$d=10$	$d=11$	$d=12$	$d=13$	$d=14$	$d=15$
	NonRedundantGuess	TL	TL	TL	TL	TL	TL
	LargeGuess	2.6s	9.6s	13s	54s	6.7m	29m
	2-string	< 0.05s	< 0.05s	< 0.05s	< 0.05s	< 0.05s	< 0.05s
	3-string	< 0.05s	< 0.05s	< 0.05s	< 0.05s	< 0.05s	< 0.05s
p	TraverString	< 0.05s	< 0.05s	< 0.05s	< 0.05s	< 0.05s	< 0.05s
r	qPMS9	12s	12s	12s	12s	12s	12s
o	Program	$d=82$	$d=83$	$d=84$	$d=85$	$d=86$	$d=87$
t	2-string	< 0.05s	< 0.05s	< 0.05s	< 0.05s	< 0.05s	< 0.05s
e	3-string	< 0.05s	< 0.05s	< 0.05s	< 0.05s	< 0.05s	< 0.05s
i	TraverString	< 0.05s	< 0.05s	< 0.05s	< 0.05s	3.6s	< 0.05s
n	qPMS9	10m	53m	21m	TL	1.2h	TL
	Program	$d=89$	$d=92$	$d=95$	$d=98$	$d=101$	
	2-string	0.2s	1.6s	2.0s	2.2s	22s	
	3-string	0.2s	2.2s	2.1s	2.1s	25s	
	TraverString	1.6s	1.5m	2.0m	2.2m	TL	
b	Program	$d=28$	$d=29$	$d=30$	$d=31$	$d=32$	$d=33$
i	New	11m	19m	32m	1h	1.9h	TL
n	2-string	< 0.05s	< 0.05s	< 0.05s	< 0.05s	< 0.05s	< 0.05s
a	3-string	< 0.05s	< 0.05s	< 0.05s	< 0.05s	< 0.05s	< 0.05s
r	TraverString	< 0.05s	< 0.05s	< 0.05s	< 0.05s	< 0.05s	< 0.05s
y	qPMS9	3.1s	3.1s	3.1s	3.1s	3.1s	3.1s

of the five instances within the time limit, we calculate the average running time
and show it as is in a table; otherwise, we put a TL symbol in the corresponding
cell of the table, where TL stands for "time limit".

Table 1 shows the comparison of the programs for DNA, protein, or binary
strings. As seen from the table, NonRedundantGuess and LargeGuess are slower
than the other algorithms even though NonRedundantGuess and LargeGuess
have better theoretical time-bounds. In the table, New means our new algorithm
in Sect. 3. We exclude the experimental results for NonRedundantGuess and
LargeGuess from the table, because both failed to solve a single instance within
the time limit. As seen from the table, our new algorithm in Sect. 3 is much

faster than *NonRedundantGuess* and *LargeGuess*, but is much slower than the other deterministic algorithms. The reason why the new algorithm runs slower seems to be that the pruning inequalities in Lemma 3 are much less effective in cutting unnecessary branches of a search tree for our new algorithm.

Based on our above experimental results, we conclude that the *2-string* and the *3-string* algorithms have little difference in running time and they are the stablest algorithms among the tested algorithms. In particular, for large alphabets (such as protein strings), the *2-string* algorithm is an ideal algorithm because it not only has the best-known theoretical time-bound but also can be implemented into a program that outperforms the other known programs for CSP in practice.

References

1. Chen, Z.-Z., Ma, B., Wang, L.: A three-string approach to the closest string problem. J. Comput. Syst. Sci. **78**, 164–178 (2012)
2. Chen, Z.-Z., Wang, L.: Fast exact algorithms for the closest string and substring problems with application to the planted (ℓ, d)-motif model. IEEE/ACM Trans. Comput. Biol. Bioinf. **8**(5), 1400–1410 (2011)
3. Chen, Z.-Z., Ma, B., Wang, L.: Randomized fixed-parameter algorithms for the closest string problem. Algorithmica **74**, 466–484 (2016)
4. Frances, M., Litman, A.: On covering problems of codes. Theoret. Comput. Sci. **30**, 113–119 (1997)
5. Gramm, J., Niedermeier, R., Rossmanith, P.: Fixed-parameter algorithms for closest string and related problems. Algorithmica **37**, 25–42 (2003)
6. Hufsky, F., Kuchenbecker, L., Jahn, K., Stoye, J., Böcker, S.: Swiftly computing center strings. BMC Bioinform. **12**, 106 (2011)
7. Lanctot, K., Li, M., Ma, B., Wang, S., Zhang, L.: Distinguishing string search problems. Inform. Comput. **185**, 41–55 (2003)
8. Ma, B., Sun, X.: More efficient algorithms for closest string and substring problems. SIAM J. Comput. **39**(4), 1432–1443 (2010)
9. Nicolae, M., Rajasekaran, S.: qPMS9: an efficient algorithm for quorum planted motif search. Nat. Sci. Rep. **5** (2015)
10. Nishimura, N., Simjour, N.: Enumerating neighbour and closest strings. In: Thilikos, D.M., Woeginger, G.J. (eds.) IPEC 2012. LNCS, vol. 7535, pp. 252–263. Springer, Heidelberg (2012). doi:10.1007/978-3-642-33293-7_24
11. Tanaka, S.: Improved exact enumerative algorithms for the planted (l, d)-motif search problem (2014). IEEE/ACM Trans. Comput. Biol. Bioinf. **11**, 361–374 (2014)

The Broken-Triangle Property
with Adjoint Values

Jian Gao[1]($\boxtimes$), Rong Chen[1]($\boxtimes$), Minghao Yin[2]($\boxtimes$), and Hui Li[1]($\boxtimes$)

[1] College of Information Science and Technology, Dalian Maritime University,
Dalian 116026, China
{gaojian,rchen,lih2002}@dlmu.edu.cn
[2] College of Computer Science, Northeast Normal University,
Changchun 130024, China
ymh@nenu.edu.cn

Abstract. Recently, the Broken Triangle Property (BTP) and its extensions have been proposed to identify hybrid tractable classes of Constraint Satisfaction Problems (CSPs). In this paper, we extend the BTP to the concept of the Broken Triangle Property with adjoint values (BTPv), and then identify a more general hybrid tractable class of binary CSPs. To prove tractability, we present a polynomial-time algorithm to solve CSP instances in the new tractable class using a novel variable selection mechanism, and show correctness of it. We also show that determining whether an instance is in the class can be achieved efficiently. Furthermore, we provide comparisons with the BTP and its extensions showing that as a generalization of the BTP, the BTPv can find novel tractable CSPs, which cannot be identified by those existing tractable classes.

Keywords: Constraint Satisfaction Problem · Broken Triangle Property · Tractable class

1 Introduction

The Constraint Satisfaction Problem (CSP) [1] has emerged as a fundamental methodology to express and solve a large variety of practical problems in our society. While it can neatly model the real-world problem as a set of variables and a set of constraints that restrict the allowed values, the CSP is accompanied with computational complexity — in general case deciding whether a given CSP has a solution is an NP-complete problem.

CSP instances in the tractable classes can be solved (and often recognized) in polynomial time. Over the past three decades, a great effort has been made to find tractable classes by restricting the allowed constraint forms, or by identifying

This work was supported by the National Natural Science Foundation of China (No. 61402070, No. 61370156, No. 61503074, No. 61672122 and No. 61602077), and Natural Science Foundation of Liaoning Province (No. 2015020023).

© Springer International Publishing AG 2017
M. Xiao and F. Rosamond (Eds.): FAW 2017, LNCS 10336, pp. 91–102, 2017.
DOI: 10.1007/978-3-319-59605-1_9

structural properties of constraint networks [2]. Recently, many hybrid tractable classes [3–7] that combine structural and relational restrictions have been identified. Among them, the Broken Triangle Property (BTP) [8] outstands due to its prefect properties that not only include several previous tractable classes but also ensure its instances to be solved in polynomial time by maintaining arc consistency with any instantiation order [3]. The BTP also has important applications such as variable elimination [9] and identifying hidden tractable classes by enforcing local consistencies [10]. So far the BTP has been extended to the joint-winner property for valued CSPs [11], the extension for quantified CSPs [12], the Extendable-Triple Property (ETP) [13], and the dual broken triangle property for non-binary CSPs [14]. In addition, recently Naanaa [15] has proposed rank CSPs using independent set based on set theory. If a binary CSP instance has a fixed rank bounded by a value κ (the maximum independent set does not exceed κ) with respect to a fixed variable ordering and satisfies $\kappa + 1$ consistency, the problem is globally consistent.

The hybrid tractable classes of the BTP and of its extensions have a common property which guarantees tractability. Namely tractability is ensured during assigning variables, where for any current partial assignment, all unassigned variables have at least one consistent value to be assigned, and thus all variables can be assigned so as to construct a solution. To solve CSP instances in the BTP class and others, variables should be assigned with a pre-computed variable ordering. Furthermore, it can be seen that if there exists a right variable to assign for any partial assignment keeping the tractability property and if the right variable can be computed efficiently according to values assigned in the current partial assignment, the CSP can also be solved in a backtrack-free manner. So it is interesting to investigate required properties for computing the right variable efficiently to ensure the tractability and thus we can obtain novel tractable classes. Motivated by this, we propose the concept of adjoint values as requirements for computing dynamical variable orderings, and then we generalize the BTP by introducing the BTP with adjoint values. In doing so, we identify a novel tractable class as well as present an algorithm to solve CSPs in that class. Moreover, we prove determining whether a CSP is in the class can be achieved in polynomial time. Finally, we compare this new class with the BTP class and other existing hybrid tractable classes, and indicate that using the BTPv new tractable cases of CSPs can be identified.

2   Background

2.1   The Constraint Satisfaction Problem

Formally, a *CSP* is defined as a triple $P = (V, D, C)$, where $V = \{x_1, x_2, \ldots, x_n\}$ is a set of n variables. $D = \{D_1, D_2, \ldots, D_n\}$ is a set of domains of values, and each variable $x_i \in V$ can take its values in the finite domain $D_i \in D$. $C = \{C_1, \ldots, C_e\}$ is a set of constraints, which specifies the allowable combinations of values. A *constraint* $C_i \in C$ is a pair $(vars(C_i), rel(C_i))$, where $vars(C_i)$ is a subset of V called the constraint scope and $rel(C_i)$ is a subset of the *Cartesian* product $D_{i_1} \times \ldots \times D_{i_k}$. The constraint C_i is binary if and only if there are only two variables involved in C_i. A CSP is binary if all constraints are binary. For the sake of simplicity, we only consider binary constraints in this paper, and use the notation C_{ij} to denote the constraint between x_i and x_j, R_{ij} to denote its relations. Note that C_{ij} and C_{ji} are different constraints, and universal constraints should be added if there is no constraint between two variables.

A *partial assignment* to a set of variables $V' \subseteq V$ is a sequence of pairs of $\langle x_i, a_i \rangle$ where $x_i \in V'$ and $a_i \in D_i$. A partial assignment is *consistent* if it satisfies all constraints whose scope is a subset of the variables involved in the assignment. A *solution* to a CSP is a consistent partial assignment that includes all variables in V. Solving a CSP is to find a solution to it (satisfiable) or to determine there is no solution (unsatisfiable).

Given a constraint C_{ij}, a value $a_j \in D_j$ is a *support* of $a_i \in D_i$ if and only if $(a_i, a_j) \in R_{ij}$; support set $D_j(x_i = a_i)$ is a subset of D_j that is composed of all supports of a_i; we use $D_j(a_i)$ for short if x_i is specified in the context. Moreover, given a subnetwork including three variables x_i, x_j and x_k, $D_k(a_i, a_j) = D_k(a_i) \bigcap D_k(a_j)$ where $a_i \in D_i, a_j \in D_j$.

To ensure tractability for CSPs, consistencies are always required. A binary CSP $P = (V, D, C)$ is *arc consistent* if and only if for all ordered pairs of variables (x_i, x_j), for each value $a_i \in D_i$, there exists at least one value $a_j \in D_j$ such that $(a_i, a_j) \in R_{ij}$.

A binary CSP $P = (V, D, C)$ is *path consistent* if and only if for all ordered triples of variables (x_i, x_j, x_k), for each pair of $(a_i, a_j) \in R_{ij}$, there exists at least one value $a_k \in D_k$ such that $(a_i, a_k) \in R_{ik}$ and $(a_j, a_k) \in R_{jk}$. P is *Strong Path Consistent* (SPC) if and only if it is arc consistent and path consistent.

Given a binary CSP, enforcing arc consistency can be achieved in $O(ed^2)$ time [16], where e is the number of constraints and d is the domain size. On the other hand, enforcing path consistency on a binary CSP can be achieved in $O(n^3 d^3)$ time [17].

2.2 The Broken Triangle Property

Cooper et al. [8] defined the BTP on binary CSPs. The BTP generalizes tree structure, and identifies a new tractable class of CSPs. To extend the BTP by adjoint values, it requires deeper investigation on microstructure of a CSP (e.g. the BTP between two values and one variable). So we give a definition about the BTP on triples of two values and a variable.

Given a binary CSP $P = (V, D, C)$, where $a_i \in D_i$, $a_j \in D_j$ such that $(a_i, a_j) \in R_{ij}$, and $x_k \in V$ $(i, j \neq k)$. The ordered triple of (a_i, a_j, x_k) satisfies the BTP if and only if for each pair of values (a_{k_1}, a_{k_2}) such that $(a_i, a_{k_1}) \in R_{ik}$ and $(a_j, a_{k_2}) \in R_{jk}$, either $(a_i, a_{k_2}) \in R_{ik}$ or $(a_j, a_{k_1}) \in R_{jk}$. We use the notation $\mathcal{BTP}$ to denote the set including all triples of values and variables that satisfies the BTP, so the ordered triple (a_i, a_j, x_k) satisfies the BTP can be described as $(a_i, a_j, x_k) \in \mathcal{BTP}$. Also, it is noted that $(a_j, a_i, x_k) \in \mathcal{BTP}$ if $(a_i, a_j, x_k) \in \mathcal{BTP}$ because the order a_i and a_j can be exchanged in the BTP definition.

Moreover, let $x_i, x_j, x_k \in V$ ($i \neq j$, and $i, j \neq k$), the ordered triple of variables $(x_i, x_j, x_k) \in \mathcal{BTP}$ if and only if for each pair of $(a_i, a_j) \in R_{ij}$, the triple of $(a_i, a_j, x_k) \in \mathcal{BTP}$.

Let $\mathcal{A}$ be a consistent partial assignment and x_k be a variable that is not involved in $\mathcal{A}$. The ordered pair of $(\mathcal{A}, x_k) \in \mathcal{BTP}_{\mathcal{A}}$ if and only if either there is only one assignment in $\mathcal{A}$ or for each pair of (a_i, a_j) such that $\langle x_i, a_i \rangle \in \mathcal{A}$ and $\langle x_j, a_j \rangle \in \mathcal{A}$, the ordered triple of $(a_i, a_j, x_k) \in \mathcal{BTP}$. It is also said that $\mathcal{A}$ satisfies the BTP with respect to x_k.

Based on the definitions above, the BTP of a CSP is defined as follows.

Definition 1. *Given a binary CSP $P = (V, D, C)$, P satisfies the BTP with respect to a variable ordering $\prec$ if and only if for all triples of variables (x_i, x_j, x_k) such that $x_i \prec x_j \prec x_k$, the triple $(x_i, x_j, x_k) \in \mathcal{BTP}$.*

According to Lemma 2.4 in [3], the BTP can also be described by set inclusion. Given $(a_i, a_j) \in R_{ij}$ and a variable x_k ($i, j \neq k$), $(a_i, a_j, x_k) \in \mathcal{BTP}$ if and only if either $D_k(a_i) \subseteq D_k(a_j)$ or $D_k(a_i) \supseteq D_k(a_j)$.

A binary CSP that satisfies the BTP with a variable ordering $\prec$ can be solved in polynomial time. Moreover, there exists an algorithm that runs in polynomial time to find the variable ordering if it exists or to determine that there is no such ordering. The BTP also has some nice properties that distinguish the BTP class from other tractable classes. For example, the BTP with respect to any fixed variable ordering is conservative, and this property ensures that it is closed under domain restrictions, so a CSP satisfying the BTP retains the property after enforcing arc consistency.

3 Adjoint Values of the BTP

3.1 Definitions

First, we define adjoint values with respect to the BTP.

Definition 2. *Given a binary CSP $P = (V, D, C)$, a value $a_j \in D_j$ is adjoint to $a_i \in D_i$ ($i \neq j$) with respect to the BTP if and only if $(a_i, a_j) \in R_{ij}$, and for each value $a_k \in D_k$ and for each variable x_l such that $(a_i, a_k) \in R_{ik}$, $(a_j, a_k) \in R_{jk}$ where x_i, x_j, x_k and x_l are different variables, $(a_i, a_k, x_l) \in \mathcal{BTP}$ implies $(a_j, a_k, x_l) \in \mathcal{BTP}$.*

An adjoint value a_j is included in at least as many BTP triples as a_i, which a_j is adjoint to. So assigning a_j together with a_i will not break the BTP held by a_i. For the simplicity, we always say a_j is adjoint to a_i for short instead of a_j is adjoint to a_i with respect to the BTP.

A *BTP-adjoint value set* of a value a_i with respect to x_j ($a_i \notin D_j$) is a subset of D_j such that each a_j in the set is adjoint to a_i. $D_j^{adj}(a_i)$ denotes the BTP-adjoint value set of a_i.

Based on adjoint values, we introduce the notion *Broken-Triangle Property with adjoint values* (BTPv for short). Given a_i, a_j such that $(a_i, a_j) \in R_{ij}$ and a variable x_k $(i,j \neq k)$, (a_i, a_j, x_k) satisfies the *BTP-adjoint property* if and only if for each $a_k \in D_k(a_i, a_j)$, both $a_k \in D_k^{adj}(a_i)$ and $a_k \in D_k^{adj}(a_j)$. The BTP-adjoint property requires that any consistent value in D_k must be adjoint to both a_i and a_j. Same as the BTP, notation $\mathcal{BTP_V}$ is defined as follows: $(a_i, a_j, x_k) \in \mathcal{BTP_V}$ if and only if either $(a_i, a_j, x_k) \in \mathcal{BTP}$ or (a_i, a_j, x_k) satisfies the BTP-adjoint property, so it can be seen that the $\mathcal{BTP_V}$ includes the $\mathcal{BTP}$. In addition, given three distinct variables x_i, x_j and x_k, $(x_i, x_j, x_k) \in \mathcal{BTP_V}$ if and only if for each pair of $(a_i, a_j) \in R_{ij}$, the triple $(a_i, a_j, x_k) \in \mathcal{BTP_V}$.

A binary CSP satisfies the BTPv is defined as follows.

Definition 3. *A CSP $P = (V, D, C)$ satisfies the BTPv with respect to a variable ordering $\prec$ if and only if for each triple of variables (x_i, x_j, x_k) such that $x_i \prec x_j \prec x_k$, $(x_i, x_j, x_k) \in \mathcal{BTP_V}$.*

Next, we analyze the computational complexity. Given two values a_i and a_j, testing whether a_j is in $D_j^{adj}(a_i)$ can be decided in $O(n^2d^2)$ time. This task can be achieved by testing whether each triple of $(a_i, a_l, x_k) \in \mathcal{BTP}$ implies $(a_j, a_l, x_k) \in \mathcal{BTP}$. Since there are $O(n^2d)$ triples to be tested, while testing whether a support set includes another one requires $O(d)$ time, so the testing is in $O(n^2d^2)$ time. Moreover, to compute all adjoint values of each value in a CSP, it requires $O(n^4d^4)$ time as there are $O(n^2d^2)$ pairs of values.

Finally, we give an example of the BTPv.

Example 1. Figure 1 shows a microstructure of a constraint network. The triple (x_1, x_2, x_3) satisfies the BTP, while triples involving variable x_4 does not. It can be seen that $(x_1, x_2, x_4) \notin \mathcal{BTP}$ as $D_4(x_1 = 1)$ and $D_4(x_2 = 1)$ cannot include each other, whereas $(x_1, x_2, x_4) \in \mathcal{BTP_V}$ as $x_4 = 1, 2, 3$ are adjoint to both $x_1 = 1$ and $x_2 = 1$, respectively.

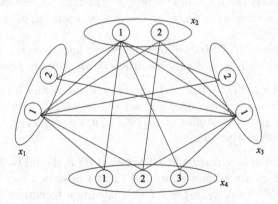

Fig. 1. The microstructure of Example 1

3.2 Conflicts in the BTPv

This subsection discusses a special case of the BTPv. Consider a subnetwork with four variables x_i, x_j, x_k and x_l, which is shown as Fig. 2, where $D_i = D_j = \{1\}$, $D_k = \{2\}$, $D_l = \{1, 2, 3, 4\}$, and lines between values are binary relations, for example, $D_l(x_i = 1) = \{1, 2, 3\}$. Assume $x_k = 2$ is adjoint to $x_i = 1$ and $x_j = 1$; $x_l = 1, 2, 3$ is adjoint to $x_i = 1$; $x_l = 2, 3, 4$ is adjoint to $x_j = 1$; and $x_l = 1, 4$ is adjoint to $x_k = 2$, so $(x_i = 1, x_j = 1, x_k) \in \mathcal{BTP}_\mathcal{V}$ and $(x_i = 1, x_j = 1, x_l) \in \mathcal{BTP}_\mathcal{V}$. Under the current partial assignment $\{\langle x_i, 1\rangle, \langle x_j, 1\rangle, \langle x_k, 2\rangle\}$, the next variable to be assigned is x_l, but this partial assignment leads to no value that can be selected for x_l due to the empty set $D_l(x_i = 1) \bigcap D_l(x_j = 1) \bigcap D_l(x_k = 2)$, though the BTPv can be satisfied with respect to the ordering $x_i \prec x_j \prec x_k \prec x_l$.

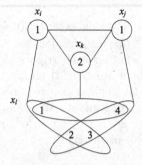

Fig. 2. Relations of a subnetwork with four variables

The case we describe above is a conflict in adjoint situations, where x_k and x_l are adjoint to $x_i = 1$ and $x_j = 1$ at the same time. Formally, 4-tuple (a_i, a_j, x_k, x_l) has a *BTPv-conflict* with respect to x_l if and only if (a_i, a_j, x_k) and (a_i, a_j, x_l) satisfy the BTP-adjoint property but they do not satisfy the BTP, and there exists a value $a_k \in D_k(a_i, a_j)$ such that both $(a_i, a_k, x_l) \notin \mathcal{BTP}$ and $(a_j, a_k, x_l) \notin \mathcal{BTP}$.

To avoid this case in CSPs we investigate, we introduce the concept of no-conflict BTPv, which is a necessary condition to identify our tractable class.

Definition 4. *A CSP $P = (V, D, C)$ satisfies the no-conflict BTPv with respect to a variable ordering $\prec$ if and only if P satisfies the BTPv with $\prec$, and for each 4-tuple of (a_i, a_j, x_k, x_l) such that $(a_i, a_j) \in R_{ij}$ and $x_i, x_j, x_k \prec x_l$, there are no BTPv-conflict with respect to x_l.*

The no-conflict BTPv restricts that external BTP should be satisfied if values from x_k and x_l are adjoint to a_i and a_j at the same time. In fact, either $(a_i, a_k, x_l) \in \mathcal{BTP}$ or $(a_j, a_k, x_l) \in \mathcal{BTP}$ is adequate to ensure that x_l can be assigned by a consistent value if the problem is path consistent, so it is not a conflict if one of the triples (a_i, a_k, x_l) and (a_j, a_k, x_l) satisfies the BTP. The restriction of no-conflict BTPv is a necessary condition for the tractable class, which will be proved in the next section.

4   The Tractable Class

In this section, we identify a new tractable class of binary CSPs and show the main tractable result with the no-conflict BTPv, and also present an algorithm to solve CSPs in the class with a dynamic variable ordering.

Before presenting the tractable class, some lemmas are introduced first. The following one concerns the adjoint values.

Lemma 1. *Let $P = (V, D, C)$ be a binary CSP, $\mathcal{A}$ be a consistent partial assignment and x_k be a variable such that $(\mathcal{A}, x_k) \in \mathcal{BTP}_\mathcal{A}$. If a_l is adjoint to a_i such that $\langle x_i, a_i \rangle \in \mathcal{A}$, and if $(a_i, a_l, x_k) \in \mathcal{BTP}$, then $(\mathcal{A} \bigcup \{\langle x_l, a_l \rangle\}, x_k) \in \mathcal{BTP}_\mathcal{A}$.*

Proof. It is clear the lemma holds if $\langle x_l, a_l \rangle \in \mathcal{A}$. Otherwise, it suffices to show for each $\langle x_j, a_j \rangle \in \mathcal{A}$, $(a_j, a_l, x_k) \in \mathcal{BTP}$. If $x_j \neq x_i$, then $(a_j, a_i, x_k) \in \mathcal{BTP}$ as $(\mathcal{A}, x_k) \in \mathcal{BTP}_\mathcal{A}$. Since a_l is adjoint to a_i, $(a_j, a_l, x_k) \in \mathcal{BTP}$. If $x_j = x_i$, then $(a_j, a_l, x_k) \in \mathcal{BTP}$ since $(a_i, a_l, x_k) \in \mathcal{BTP}$. Therefore, $(\mathcal{A} \bigcup \{\langle x_l, a_l \rangle\}, x_k) \in \mathcal{BTP}_\mathcal{A}$. $\square$

The second lemma is an extended result of the BTP. From [3], we can learn that the BTP ensures that one can always assign consistent values to variables so as to find a solution. Lemma 2 extends the case of the BTP with the help of bipartite graph.

Here we employ a graph to describe relations of the BTP in a partial assignment. Given a variable x_k and a consistent partial assignment $\mathcal{A}$ (x_k is not involved in $\mathcal{A}$), we can construct a graph according to the BTP relation of $\mathcal{A}$ with respect to x_k, The assigned values in $\mathcal{A}$ are nodes of the graph, where an edge will be added between two nodes a_i and a_j if $(a_i, a_j, x_k) \notin \mathcal{BTP}$. Such BTP-relation graph constructed above is called a *BTP-graph of $\mathcal{A}$* with respect to x_k. Lemma 2 shows a case that x_k can take a value from its domain such that the partial assignment is also consistent after involving the value.

Lemma 2. *Given a path consistent binary CSP $P = (V, D, C)$, let x_k be a variable and $\mathcal{A}$ be a consistent partial assignment. If the BTP-graph of $\mathcal{A}$ with respect to x_k is a bipartite graph, then there exists a value a_k from D_k such that $\mathcal{A} \bigcup \{\langle x_k, a_k \rangle\}$ is consistent.*

Proof. Since the BTP-graph is a bipartite graph, nodes in the graph can be divided into two subsets V_1 and V_2. Without loss of generality, suppose the partial assignment $\mathcal{A}$ is $\{\langle x_{i_1}, a_1 \rangle, \ldots, \langle x_{i_f}, a_f \rangle, \langle x_{j_1}, b_1 \rangle, \ldots, \langle x_{j_g}, b_g \rangle\}$, where $x_{i_1}, \ldots, x_{i_f} \in V_1$ and $x_{j_1}, \ldots, x_{j_g} \in V_2$. It can be seen that for each pair of values (a_i, a_j) that are assigned to variables in V_1, the triple $(a_i, a_j, x_k) \in \mathcal{BTP}$, so there exists a minimal support set among $D_k(a_1), \ldots, D_k(a_f)$, according to the proof of Theorem 3.1 in [3]. Also, there exists a minimal support set for V_2. Assume the minimal support set for V_1 is $D_k(a)$ and the minimal support set for V_2 is $D_k(b)$. Hence there are two cases: the first one is that there is no edge between nodes a and b, so the triple $(a, b, x_k) \in \mathcal{BTP}$, i.e., one of $D_k(a_l)$ and $D_k(b_r)$ includes the other, and thus there exists a value $a_k \in D_k(a) \bigcap D_k(b)$ such that $\mathcal{A} \bigcup \{\langle x_k, a_k \rangle\}$ is consistent; the second case is that there is an edge between nodes a and b, but $D_k(a) \bigcap D_k(b)$ is not empty as the CSP is path consistent, so there also exists a value $a_k \in D_k(a) \bigcap D_k(b)$ such that the partial assignment $\mathcal{A} \bigcup \{\langle x_k, a_k \rangle\}$ is consistent. □

In the following, we present the tractable class characterizing by the no-conflict BTPv.

Theorem 1. *A binary CSP $P = (V, D, C)$ that is strong path consistent is satisfiable if there exists a variable ordering $\prec$ such that P satisfies the no-conflict BTPv.*

Proof. To prove this theorem, we divide the proof into two parts: in the first part, it is to propose a polynomial-time algorithm for assigning all variables in a CSP, and in the second part we prove correctness of the algorithm, i.e., it can always construct a solution if the input CSP is included in the class. The detailed procedure of the algorithm is shown as follows.

Algorithm 1. Solve-BTPv

Require:
  A binary CSP $P = (V, D, C)$ with a variable ordering $\prec$
Ensure:
  A solution $\mathcal{A}$
1: let x_i be the first variable of the ordering $\prec$, and a_i be a value from D_i
2: $\mathcal{A} \leftarrow \{\langle x_i, a_i \rangle\}$
3: $V_u \leftarrow V - \{x_i\}$; $V_{mark} \leftarrow \{x_i\}$
4: **while** $V_u \neq \emptyset$ **do**
5:   let x_i be the first variable in V_u of the ordering $\prec$
6:   select a value a_i from $\bigcap_{\langle x_j, a_j \rangle \in \mathcal{A}} D_i(a_j)$
7:   $\mathcal{A} \leftarrow \mathcal{A} \bigcup \{\langle x_i, a_i \rangle\}$
8:   $V_u \leftarrow V_u - \{x_i\}$; $V_{mark} \leftarrow V_{mark} \bigcup \{x_i\}$
9:   $V_{adj} \leftarrow$ AdjointVariables($V_u, \mathcal{A}, V_{mark}, x_i$)
10: $V_u \leftarrow V_u - V_{adj}$
11: **while** $V_{adj} \neq \emptyset$ **do**
12:    let x_k be the first variable in V_{adj} of the ordering $\prec$
13:    select a value a_k from $\bigcap_{\langle x_j, a_j \rangle \in \mathcal{A}} D_k(a_j)$
14:    $\mathcal{A} \leftarrow \mathcal{A} \bigcup \{\langle x_k, a_k \rangle\}$
15:    $V_{adj} \leftarrow V_{adj} - \{x_k\}$.
16: **return** $\mathcal{A}$

17: **Function** AdjointVariables($V_u, \mathcal{A}, V_{mark}, x_i$)
18: $V_{adj} \leftarrow \emptyset$
19: **for** each variable $x_k \in V_u$ **do**
20:   **if** there exists a pair of $(\langle x_i, a_i \rangle, \langle x_j, a_j \rangle)$ in $\mathcal{A}$ ($x_j \in V_{mark}, i \neq j$) s.t. $(a_i, a_j, x_k) \notin \mathcal{BTP}$ **then**
21:     $V_{adj} \leftarrow V_{adj} \bigcup \{x_k\}$
22: **return** V_{adj}

Algorithm 1 assigns variables in a backtrack-free manner, and determines the next variable dynamically. It starts from assigning a value to the first variable, and assigns other variables according to the ordering $\prec$. V_{mark} is used to record the assigned variables that are not included in any V_{adj}. However, in each iteration, variables in V_{adj} will be assigned immediately after x_i is assigned, where the ordering for assigning those variables is also according to $\prec$ if there is more than one variable in V_{adj}. Function AdjointVariables computes V_{adj} for each iteration by checking whether triples satisfy the BTP, where an unassigned variable x_k will be included in V_{adj} if there exists a pair of values (a_i, a_j) assigned to variables in the current V_{mark} such that $(a_i, a_j, x_k) \notin \mathcal{BTP}$.

Time complexity of Algorithm 1 can be computed as follows: selecting a value from its domain requires intersection of $O(n)$ support sets, so computing $\bigcap D_i(a_j)$ (Line 6 or Line 12) can be achieved in $O(nd)$ time; and to compute V_{adj}, there are at most $O(n)$ pairs to check for each variable in V_u, whereas for each triple, checking satisfaction of the BTP needs $O(d)$, so time complexity of Function AdjointVariables is $O(n^2 d)$. As a result, because there are n variables to be assigned for constructing a solution, Algorithm 1 runs in $O(n^3 d)$ time.

We are now turning to the proof of Theorem 1. The idea is to show a path consistent CSP that satisfies the no-conflict BTPv can be solved by Algorithm 1. State in another way, the partial assignment including all variables constructed by Algorithm 1 is a solution to such a CSP.

The key condition that guarantees Algorithm 1 works is that in each assignment step there is a consistent value that can be assigned to the current variable under any consistent partial assignment. To prove this, we discuss the assignment for each variable. Clearly, consistent values can be assigned to the first and the second variables as the problem is arc consistent. For the other variables, the trick of the proof is to show that for each variable, the BTP-graph of the current partial assignment with respect to it is a bipartite graph.

Without loss of generality, suppose x_k is a variable to be assigned, and $\mathcal{A}$ is the current consistent partial assignment. The bipartite graph can be constructed by considering two cases: $x_k \in V_{adj}$ of a variable x_p and otherwise. In the first case, it is obvious that $\langle x_p, a_p \rangle \in \mathcal{A}$, and all values in $\mathcal{A}$ assigned before a_p are put into the first subset, and a_p is put into the second subset. Moreover, it is obvious that all variables in $\mathcal{A}$ assigned after a_p are included in V_{adj} of x_p from the description of Algorithm 1. Hence, for each $x_i \in V_{adj}$ of x_p ($i \neq k$), its value a_i is put into the first subset if $(a_p, a_i, x_k) \notin \mathcal{BTP}$, otherwise it is put into the second subset. In the second case, we put all assigned values into the first subset with the second subset empty.

Next, we will prove the BTP-graph with the above subsets is a bipartite graph.

For the second subset, there is no edge if only one node is in the subset or if the subset is empty. If there are two or more nodes, we will prove that for each pair of $\{\langle x_i, a_i \rangle, \langle x_j, a_j \rangle\}$ within the subset, triple $(a_i, a_j, x_k) \in \mathcal{BTP}$. If a_i is a_p, then $(a_i, a_j, x_k) \in \mathcal{BTP}$ because a_j would be put into the first subset with the case that $(a_p, a_j, x_k) \notin \mathcal{BTP}$. Also, $(a_i, a_j, x_k) \in \mathcal{BTP}$ if a_j is a_p. Furthermore, we discuss the case that a_i and a_j are not a_p. It is clear that there exists a pair of $\{\langle x_g, a_g \rangle, \langle x_h, a_h \rangle\}$ in $\mathcal{A}$ such that $x_g, x_h \in V_{mark}$ and $(a_g, a_h, x_i) \notin \mathcal{BTP}$, and it is easy to see one of a_g and a_h is a_p because if not x_i would be in a V_{adj} before x_p. Suppose x_h is x_p, so (a_g, a_p, x_i) satisfies the BTP-adjoint property and a_i is adjoint to a_p. Also, $(a_p, a_j, x_k) \in \mathcal{BTP}$, and it follows that $(a_i, a_j, x_k) \in \mathcal{BTP}$.

For the first subset, let $\mathcal{A}' \subseteq \mathcal{A}$ be the partial assignment composed of all assignments whose variables are in V_{mark}. Clearly, $(\mathcal{A}', x_k) \in \mathcal{BTP_A}$, since if not, x_k would be in a V_{adj} before x_p. It is obvious that there exist at least two variables in $\mathcal{A}'$. Next, we add other variables in the first subset to $\mathcal{A}'$ iteratively keeping $(\mathcal{A}', x_k) \in \mathcal{BTP_A}$. Suppose a_i is in the first subset but not in $\mathcal{A}'$, so according to Function AdjointVariables in Algorithm 1 there exist $\langle x_g, a_g \rangle$ and $\langle x_h, a_h \rangle$ in $\mathcal{A}'$ such that a_i is adjoint to both a_g and a_h, where x_g and x_h are included in V_{mark}. Moreover, there is at least one of a_g and a_h is in the first subset according to the partition method mentioned above, and thus there are two cases to discuss:

(a) If both a_g and a_h are in the first subset, then a_g and a_h are in $\mathcal{A}'$, so we have $(a_g, a_h, x_k) \in \mathcal{BTP}$, and thus $(a_g, a_i, x_k) \in \mathcal{BTP}$. Hence new $\mathcal{A}'$ after adding $\langle x_i, a_i \rangle$ satisfies the BTP with respect to x_k according to Lemma 1.

(b) If only one of a_g and a_h is in the first subset, suppose a_g is in the first subset, and thus from the partition method it can be seen that a_h is a_p. Suppose $(a_g, a_p, x_k) \in \mathcal{BTP}$, then we have $(a_i, a_p, x_k) \in \mathcal{BTP}$, while this is contrary to the fact that $(a_i, a_p, x_k) \notin \mathcal{BTP}$ because a_i is put into the first subset, so $(a_g, a_p, x_k) \notin \mathcal{BTP}$ and thus (a_g, a_p, x_k) satisfies the BTP-adjoint property. Moreover, (a_g, a_p, x_i) also satisfies the BTP-adjoint property. In addition, both x_i and x_k are in V_{adj} of x_p, so it follows that $x_i \prec x_k$. According to the satisfaction of the no-conflict BTPv between 4-tuple (a_g, a_p, x_i, x_k), and $(a_i, a_p, x_k) \notin \mathcal{BTP}$ so $(a_g, a_i, x_k) \in \mathcal{BTP}$. So new $\mathcal{A}'$ after adding $\langle x_i, a_i \rangle$ also satisfies the BTP with respect to x_k according to Lemma 1.

Clearly, variables can be added into $\mathcal{A}'$ iteratively, and $\mathcal{A}'$ including all variables in the first subset satisfies the BTP with respect to x_k, so it can be seen that there is no edge in the first subset.

From the above discussions, we can see that the BTP-graph is a bipartite graph. Furthermore, as the CSP is path consistent, according to Lemma 2, there exists a value a_k from D_k such that $\mathcal{A} \bigcup \{\langle x_k, a_k \rangle\}$ is consistent. As x_k is any variable, Algorithm 1 can find a solution to the CSP after assigning all variables. Therefore the CSP is satisfiable.   □

According to discussions in [18], a CSP is tractable if both its search problem and its identification problem are tractable, i.e., there exists a polynomial-time algorithm to find a solution of the problem, as well as there exists a polynomial-time algorithm to determine a CSP is in the tractable class. So it is also needed to find the variable ordering $\prec$ in polynomial time.

Theorem 2. *Given a binary CSP $P = (V, D, C)$, determining whether there exists a variable ordering $\prec$ such that P satisfies the no-conflict BTPv or determining there is no such ordering can be achieved in polynomial time.*

Proof. The proof of this result is similar to that given in Theorem 3.2 in [3] by constructing an indicator problem whose solution is a suitable variable ordering. For each triple of variables (x_i, x_j, x_k), if $(x_i, x_j, x_k) \notin \mathcal{BTP_V}$, then a disjunctive constraint $x_k < x_i \lor x_k < x_j$ is added into the indicator problem.

To ensure the no-conflict BTPv, for each 4-tuple of variables (x_i, x_j, x_k, x_l), if a 4-tuple (a_i, a_j, x_k, x_l) such that $(a_i, a_j) \in R_{ij}$ has a BTPv-conflict with respect to x_l, then a disjunctive constraint $x_l < x_i \lor x_l < x_j \lor x_l < x_k$ will be added. It is clear that all constraints added into the indicator problem are max-closed [19], so the problem can be solved in polynomial time.

To construct the indicator problem, we should first check that for each pair of values whether one is adjoint to the other, and obtain all adjoint values of each value. This can be achieved in $O(n^4 d^4)$ time as mentioned above. Next, we should check whether all triples of variables satisfy the BTPv, which will cost at most $O(n^3 d^3)$ time for examining all triples. Finally, the no-conflict BTPv of each 4-tuple will be considered, and this check will take $O(n^4 d^4)$ time. In a word, time complexity for constructing the indicator problem can be achieved in $O(n^4 d^4)$ time.

As both constructing the indicator problem and solving the problem can be achieved in polynomial time, the theorem is proved.   □

To sum up, given a binary CSP, checking whether it is strong path consistent can be achieved in polynomial time, and if it is strong path consistent and there exists a variable ordering such that it satisfies the no-conflict BTPv, then a solution can be found in polynomial time. It is noted that though a static variable ordering should be computed, the actual ordering for assigning variables depends on values assigned currently, since variables in V_{adj} will be assigned before their positions in the ordering.

Clearly, CSP instances satisfying the BTP satisfy the BTPv, so the new class is an extension of the BTP. Moreover, the rank CSP includes many other known tractable classes. We denote the class including all strong path consistent CSP instances of rank 2 discussed in [15] by Rank₂-SPC and the class identified by Theorem 1 by BTPv-SPC, then we have that BTPv-SPC and Rank₂-SPC cannot include each other (the proof is omitted here), so BTPv-SPC can identify new tractable cases not included in known tractable classes.

5   Conclusions

In this paper, we discuss a new method to generalize the BTP class by proposing the notion of adjoint values and the no-conflict BTPv, and then identify a novel tractable class using those concepts. We also prove that determining whether a CSP instance is in the new class can be checked efficiently. Besides, we show that the new generalized class is incomparable with the class of rank CSPs. A possible further work would be to investigate extending other hybrid tractable class by using the concept of adjoint values.

References

1. Dechter, R.: Constraint Processing. Morgan Kaufmann, Burlington (2003)
2. Carbonnel, C., Cooper, M.C.: Tractability in constraint satisfaction problems: a survey. Constraints **21**(2), 115–144 (2016)
3. Cooper, M.C., Jeavons, P.G., Salamon, A.Z.: Generalizing constraint satisfaction on trees: hybrid tractability and variable elimination. Artif. Intell. **174**(9–10), 570–584 (2010)
4. Cohen, D.A., Cooper, M.C., Creed, P., Marx, D., Salamon, A.Z.: The tractability of CSP classes defined by forbidden patterns. J. Artif. Intell. Res. (JAIR) **45**, 47–78 (2012)
5. Cohen, D.A., Cooper, M.C., Jeavons, P.G., Zivny, S.: Tractable classes of binary CSPs defined by excluded topological minors. In: Proceedings of the Twenty-Fourth International Joint Conference on Artificial Intelligence, IJCAI 2015, Buenos Aires, Argentina, 25–31 July 2015, pp. 1945–1951 (2015)
6. Thorstensen, E.: Hybrid tractability of constraint satisfaction problems with global constraints. Ph.D. thesis, University of Oxford (2013)
7. Cooper, M.C., Duchein, A., Mouelhi, A.E., Escamocher, G., Terrioux, C., Zanuttini, B.: Broken triangles: from value merging to a tractable class of general-arity constraint satisfaction problems. Artif. Intell. **234**, 196–218 (2016)
8. Cooper, M.C., Jeavons, P.G., Salamon, A.Z.: Hybrid tractable CSPs which generalize tree structure. In: Proceedings of the ECAI 2008, pp. 530–534. IOS Press (2008)
9. Cohen, D.A., Cooper, M.C., Escamocher, G., Zivny, S.: Variable elimination in binary CSP via forbidden patterns. In: Rossi, F. (ed.) Proceedings of the IJCAI, IJCAI 2013. AAAI, Menlo Park (2013)
10. Mouelhi, A.E., Jégou, P., Terrioux, C., Classes, H.T.: From theory to practice. In: Proceedings of the ICTAI 2014, pp. 437–445 (2014)
11. Cooper, M.C., Zivny, S.: Hybrid tractability of valued constraint problems. Artif. Intell. **175**(9–10), 1555–1569 (2011)
12. Gao, J., Yin, M., Zhou, J.: Hybrid tractable classes of binary quantified constraint satisfaction problems. In: Burgard, W., Roth, D. (eds.) Proceedings of the AAAI 2011. AAAI Press (2011)
13. Jégou, P., Terrioux, C.: The extendable-triple property: a new CSP tractable class beyond BTP. In: Proceedings of the AAAI 2015 (2015)
14. Mouelhi, A.E., Jégou, P., Terrioux, C.: A hybrid tractable class for non-binary CSPs. In: Proceedings of the ICTAI 2013, pp. 947–954. IEEE Computer Society (2013)
15. Naanaa, W.: Unifying and extending hybrid tractable classes of CSPs. J. Exp. Theor. Artif. Intell. **25**(4), 407–424 (2013)
16. Bessière, C., Régin, J.: Refining the basic constraint propagation algorithm. In: Nebel, B. (ed.) Proceedings of the IJCAI 2001, pp. 309–315. Morgan Kaufmann (2001)
17. Singh, M.: Path consistency revisited. Int. J. Artif. Intell. Tools **5**(1–2), 127–142 (1996)
18. Green, M.J., Cohen, D.A.: Domain permutation reduction for constraint satisfaction problems. Artif. Intell. **172**(8–9), 1094–1118 (2008)
19. Jeavons, P., Cooper, M.C.: Tractable constraints on ordered domains. Artif. Intell. **79**(2), 327–339 (1995)

Online Knapsack Problem Under Concave Functions

Xin Han[1](✉), Ning Ma[1], Kazuhisa Makino[2], and He Chen[1]

[1] Software School, Dalian University of Technology, Dalian 116620, China
hanxin.mail@gmail.com, 912583307@qq.com, ankachan@126.com
[2] Research Institute for Mathematical Sciences,
Kyoto University, Kyoto, Japan
makino@kurims.kyoto-u.ac.jp

Abstract. In this paper, we address an online knapsack problem under concave function $f(x)$, i.e., an item with size x has its profit $f(x)$. We first obtain a simple lower bound $\max\{q, \frac{f'(0)}{f(1)}\}$, where $q \approx 1.618$, then show that this bound is not tight, and give an improved lower bound. Finally, we find the online algorithm for linear function [8] can be employed to the concave case, and prove its competitive ratio is $\frac{f'(0)}{f(1/q)}$, then we give a refined online algorithm with a competitive ratio $\frac{f'(0)}{f(1)} + 1$. And we also give optimal algorithms for some piecewise linear functions.

1 Introduction

Knapsack problem is one of the most classical and studied problems in combinatorial optimization field and has a lot of applications in the real world [11]. The (classical) knapsack problem is defined as below: given a set of items with weights (here, we call it *profit*) and sizes, and the capacity of a knapsack, to maximize the total weight (profit) of selected items in the knapsack satisfying the capacity constraint.

In this paper, we study the online knapsack problem: (i) items are given one by one over time, i.e., after a decision is made on a given item, the next one is then known or given, (ii) in order to accept a new item, it is allowed to remove old items accepted in the knapsack, (iii) the objective is the same as the offline version, i.e., to maximize the total profit in the knapsack. If the profit function is a linear function, then there is an optimal online algorithm in [8], i.e., the problem admits an online algorithm with a competitive ratio matching the lower bound of the problem. Motivated by the above results, in this paper, we consider the case in which the profit function is a concave curve. Require that function $f(x)$ has the following properties: Concave; Monotone increasing; $f(0) = 0$; $f'(0)$ exists and is bounded.

Related Work: For the offline case, there are many papers working on it, refer to [11]. As for the online maximization knapsack problem, it was first studied on average case analysis by Marchetti-Spaccamela and Vercellis [13]. They proposed

© Springer International Publishing AG 2017
M. Xiao and F. Rosamond (Eds.): FAW 2017, LNCS 10336, pp. 103–114, 2017.
DOI: 10.1007/978-3-319-59605-1_10

a linear time approximation algorithm such that the expected difference between the optimal and the approximation value is $O(\log^{3/2} n)$ under the condition that the capacity of the knapsack grows proportionally to n, the number of items. Lueker [12] further improved the expected difference to $O(\log n)$ under a fairly general condition on the distribution. In 2002 Iwama and Taketomi [8] studied the problem on worst case analysis. They obtained an 1.618-competitive algorithm for the online Max-Knapsack under the *removable* condition, if each item has its size equal to its profit. Here the *removable* condition means that it is allowed to remove some items accepted in the knapsack in order to accept a new item. They also showed that this is best possible by providing a lower bound 1.618 for this case. For the general case, Iwama and Zhang [9] showed that no algorithm for online Max-Knapsack has a bounded competitive ratio, even if the removal condition is allowed. Iwama and Zhang gave some tight bounds for the online Max-Knapsack problem with resource augmentation [9]. Noga and Sarbua studied an online partially fractional knapsack problem with resource augmentation, in which items are only allowed to be cut just before they are packed into the knapsack. They gave an upper bound $2/m$ and proved the bound is the best possible, where $m \geq 1$ is the capacity of the knapsack used by online algorithms while the optimal offline algorithm uses a unit capacity knapsack. Han and Makino studied an online fractional knapsack problem with limited cuts, where items are allowed to be cut at most $k (\geq 1)$ times, they obtained an optimal algorithm with a competitive ratio $\frac{k+1}{k}$ [5]. For other results, refer to papers in [1–4, 6, 7, 10, 14].

Our Contributions: For the online knapsack problem with concave function $f(x)$, we first obtain a lower bound $\max\{q, \frac{f'(0)}{f(1)}\}$, where $q \approx 1.618$, then prove that this bound is not tight, and give an improved for some specific functions. Finally, we find the online algorithm for linear function [8] can be employed to the concave case, and prove its competitive ratio is $\frac{f'(0)}{f(1/q)}$, then we give a refined online algorithm with a competitive ratio $\frac{f'(0)}{f(1)} + 1$. The main ideas in the improved algorithm are below: (i) if there are large items, then we accept the largest one, (ii) otherwise we use the policy: *the smaller the first* to select small items. And we also give optimal algorithms for some piecewise linear functions.

2   Preliminary

Knapsack Problem Under Concave Function: The input is a unit size of knapsack and a set of items associated with a size-profit function $f(x)$, where function $f(x)$ is concave. The output is to select a subset of items to maximize the total profit of all the selected items without exceeding the capacity of the knapsack.

Online Knapsack Problem with Concave Function: In this paper, we study an online knapsack problem under a concave size-profit function. The capacity of the knapsack and the function $f(x)$ are known before packing.

The word "Online" means that (i) items are given one by one over time, i.e., after a decision is made on the current item, then the next one is known, (ii) in order to accept a new item, it is allowed to remove old items in the knapsack. During selection, in order to accept a new item, some old items are allowed to be discarded or removed. The objective of the online knapsack is the same as the offline version, i.e., to maximize the profit under the capacity constraint. Let $f(x)$ be the concave profit function, i.e., for each item with size x, its profit is $f(x)$. Here we require that function $f(x)$ has the following properties:

- Concave;
- Monotone increasing;
- $f(0) = 0$; ·
- $f'(0)$ exists and is bounded.

If $f'(0)$ is unbounded, then following the proof in [9], we can prove that there is no online algorithm with a bounded competitive ratio (we will give the definition of the competitive ratio later). So, in the following, we assume that $f'(0)$ exists and is bounded.

Competitive Ratio: We analyze online algorithms by using one of the standards: the competitive ratio. Given an input sequence L and an online algorithm A, the *competitive ratio* of algorithm A is defined as follows:

$$R_A = \sup_L \frac{OPT(L)}{A(L)},$$

where $OPT(L)$ and $A(L)$ denote the profits obtained by an optimal algorithm and algorithm A, respectively.

3   Lower Bounds

We first give a simple lower bound, then show the bound is not tight, finally give an improved bound.

3.1   A Simple Lower Bound

Theorem 1. *There is no online algorithm with a competitive ratio strictly less than* $\max\{q, \frac{f'(0)}{f(1)}\}$, *where* $q \approx 1.618$ *is the golden ratio.*

Proof. We first prove that the lower bound of the competitive ratio cannot be strictly less than $q \approx 1.618$. The proof is similar with the one in [8]. Find a constant $x_0 > 0.5$ such that

$$\frac{f(x_0) + f(1 - x_0)}{f(x_0)} = \frac{f(x_0)}{f(1 - x_0)} \quad \Rightarrow \quad \frac{f(x_0)}{f(1 - x_0)} = q \approx 1.618.$$

We can prove that such $x_0 (> 0.5)$ must exist, by constructing the following function:

$$F(x) = f(x) - q \times f(1 - x).$$

Observe that $F(1) = f(1) - 0 > 0$ and $F(0.5) = -(q-1) \cdot f(0.5) < 0$. $F(x)$ is continuous, hence there is $x_0 > 0.5$ such that $F(x_0) = 0$, i.e., $f(x_0) = q \times f(1 - x_0)$.

Assume there is an online algorithm with a competitive ratio r which is strictly less than $q \approx 1.618$. Next we construct an instance to prove that such online algorithm does not exist. In the instance, the first item a_1 is of size $1 - x_0$ and the second item a_2 is of size $x_0 + \epsilon$, where we require that $\frac{f(x_0) + f(1-x_0)}{f(x_0 + \epsilon)} > r$ (such that ϵ must exist by the same approach of proving x_0 must exist). Note that items a_1 and a_2 cannot be accepted together. Before the third item a_3 is given, in order to achieve the competitive ratio r, the online algorithm has to accept item a_2 and discard item a_1. Otherwise the competitive ratio is at least $\frac{f(x_0 + \epsilon)}{f(1-x_0)} > q > r$. Finally item a_3 of size x_0 is given and the input stops. Either a_3 is discarded or is accepted, we have that the competitive ratio of the online algorithm is at least $\frac{f(x_0) + f(1-x_0)}{f(x_0 + \epsilon)} > r$. Hence, there is no online algorithm with a competitive ratio strictly less than q.

Next we prove that the lower bound is at least $\frac{f'(0)}{f(1)}$. With the similar idea, we construct an instance to prove that there is no online algorithm with a competitive ratio strictly less than $\frac{f'(0)}{f(1)}$. Find a sufficiently small real $\xi > 0$ such that

$$\frac{f(1)}{f(\xi)} > \frac{f'(0)}{f(1)}.$$

Since $f'(0)$ is bounded, $f(0) = 0$ and $f(x)$ is monotone increasing, such ξ must exist (we skip the proof here). Then we consider the following instance:

$$1, \xi, \xi, \xi, \dots,$$

the first item is of size 1 and all the others have the same size ξ. Observe that any time once an online algorithm discards the first item then the input stops immediately, which will cause the online algorithm has a competitive ratio at least $\frac{f(1)}{f(\xi)}$ since there is only one item of size ξ accepted by the online algorithm and the optimal value is at least $f(1)$. Also find if an online algorithm keeps holding the item with size 1 in the knapsack all the times and discards all the others, the competitive ratio is at least $\frac{\lfloor \frac{1}{\xi} \rfloor f(\xi)}{f(1)}$, since there is a feasible solution of keeping $\lfloor \frac{1}{\xi} \rfloor$ items of size ξ in the knapsack. If we can prove that

$$\lim_{\xi \to 0} \lfloor \frac{1}{\xi} \rfloor f(\xi) = f'(0),$$

then we are done. Since

$$(\frac{1}{\xi} - 1)f(\xi) \le \lfloor \frac{1}{\xi} \rfloor f(\xi) \le \frac{f(\xi)}{\xi},$$

observe that

$$\lim_{\xi \to 0} \frac{f(\xi)}{\xi} = f'(0),$$

and

$$\lim_{\xi \to 0}(\frac{1}{\xi} - 1)f(\xi) = f'(0) - f(0) = f'(0).$$

Hence we have $\lim_{\xi \to 0}\lfloor \frac{1}{\xi} \rfloor f(\xi) = f'(0)$, and this theorem holds. □

3.2   An Improved Lower Bound

It seems that the lower bound $\max\{q, \frac{f'(0)}{f(1)}\}$ is tight for all concave functions. However this is not true. We first give an evidence to show that for some specific function the lower bound $\max\{q, \frac{f'(0)}{f(1)}\}$ can be improved. Then we generalize the approach and obtain an improved lower bound.

Lemma 1. *The lower bound* $\max\{q, \frac{f'(0)}{f(1)}\}$ *is not tight.*

Proof. Consider the following function f(x), refer to Fig. 1:

$$f(x) = \begin{cases} x, & \text{if } 0 \le x \le 1/3, \\ (\sqrt{3} - 1)(x - \frac{1}{3}) + \frac{1}{3}, & \text{else } 1/3 < x \le 1. \end{cases}$$

Observe that

$$f(\frac{1}{3}) = \frac{1}{3}, \quad f(\frac{2}{3} + \epsilon) = \frac{\sqrt{3}}{3} + O(\epsilon), \quad f'(0) = 1.$$

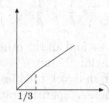

Fig. 1. An instance of $f(x)$

Next we prove that there is no online algorithm with a competitive ratio strictly less than $\sqrt{3}$, which is larger than $\max\{q, \frac{f'(0)}{f(1)}\} \approx 1.618$. The main idea is similar with the one in Theorem 1. We construct the following instance:

$$\frac{2}{3} + \epsilon, \frac{1}{3}, \frac{1}{3}, \frac{1}{3}, \dots,$$

where ϵ is sufficiently small. Observe that an item of size $\frac{2}{3} + \epsilon$ and an item of size $\frac{1}{3}$ can not be accepted together. If there is an online algorithm with a competitive ratio strictly less than $\sqrt{3}$, then before the third item is given the online algorithm has to keep the first item in the knapsack, i.e., discard the second item. Otherwise if the first item is discarded, then the competitive ratio is at least $\frac{f(\frac{2}{3}+\epsilon)}{\frac{1}{3}} > \sqrt{3}$. However if the online algorithm keeps holding the first item after the fourth item is given, then the competitive ratio is at least $\frac{3f(\frac{1}{3})}{f(\frac{2}{3}+\epsilon)}$, which will be close to $\sqrt{3}$ when ϵ approaches to zero.

For the function $f(x)$ we defined above, there is no online algorithm with a competitive ratio less than $\sqrt{3}$. Hence the lower bound $\max\{q, \frac{f'(0)}{f(1)}\}$ is *not* tight.   □

Theorem 2. *Given a concave function* $f(x)$*, find a maximum integer* $k \geq 2$ *such that*

$$\frac{f(1-\frac{1}{k})}{f(\frac{1}{k})} = \frac{k \cdot f(\frac{1}{k})}{f(1-\frac{1}{k})}, \tag{1}$$

If such integer k *exists, then there is no online algorithm with a competitive ratio less than* $\sqrt{k}$*.*

Proof. It is not difficult to see that

$$\frac{f(1-\frac{1}{k})}{f(\frac{1}{k})} = \frac{k \cdot f(\frac{1}{k})}{f(1-\frac{1}{k})} \quad \Rightarrow \quad \frac{f(1-\frac{1}{k})}{f(\frac{1}{k})} = \sqrt{k}. \tag{2}$$

Similar with the instance used in Lemma 1, we construct the following instance, where $\epsilon > 0$ is sufficiently small,

$$1 - \frac{1}{k} + \epsilon, \frac{1}{k}, \frac{1}{k}, \frac{1}{k}, \dots, \frac{1}{k},$$

i.e., the first item has size $1 - \frac{1}{k} + \epsilon$ and all the others have size $\frac{1}{k}$ (once the first item is discarded or removed, the input stops).

By Eq. (2) and the analysis framework used in Lemma 1, we can prove that there is no online algorithm with a competitive ratio less than $\sqrt{k}$.   □

4   Upper Bounds

First we observe that the algorithm [8] for linear online knapsack problem can be exploited to the concave case, then prove the competitive ratio is $\approx 1.618 \cdot \frac{f'(0)}{f(1)}$. Finally we give a new online algorithm and prove that its competitive ratio is $\frac{f'(0)}{f(1)} + 1$. The idea in the new algorithm is very simple: (i) if there are large items, then we accept the largest one, (ii) otherwise we use the policy: *the smaller the first* to select small items.

Lemma 2. *If function* $f(x)$ *is concave and monotone increasing,* $f(0) = 0$*, then we have* $\frac{f(x)}{x}$ *is non-increasing for any* $x > 0$*.*

Proof. Given two variables $0 < x_1 < x_2 \leq 1$, we will prove that $\frac{f(x_1)}{x_1} \geq \frac{f(x_2)}{x_2}$. Since

$$x_1 = 0 \cdot (1 - \frac{x_1}{x_2}) + x_2 \cdot \frac{x_1}{x_2},$$

and function $f(\cdot)$ is concave and $f(0) = 0$, we have

$$f(x_1) \geq (1 - \frac{x_1}{x_2})f(0) + f(x_2) \cdot \frac{x_1}{x_2} = f(x_2) \cdot \frac{x_1}{x_2}. \cdot$$

Hence we have this lemma.   □

By the above lemma, we have the following result.

Corollary 1. *If the total size in the knapsack is x then the profit in the knapsack is at least $f(x)$.*

4.1   Upper bound $\frac{f'(0)}{f(1/q)}$

Our algorithm generalizes the idea in [8]. Let $q \approx 1.618$ be the golden ratio. We group items into three classes: *large*, *medium* and *small*. If an item has size larger than $\frac{1}{q}$, then we say it is large, if its size is in $(\frac{1}{q^2}, \frac{1}{q}]$, we say it is medium, otherwise we call it small. The main idea of our algorithm is given as below:

- If there is a large given item, then we select one of them and keep it in the knapsack, discard all the others.
- Else if there are medium items given, we use the policy: *the smaller the first* to select medium items, if there are two mediums selected then we can discard all the others, else keep the minimal medium item in the knapsack.
- Else we first keep the smallest medium item in the knapsack if possible, then remove small items one by one until all the remaining items can be accommodated in the knapsack.

Lemma 3. *At line 7 in algorithm A_1, item u must be medium.*

Proof. If items u is a small item, then the total size in the knapsack is at least

$$1 - \frac{1}{q^2} = \frac{1}{q},$$

where the equality holds from the definition of the golden ratio q. Then by Corollary 1, we have the profit in the knapsack is at least $f(1/q)$, which means that item u would have been discarded at line 2 in the algorithm. Hence we have item u must be a medium item.   □

Theorem 3. *The competitive ratio of the above online algorithm is $\frac{f'(0)}{f(1/q)}$, where $q \approx 1.618$ is the golden ratio.*

Algorithm 1. A_1

Input: the set B of items accepted in knapsack, and the new arriving u-item
Output: the new knapsack set: B'
1: **if** the profit in set B is at least $f(1/q)$ **then**
2:      discard u
3: **else if** item u can be accepted in the knapsack without removing any item **then**
4:      $B' = B \cup \{u\}$
5: **else if** u-item is a large item **then**
6:      keep u-item and discard all others
7: **else**(u-item must be medium item, refer to Lemma 3.)
8:      **if** B only contains small items **then**
9:          holds u-item and removes small items until the total size of items is at most
            1.
10:     **else**(B contains one medium v-item)
11:         **if** $u + v \leq 1$ **then**
12:             keep u-item and v-item, $B' = \{u, v\}$.
13:         **else**
14:             keep $\min\{u, v\}$-item and discard the bigger item.
15:         **end if**
16:     **end if**
17: **end if**
18: **return** B'

Proof. For an input L, let $A(L)$, $OPT(L)$ be the profit by our online algorithm and an optimal algorithm, respectively. We are going to prove that

$$\frac{f'(0)}{f(1/q)} \cdot A(L) \geq OPT(L).$$

Case 1: the total size in the knapsack is at least $1/q$. Then we have $A(L) \geq f(1/q)$ by Corollary 1. And $OPT(L) \leq f'(0)$ by Lemma 2. Therefore in this case $R_A \leq \frac{f'(0)}{f(1/q)}$.

Case 2: the total size in the knapsack is less than $1/q$. Then we have the following facts:

- there is no large item in L;
- no small item is discarded or removed;
- in the knapsack there is at most one medium item.

Otherwise the total size in the knapsack would have been at least $1/q$.

Case 2.1: there is no medium item in L. Then we have $A(L) = OPT(L)$, since our algorithm accepts all the small items.

Case 2.2: there is a medium item in L. And we have the following important fact:

- any two medium items cannot be packed in the knapsack if there are at least two medium items in L.

The reason is that: refer to line 14 in A_1, we always select the minimal medium item in the knapsack. So if any two medium items can be packed in the knapsack then algorithm A_1 would have kept two medium items in the knapsack, which contradicts with the fact: the total size in the knapsack is less than $1/q$. Then we have

$$A(L) \geq f(S) + f(u), \quad OPT(L) \leq f(S) + f(m),$$

where set S is all the small items in L and item u is the minimal medium item in L and item m is the maximal medium item in L. By Lemma 2, we have $\frac{f(m)}{f(u)} \leq \frac{m}{u} \leq q$. Then in this case $R_A \leq q$. Observe that

$$\lim_{\xi \to 0} \frac{f(\xi)}{\xi} = f'(0),$$

and $\frac{f(\xi)}{\xi}$ is a monotonically decreasing function by Lemma 2. For some $\epsilon > 0$,

$$\frac{f'(0)}{f(1/q)} \geq \frac{f(\epsilon)}{\epsilon f(1/q)} \geq \frac{\epsilon}{\epsilon/q} = q,$$

where the last inequality holds by Lemma 2. $\qquad\square$

4.2   Upper bound $\frac{f'(0)}{f(1)} + 1$

Let $\alpha = \frac{f'(0)}{f(1)}$. We find if α is very large, the gap between the lower bound α and $\frac{f(1)}{f(1/q)} \cdot \alpha$ may be very large. So we need a new algorithm for the case when α is very large.

Theorem 4. *There is an online algorithm with a competitive ratio $\alpha + 1$ for the problem.*

Proof. We first give the online algorithm and analyze its competitive ratio. If an item has size at least $\frac{\alpha}{\alpha+1}$, it is called *large*, otherwise *small*. Our algorithm works as below:

– If there are large items, then we select one of them and keep it in the knapsack, discard all the others.
– Else we use a greedy algorithm (*the smallest the first*) for small items, i.e., if all the small items can be accepted then hold all of them, otherwise discard the largest one until all the remaining items can be accommodated in the knapsack.

Again, for an input L, let $A(L), OPT(L)$ be the profit by our online algorithm and an optimal algorithm, respectively. Next we are going to prove that $(\alpha + 1)$ $A(L) \geq OPT(L)$. There are two cases.

Case 1: there is one large item in the input L. We have

$$A(L) \geq f(\frac{\alpha}{\alpha+1}) \geq \frac{\alpha}{\alpha+1} \cdot f(1),$$

where the last inequality holds from Lemma 2. Again, by Lemma 2 it is not difficult to see that

$$OPT(L) \leq f'(0).$$

Hence in this case, we have $(\alpha + 1)A(L) \geq OPT(L)$, where $\alpha = \frac{f'(0)}{f(1)}$.

Case 2: there is no large item in the input. If there is no item discarded, then we are done. Observe that if some small items are discarded, then the total size in the knapsack is at least $\frac{1}{\alpha+1}$ since a small item has size less than $\frac{\alpha}{\alpha+1}$. By Lemma 2, we know a smaller item has a higher density (the ratio between the profit and size). Since in our algorithm, the greedy algorithm (the smallest the first) is used for small items, we have $A(L) \geq \frac{OPL(L)}{\alpha+1}$.

4.3   Tight Upper Bounds for Piecewise Linear Functions

In this subsection, we give some optimal online algorithms for some piecewise linear functions, refer to Figs. 2 and 3. Given $0 \leq c_2 < c_1$, define piecewise linear function $f(x)$ as below:

$$f(x) = \begin{cases} c_1 \cdot x, \text{ if } 0 \leq x \leq p; \\ c_2 \cdot x, p \leq x \leq 1. \end{cases}$$

Similar with the proof in Theorem 4, we have the following result.

Lemma 4. *The competitive ratio of algorithm A_1 for function $f(x)$ is $\frac{f'(0)}{f(1/q)} = q$, which meets the lower bound q, where $p \geq 1/q$.*

The above result also holds for the case: the second line-segment is changed by any concave function. Next we focus on the case: $p < 1/q$.

Lemma 5. *There is an optimal online algorithm with competitive ratio $\frac{1}{p}$ for function $f(x)$ with $c_2 = 0$ and $p < 1/q$.*

Proof. Case 1: $0.5 < p < 1/q$. We group items into three classes: *large*, *medium* and *small*. If an item has size larger than p, then we say it is large, if its size is in $(p^2, p]$, we say it is medium, otherwise we call it small. Then we run the same algorithm as A_1 by just changing the threshold $f(1/q)$ into $f(p)$. By the same proof in Theorem 4 and $0.5 < p < q$, we can prove that the competitive ratio is $1/p$, which matches the lower bound $1/p$ by Theorem 1.

Case 2: $p \leq 0.5$. We group items into two classes: *large*, and *small*. If an item has size larger than p, then we say it is large, otherwise we call it small. Then we run the following algorithm which is similar with A_1. Observe that at line 5 in algorithm A_2 item-u must be a large item, otherwise the total size in the knapsack is at least $1 - p \geq p$ since $p \leq 0.5$. Then the total profit in the knapsack would have been at least $f(p)$ by Corollary 1. By the same proof in Theorem 4 and $p \leq 0.5$, we can prove that the competitive ratio is $1/p$, which matches the lower bound $1/p$ by Theorem 1. □

Algorithm 2. A_2

Input: the set B of items accepted in knapsack, and the new arriving u-item
Output: the new knapsack set: B'
1: **if** the profit in set B is at least $f(p)$ **then**
2:     discard u
3: **else if** item u can be accepted in the knapsack without removing any item **then**
4:     $B' = B \cup \{u\}$
5: **else**(u-item must be a large item)
6:     keep u-item and discard all others
7: **end if**
8: **return** B'

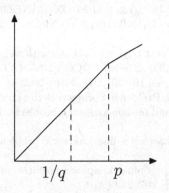

Fig. 2. $p \geq 1/q$

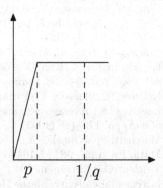

Fig. 3. $c_2 = 0$ and $p < 1/q$

5   Concluding Remarks

The gap between the lower bound and upper bound is at most $(\min\{\frac{f'(0)}{f(1)} + 1, \frac{f'(0)}{f(1/q)}\} - \max\{q, \frac{f'(0)}{f(1)}\})$. It will be very interesting to reduce the gap as an open question.

Acknowledgment. This research was partially supported by NSFC (11101065), RGC (HKU716412E).

References

1. Cygan, M., Jez, L., Sgall, J.: Online knapsack revisited. Theory Comput. Syst. **58**(1), 153–190 (2016)
2. Han, X., Kawase, Y., Makino, K.: Online unweighted knapsack problem with removal cost. Algorithmica **70**(1), 76–91 (2014)
3. Han, X., Kawase, Y., Makino, K.: Randomized algorithms for online knapsack problems. Theor. Comput. Sci. **562**, 395–405 (2015)
4. Han, X., Kawase, Y., Makino, K., Guo, H.: Online removable knapsack problem under convex function. Theor. Comput. Sci. **540**, 62–69 (2014)

5. Han, X., Makino, K.: Online removable knapsack with limited cuts. Theor. Comput. Sci. **411**(44–46), 3956–3964 (2010)
6. Han, X., Makino, K.: Online minimization knapsack problem. Theor. Comput. Sci. **609**, 185–196 (2016)
7. Horiyama, T., Iwama, K., Kawahara, J.: Finite-state online algorithms and their automated competitive analysis. In: Asano, T. (ed.) ISAAC 2006. LNCS, vol. 4288, pp. 71–80. Springer, Heidelberg (2006). doi:10.1007/11940128_9
8. Iwama, K., Taketomi, S.: Removable online knapsack problems. In: Widmayer, P., Eidenbenz, S., Triguero, F., Morales, R., Conejo, R., Hennessy, M. (eds.) ICALP 2002. LNCS, vol. 2380, pp. 293–305. Springer, Heidelberg (2002). doi:10.1007/3-540-45465-9_26
9. Iwama, K., Zhang, G.: Optimal resource augmentations for online knapsack. In: Charikar, M., Jansen, K., Reingold, O., Rolim, J.D.P. (eds.) APPROX/RANDOM-2007. LNCS, vol. 4627, pp. 180–188. Springer, Heidelberg (2007). doi:10.1007/978-3-540-74208-1_13
10. Kawase, Y., Han, X., Makino, K.: Proportional cost buyback problem with weight bounds. In: Lu, Z., Kim, D., Wu, W., Li, W., Du, D.-Z. (eds.) COCOA 2015. LNCS, vol. 9486, pp. 794–808. Springer, Cham (2015). doi:10.1007/978-3-319-26626-8_59
11. Kellerer, H., Pferschy, U., Pisinger, D.: Knapsack Problems. Springer, Berlin (2004)
12. Lueker, G.S.: Average-case analysis of off-line and on-line knapsack problems. In: SODA, pp. 179–188 (1995)
13. Marchetti-Spaccamela, A., Vercellis, C.: Stochastic on-line knapsack problems. Math. Program. **68**, 73–104 (1995)
14. Thielen, C., Tiedemann, M., Westphal, S.: The online knapsack problem with incremental capacity. Math. Methods OR **83**(2), 207–242 (2016)

Fluctuated Fitting Under the ℓ_1-metric

Kai Jin[✉]

University of Hong Kong, Pokfulam, Hong Kong SAR, China
cscjjk@gmail.com

Abstract. We consider the problem of fitting a given sequence of integers by an (α, β)-fluctuated one. For a sequence of numbers, those elements which are larger than their direct precursors are called *ascends*, those elements which are smaller than their direct precursors are called *descends*. A sequence is said to be (α, β)-fluctuated if there is a descend between any $\alpha + 1$ ascends and an ascend between any $\beta + 1$ descends; or equivalently, if it has at most α consecutive ascends and at most β consecutive descends, when adjacent equal values are ignored.

Given a sequence of integers $\mathbf{a} = (a_1, \ldots, a_n)$ and two parameters α, β in $[1, n]$, we compute (1) a sequence $\mathbf{b} = (b_1, \ldots, b_n)$ of integers that is (α, β)-*fluctuated* and is closest to $\mathbf{a}$ among all such sequences; (2) a sequence $\mathbf{b'} = (b'_1, \ldots, b'_n)$ of integers that is (α, β)-*fluctuated* and is bounded by $\mathbf{a}$ (i.e. $b'_i \leq a_i$ for all i) and is closest to $\mathbf{a}$ among all such sequences. We measure the distance between sequences under ℓ_1 metric.

Our algorithm runs in $O((\alpha + \beta) \cdot n)$ time, which is linear when α, β are considered as constants. We also show that a variation of our problem can be solved in the same time complexity. We achieve our result mainly by exploiting and utilizing the property of the closest sequence.

Keywords: Curve approximation · Histogram · Fitting

1 Introduction

Curve fitting and polygonal regression are fundamental problems in statistics, which aim to approximate a set of points, discrete or continuous, by a function that satisfies certain constraints. They have received much attention since their numerous applications, including operations research, signal processing, image compression, data analysis, and information retrieve [3,7,8,16,20].

In this paper, we consider a new shape-constrained regression problem, which aims to approximate a sequence of integers (which can be regarded as a histogram or a set of points) by another sequence of integers that is "fluctuated". We define the notion of (α, β)-fluctuated as follows; see Fig. 1 for an illustration.

Definition 1. *Given two parameters $\alpha \geq 1, \beta \geq 1$ and a sequence of integers $\mathbf{x} = (x_1, \ldots, x_n)$. An element x_i is called an* ascend *if $i > 1$ and $x_i > x_{i-1}$; a* descend *if $i > 1$ and $x_i < x_{i-1}$. The sequence $\mathbf{x}$ is called (α, β)-fluctuated if there is a descend between any $\alpha + 1$ ascends and an ascend between any $\beta + 1$ descends; or equivalently, if it has at most α consecutive ascends and at most β consecutive descends, when adjacent equal values are ignored.*

© Springer International Publishing AG 2017
M. Xiao and F. Rosamond (Eds.): FAW 2017, LNCS 10336, pp. 115–126, 2017.
DOI: 10.1007/978-3-319-59605-1_11

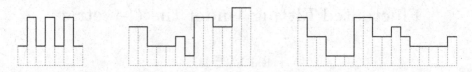

Fig. 1. Illustration of (α, β)-fluctuated. The sequences represented by the left and middle pictures are $(1,1)$-fluctuated. The right one is $(1,2)$-fluctuated

A sequence $\mathbf{b} = (b_1, \ldots, b_n)$ is *bounded by* $\mathbf{a} = (a_1, \ldots, a_n)$ if $b_i \leq a_i$ for all i.

We measure the distance between two sequences by ℓ_1 metric; so the distance between $\mathbf{a}, \mathbf{b}$ is given by $\Sigma_{i=1}^n |a_i - b_i|$.

Our problems and results are stated as follows.

Unbounded version: Given a sequence $\mathbf{a} = (a_1, \ldots, a_n)$, find a sequence $\mathbf{b} = (b_1, \ldots, b_n)$ of integers which is (α, β)-fluctuated and is closest to $\mathbf{a}$ in ℓ_1 metric among all such sequences.

Bounded version: Given a sequence $\mathbf{a} = (a_1, \ldots, a_n)$, find a sequence $\mathbf{b} = (b_1, \ldots, b_n)$ of integers which is (α, β)-fluctuated and is bounded by $\mathbf{a}$ and is closest to $\mathbf{a}$ in ℓ_1 metric among all such sequences.

Theorem 1. *Given a sequence of integers* $\mathbf{a} = (a_1, \ldots, a_n)$, *we can compute a solution to the bounded or unbounded version in* $O((\alpha + \beta) \cdot n)$ *time.*

1.1   Related Works

Fitting a function to a data set has been extensively studied and widely applied. Given a data set $(x_1, y_1), \ldots, (x_n, y_n)$, perhaps with nonnegative weights $w_1, \ldots, w_n$, the task is to find a function f in a group of candidates $\mathcal{F}$, so that it has the minimum error to represent the data, in which the most common criterions for error are

$$\ell_p : (\sum_{i=1}^n w_i \cdot |y_i - f(x_i)|^p)^{1/p}; \text{ and}$$

$$\ell_\infty : \max_{i=1}^n w_i \cdot |y_i - f(x_i)|.$$

The famous *V-optimum histogram problem* considers the case where $\mathcal{F}$ are k-step functions (i.e. histograms) and measure the error by the ℓ_2 criterion. A naive dynamic programming approach is widely known, which costs $O(kn^2)$ time, and no better results have been found for computing the best fitting; yet [13] manage to compute an $O(1 + \epsilon)$ approximation in linear time and poly-logarithmic space. Another instance is the *maximum error histogram problem*, which concerns the same set $\mathcal{F}$ but under the ℓ_∞ criterion. [12] solve the unit weighted case in $O(n + k^2 \log^3 n)$ time, and the general case in $O(n \log n + k^2 \log^6 n)$ time. Later, [10] solve the unit case and general case in $O(n)$ time and $O(n \log^4 n)$ time respectively. This result has further been improved in several papers [2,9].

Some researchers study the case where $\mathcal{F}$ are the k-link piece-wise linear functions. [11] presents an $O(n \log n)$ time algorithm for the ℓ_∞ criterion, and [1] present a FPTAS for the ℓ_1/ℓ_2 criterion.

In works more related to ours, the constraints on the candidate functions is not on the number of links but on the morphology. For example, the *monotonic regression* demands the candidate functions to be monotonic, and the *unimodal regression* asks them to be unimodal, which is increasing and then decreasing. For these regressions, [17] develop $\Theta(n), \Theta(n \log n), \Theta(n)$ algorithms for the ℓ_2, ℓ_1 and unit weighted ℓ_∞ case respectively. A generalization of the monotonic regression is the *isotonic regression*, and part of its results can be found in [18, 19]. Sometimes, the constraints can be both on the number of links and on the morphology. [14] study a variant of the V-optimal histogram problem, in which the candidate functions are required to be k-step histograms that are monotonic or unimodal. They solve this variant in $O(n^2 k)$ time, which is the same as the time required to solve the V-optimal histogram problem, i.e. the problem without the constraints on morphology.

Other researchers study the case where the data set is a curve rather than a set of discrete points. Most commonly, the data curve is an n-link piece-wise linear function (i.e. a polygonal curve); and the error criterion is defined in the same manner as the discrete case shown above. A survey for this case is given by [15]. In particular, [5] consider approximating the curve by a unimodal curve, and they solve it in $\Theta(n)$ time for the ℓ_2 criterion. [6] consider approximating the curve by a curve with at most k peaks. Under the ℓ_p $(p < \infty)$ criterion, they solve it in $\Theta(n \log n)$ time for $k = 1$ and $\Theta(kh^2 + hn \log n)$ time for $k > 1$, where h denotes the number of peaks of the given curve. [4] consider the same problem under the ℓ_∞ criterion, and their algorithm works in $\Theta(n)$ time.

In this paper, our restriction to $\mathcal{F}$ is not on the number of peaks, but the morphology of each peak.

1.2   Technique Overview

We first define four terms, namely *good, bounded-good, superb, bounded-superb.*

Definition 2. *For a sequence* $\mathbf{b} = (b_1, \ldots, b_n)$, *we define its* distance $\mathbb{D}(\mathbf{b})$, length $\mathbb{L}(\mathbf{b})$, *and* height $\mathbb{H}(\mathbf{b})$ *as follows:*

$$\mathbb{D}(\mathbf{b}) = \sum_{i=1}^{n} |b_i - a_i|, \quad \mathbb{L}(\mathbf{b}) = \sum_{i=2}^{n} |b_i - b_{i-1}|, \quad \mathbb{H}(\mathbf{b}) = \sum_{i=1}^{n} b_i.$$

For sequences $\mathbf{b}, \mathbf{b}'$, *we regard that* $\mathbf{b}$ *is* better *than* $\mathbf{b}'$ *if the following is true:*

- $\mathbb{D}(\mathbf{b}) < \mathbb{D}(\mathbf{b}')$, *or*
- $\mathbb{D}(\mathbf{b}) = \mathbb{D}(\mathbf{b}')$ *and* $\mathbb{L}(\mathbf{b}) > \mathbb{L}(\mathbf{b}')$, *or*
- $\mathbb{D}(\mathbf{b}) = \mathbb{D}(\mathbf{b}')$ *and* $\mathbb{L}(\mathbf{b}) = \mathbb{L}(\mathbf{b}')$ *and* $\mathbb{H}(\mathbf{b}) > \mathbb{H}(\mathbf{b}')$.

A sequence is *good*, if it is (α, β)-fluctuated and has minimum distance among all such sequences. A sequence is *bounded-good*, if it is (α, β)-fluctuated and is bounded by **a** and has minimum distance among all such sequences. So, a *good* sequence refers to a solution to the unbounded version, and a *bounded-good* sequence refers to a solution to the bounded version.

A sequence is *superb*, if it is (α, β)-fluctuated and no such sequence is better than it. A sequence is *bounded-superb*, if it is (α, β)-fluctuated and bounded by **a** and no such sequence is better than it. Shortly, superb means good and nothing better, bounded-superb means bounded-good and nothing better.

Note that "better than" is a partial order between the sequences. So, there is at least one superb sequence and at least one bounded-superb sequence.

To find a good or bounded-good sequence, we will actually find a superb or bounded-superb sequence. Those superb or bounded-superb sequences have a special property which allows us to compute them efficiently. This property briefly states that each element of the sequence is confined to a particular set whose size is bounded by a constant. According to this property, we use dynamic programming to compute the superb and bounded-superb sequence.

The good or bounded-good sequences in general do not have the mentioned property; this is why we should consider "better than" instead of "closer than".

Outline. We describe and prove the special property of the superb and bounded-superb sequences in Sect. 2. We present our algorithm in Sect. 3. We solve a variant in Sect. 4. We discuss a future direction in Sect. 5.

2   A Property of the Superb or Bounded-Superb Sequence

A key property of the superb or bounded-superb sequences is stated in the following lemma.

Definition 3. *Given a sequence* $\mathbf{x} = (x_1, \ldots, x_n)$. *For* $i \in [1, n]$, *denote* $\delta(i) = \{j \in [1, n] : |i - j| \leq 2\}$. *An element* x_i *is shaped if there exists* $j(i)$ *in* $\delta(i)$, *such that* $|x_i - a_{j(i)}| \leq 1$. *The sequence* $\mathbf{x}$ *is shaped if all of its elements are shaped.*

Lemma 1   ‸

1. *The superb sequence(s) must be shaped.*
2. *The bounded-superb sequence(s) must be shaped.*

The proof of Lemma 1.2 is given below. The proof of Lemma 1.1 is similar but much more complicated; it is given in the full version of this paper.

To prove this lemma, we first define the *pieces* of a given sequence.

Definition 4. *Assume that* $\mathbf{x} = (x_1, \ldots, x_n)$. *For a pair of indices* j, k *such that* $j \leq k$, *let* $\langle x_j, x_k \rangle$ *denote* $x_j, \ldots, x_k$ *and we call it a segment of* $\mathbf{x}$. *We can uniquely partition* $\mathbf{x}$ *into several segments so that the elements in each segment are identical but elements in adjacent segments are distinct; and we call each of such segment a* piece *of* $\mathbf{x}$.

We define the height *of a piece to be the value of any of its element (which are all the same). If* $\langle x_j, x_k \rangle$ *is a piece of* $\mathbf{x}$, *we refer to the two neighboring pieces of* $\langle x_j, x_k \rangle$ *as its "previous piece" and "next piece", respectively.*

Proof (of Lemma 1.2). Assume **x** is bounded-superb and $\langle x_j, x_k \rangle$ is an arbitrary piece of **x**. We shall prove that all the elements in this piece are shaped.

First of all, we make some notations and conventions.

- We denote by L the absolute height difference between piece $\langle x_j, x_k \rangle$ and its previous piece, and set $L = +\infty$ if $\langle x_j, x_k \rangle$ is the first piece.
- We denote by R the absolute height difference between piece $\langle x_j, x_k \rangle$ and its next piece, and set $R = +\infty$ if $\langle x_j, x_k \rangle$ is the last piece.
- For simplicity of presentation, we regard that *the first piece is higher than its previous piece, and that the last piece is higher than its next piece.*
- For proving some properties of a sequence **x**, we frequently show that if they did not hold, then a sequence **x'** better than **x** could be constructed from **x**. In these cases, we will make sure that **x'** is (α, β)-fluctuated. But, for brevity, we do not repeatedly declare this "fluctuated" property of **x'**.

We discuss four cases depending on the relative height between $\langle x_j, x_k \rangle$ and its neighboring pieces.

1. Piece $\langle x_j, x_k \rangle$ is lower than its previous piece but higher than its next.
    1.1. $k = j$. We claim that $x_j = a_j$ and so x_j is shaped. Otherwise, $x_j < a_j$, and **x** would be improved by augmenting x_j by 1.
    1.2. $k > j$. For the same reason as Case 1.1, we have $x_i = a_i$ for $i \in [j+1, k]$. This implies that elements $x_j, \ldots, x_k$ are shaped.
2. Piece $\langle x_j, x_k \rangle$ is higher than its previous piece but lower than its next. This case is symmetric to the previous case and we omit its proof.
3. Piece $\langle x_j, x_k \rangle$ is lower than its neighboring pieces.
    3.1. $k \geq j + 2$. Similar to case 1.2, we have $x_i = a_i$ for $i \in [j+1, k-1]$. This implies that elements $x_j, \ldots, x_k$ are shaped.
    3.2. $k = j + 1$. We claim that $x_j = a_j$ or $x_k = a_k$, thus x_j, x_k are shaped. Suppose to the contrary that $x_j < a_j$ and $x_k < a_k$. If both L, R are larger than 1, we augment x_j and x_k by 1; if $L = 1$, we augment x_j by 1; if $R = 1$, we augment x_k by 1. In each case, we obtain a better sequence.
    3.3. $k = j$. When $L > 1$ and $R > 1$, we can easily get $x_j = a_j$ and so x_j is shaped. In the following we show that if $L = 1$ then x_j is shaped. Symmetrically, x_j is shaped when $R = 1$. So, x_j is always shaped. Notice that since $\langle x_j \rangle$ is lower than its previous piece, it is not the first piece.
        3.3.1. The previous piece has one element. Then, $x_{j-1} = a_{j-1}$. Otherwise **x** can be improved by augmenting x_{j-1} by 1.
        3.3.2. The previous piece has two elements. Then, $x_{j-1} = a_{j-1}$ or $x_{j-2} = a_{j-2}$. Otherwise **x** can be improved by augmenting x_{j-2}, x_{j-1} by 1.
        3.3.3. The previous piece has at least three elements. Then, $x_{j-1} = a_{j-1}$. Otherwise **x** can be improved by augmenting x_{j-1} by 1 and reducing x_{j-2} by 1. (The distance is unchanged but the length is increased.)
    In each subcase, applying $L = 1$ we can further get that x_j is shaped.
4. Piece $\langle x_j, x_k \rangle$ is higher than its neighboring pieces.
    4.1. $k \leq j + 1$. If $x_j < a_j$ and $x_k < a_k$, we can improve **x** by augmenting this piece by 1. Therefore, $x_j = a_j$ or $x_k = a_k$. So x_j, x_k are shaped.

4.2. $k \geq j+2$. We claim $x_j = a_j$. Otherwise, **x** can be improved by augmenting x_j by 1 and reducing x_{j+1} by 1. (The distance remains unchanged but the length increases.) Symmetrically, $x_k = a_k$. Therefore, x_j, x_{j+1}, x_{j+2} and x_k, x_{k-1}, x_{k-2} are shaped. Moreover, we claim that $x_i \geq a_i - 1$ for any $i \in [j+3, k-3]$. This implies that the other elements in $\langle x_j, x_k \rangle$ are also shaped. Suppose to the contrary that $x_i < a_i - 1$, then **x** can by improved by augmenting x_i by 2 and reducing x_{i-1}, x_{i+1} by 1.

Remark 1. The definition for "shaped" is tricky and noteworthy. To explain this, let us introduce a term called k-shaped. (Notice that "shaped" is the same as "2-shaped".) An element x_i of a sequence $\mathbf{x} = (x_1, \ldots, x_n)$ is k-shaped if there exists an index $j(i)$ in $\delta^k(i) := \{j \in [1, n] \mid i - k \leq j \leq i + k\}$, such that $|x_i - a_{j(i)}| \leq 1$. Moreover, a sequence **x** is said k-shaped if all of its elements are k-shaped. At first we guess that the bounded-superb sequence must be 1-shaped. However, it is not! A counter example is: $\mathbf{a} = (3, 6, 7, 7, 10)$, $\alpha = \beta = 1$. The reader can check that the bounded-superb sequence is $(3, 6, 6, 5, 10)$. Here the 4-th element 5 is not 1-shaped. Similarly, the superb sequence may not be 1-shaped.

Also note that a good or bounded-good sequence may not be 2-shaped; only the superb or bounded-superb ones are assured to be 2-shaped.

3   Algorithm for Computing a Fluctuated Fitting Sequence

In this section, we first design an algorithm for solving the following problem.

(*) Given $\alpha, \beta, \mathbf{a} = (a_1, \ldots, a_n)$, and n sets of integers $S_1, \ldots, S_n$, each of which contains at most C elements (C is some constant). Find a sequence of integers $\mathbf{b} = (b_1, \ldots, b_n)$ such that $\sum_i |a_i - b_i|$ is minimized subject to the constraints that **b** is (α, β)-fluctuated and $b_i \in S_i$ for each i in $[1, n]$.

Briefly speaking, our algorithm is based on dynamic programming and it costs $O(C^2(\alpha + \beta) \cdot n) = O((\alpha + \beta) \cdot n)$ time.

First, we introduce the "Tail Fluctuate State" $\mathsf{TFS}(\mathbf{x})$ of sequence **x**.

Definition 5. *Assume that* $\mathbf{x} = (x_1, \ldots, x_i)$. *We call each ascend and each descend in this sequence a "change", and define* $\mathsf{TFS}(\mathbf{x})$ *to be*

$$\begin{cases} 0, & \text{if } \mathbf{x} \text{ has no changes at all;} \\ j, & \begin{array}{l} \text{if the last } j \text{ changes of } \mathbf{x} \text{ are all ascends, and there are} \\ \text{no other changes or the last } (j+1)\text{-th change is a descend;} \end{array} \\ -j, & \begin{array}{l} \text{if the last } j \text{ changes of } \mathbf{x} \text{ are all descends, and there are} \\ \text{no other changes or the last } (j+1)\text{-th change is an ascend.} \end{array} \end{cases} \quad (1)$$

For $1 \leq i \leq n$, let $t(i) = |S_i|$ and denote $S_i = \{S_{i,1}, \ldots, S_{i,t(i)}\}$.

For $-\beta \le j \le \alpha$ and $1 \le k \le t(i)$, let $F_i(j,k)$ denote the following subproblem: Find $(x_1, \ldots, x_i)$ to minimize $\sum_{h=1}^{i} |x_h - a_h|$, subject to:

(i) $x_h \in S_h$ for all $h < i$, and $x_i = S_{i,k}$.
(ii) $(x_1, \ldots, x_i)$ is $(\alpha, \beta) -$ fluctuated and $\mathsf{TFS}(x_1, \ldots, x_i) = j$.

Let $f_i(j,k)$ denote the value $(\sum_{h=1}^{i} |x_h - a_h|)$ of the minimization problem $F_i(j,k)$. The *initial condition* is given by $f_1(0,k) = |S_{1,k} - a_1|$ for $1 \le k \le t(1)$. For $i > 1$, the *recurrence* is given as follows. (Here k' denotes the index such that $S_{i-1,k'} = S_{i,k}$ when $S_{i,k} \in S_{i-1}$; and $(k' = 0)$ indicates that $S_{i,k} \notin S_{i-1}$.)

$$f_i(j,k) = \begin{cases} +\infty, & \text{if } j = 0 \text{ and } k' = 0; \\ |S_{i,k} - a_i| + f_{i-1}(0,k'), & \text{if } j = 0 \text{ and } k' > 0; \\ |S_{i,k} - a_i| + \min_{h:S_{i-1,h} \le S_{i,k}} f_{i-1}(j - \Delta_h, h), & \text{if } j > 0; \\ |S_{i,k} - a_i| + \min_{h:S_{i-1,h} \ge S_{i,k}} f_{i-1}(j + \Delta_h, h), & \text{if } j < 0; \end{cases} \quad (2)$$

where

$$\Delta_h = \begin{cases} 1, S_{i-1,h} \ne S_{i,k}; \\ 0, S_{i-1,h} = S_{i,k}. \end{cases}$$

We explain the recurrence for the case $j > 0$; the other cases are analogous. Suppose that $(x_1, \ldots, x_i)$ is a solution to problem $F_i(j,k)$, we have $\mathsf{TFS}(x) = j$. Since $j > 0$, x_i can not be a descend; namely, $x_{i-1} \le x_i$. Besides, we must have $x_i = S_{i,k}$ and $x_{i-1} \in S_{i-1}$. Therefore, x_{i-1} must be an element in S_{i-1} which is smaller than or equal to $S_{i,k}$. If x_{i-1} equals $S_{i,k}$, the problem reduces to problem $F_{i-1}(j,k')$; otherwise, $x_{i-1} = S_{i-1,h}$ for some h such that $S_{i-1,h} < S_{i,k}$, and the problem reduces to problem $F_{i-1}(j-1,h)$. Thus we prove this case.

We now give the algorithm. The proof of correctness is obvious and omitted.

---

1 Compute the array f based on Eq. (2);
2 Compute (j^*, k^*) so that $f_n(j^*, k^*) = \min_{1 \le j \le t(n), -\beta \le k \le \alpha} f_n(j, k)$;
3 $(j, k) \leftarrow (j^*, k^*)$;
4 **foreach** $i = n$ *downto 2* **do**
5 $\quad b_i \leftarrow S_{i,k}$;
6 $\quad$ Find (j^*, k^*) so that $f_i(j,k)$ is achieved from $f_{i-1}(j^*, k^*)$; (Using (2))
7 $\quad$ Let $(j,k) \leftarrow (j^*, k^*)$;
8 **end**
9 $b_1 \leftarrow S_{1,k}$;
10 Output $(b_1, \ldots, b_n)$.

Algorithm 1. An algorithm for solving Problem (*).

Apply Algorithm 1 to Compute the Fluctuated Fitting Sequence. By applying this algorithm, we can solve the original problem as follows and thus prove Theorem 1. Due to Lemma 1, the superb sequence(s) and the bounded-superb sequences(s) are shaped. This means there is at least one good sequence which is shaped, and at least one bounded-good sequence which is shaped. Therefore:

– Computing a good sequence reduces to compute the sequence closest to $\mathbf{a}$ subject to the constraint that it is (α, β)-fluctuated and shaped. This reduces to solving an instance of problem (*) - $(n, \alpha, \beta, \mathbf{a}, U)$, where

$$U_i := \{a_j - 1, a_j, a_j + 1 \mid j \in \delta(i)\} \quad \text{(for } i \in [1, n]\text{)}.$$

– Computing a bounded-good sequence reduces to compute the sequence closest to $\mathbf{a}$ subject to the constraints that it is (α, β)-fluctuated, shaped and bounded. This reduces to solving an instance of problem (*) - $(n, \alpha, \beta, \mathbf{a}, U')$, where

$$U'_i := U_i \cap (-\infty, a_i] \quad \text{(for } i \in [1, n]\text{)}.$$

Note that $|U'_i| \le |U_i| \le 15$, so we can set $C = 15$.

4   Fitting by Circularly-Fluctuated Sequence

In this section, we introduce an extended notion of "fluctuated sequence" called "*circularly fluctuated sequence*" (see Definition 6) and study the following variant problem: find the sequence closest to $\mathbf{a}$ that is bounded by $\mathbf{a}$ and is circularly fluctuated. As a result, we show that this variant can be reduced to the non-circular bounded version and thus can be solved in the same time asymptotically.

Definition 6. *A sequence* $\mathbf{x} = (x_1, \ldots, x_n)$ *is circularly-(α, β)-fluctuated if the sequence constructed by repeating* $\mathbf{x}$ *is (α, β)-fluctuated. In other words,* $\mathbf{x}$ *is circularly-(α, β)-fluctuated if* $(\mathbf{x}, \mathbf{x}) = (x_1, \ldots, x_n, x_1, \ldots, x_n)$ *is (α, β)-fluctuated.*

Recall that bounded-good means "(α, β)-fluctuated, bounded by $\mathbf{a}$, and with minimum distance among such sequences". Similarly, we define *C-bounded-good* as "circularly-(α, β)-fluctuated, bounded by $\mathbf{a}$, and with minimum distance among such sequences". To show the reduction, we first give two lemmas.

Lemma 2. *Assume that* $\mathbf{a} = (a_1, \ldots, a_n)$.

(1) If sequence $\mathbf{b}$ *is bounded-good, then* $\min(b_1, \ldots, b_n) = \min(a_1, \ldots, a_n)$.
(2) If sequence $\mathbf{c}$ *is C-bounded-good, then* $\min(c_1, \ldots, c_n) = \min(a_1, \ldots, a_n)$.

Proof. Denote $\gamma = \min(a_1, \ldots, a_n)$. Since $\mathbf{b}, \mathbf{c}$ are bounded by $\mathbf{a}$, we have

$$\min(b_1, \ldots, b_n) \le \gamma, \min(c_1, \ldots, c_n) \le \gamma.$$

So, we only need to prove that

$$\min(b_1, \ldots, b_n) \ge \gamma, \min(c_1, \ldots, c_n) \ge \gamma.$$

In other words, we only need to prove that ① $\mathbf{b}$ does not have elements smaller than γ; and ② $\mathbf{c}$ does not have elements smaller than γ.

Proof of ①: Suppose to the contrary that **b** has elements smaller than γ. We construct another sequence **b**′ and argue that (i) **b**′ is (α, β)-fluctuated; (ii) it is bounded by **a**; (iii) it has distance smaller than **b**. This means that **b** is not bounded-good, which is contradictory.

We construct **b**′ as follows. For $1 \leq i \leq n$, define $b_i' = \max(b_i, \gamma)$. Briefly, **b**′ is a modification of **b**, which increases all the elements smaller than γ to γ.

Arguments (ii) and (iii) are obvious; we prove (i) in the following. For convenience, we introduce an array $\mathsf{M} = (\mathsf{M}_2, \ldots, \mathsf{M}_n)$ defined as follows.

$$\text{For } 2 \leq i \leq n, \mathsf{M}_i = \begin{cases} \mathsf{A}, & \text{if } b_i' > b_{i-1}'; \\ \mathsf{D}, & \text{if } b_i' < b_{i-1}'; \\ \mathsf{G}, & \text{if } b_i' = b_{i-1}' = \gamma; \\ \_, & \text{otherwise.} \end{cases}$$

By the construction of **b**′, we observe: (I) if $\mathsf{M}_i = \mathsf{A}$, then b_i is an ascend; if $\mathsf{M}_i = \mathsf{D}$, then b_i is a descend; if $\mathsf{M}_i = \_$, then $b_i = b_{i-1}$, i.e. b_i is neither an ascend nor a descend; and (II) if b_i is a descend, then M_i equals D or G.

To prove that **b**′ is (α, β)-fluctuated, we need to prove statements (A), (B):

(A) If $\mathsf{M}_j, \ldots, \mathsf{M}_k$ $(j < k)$ contain $\alpha + 1$ A's, they contain at least one D;
(B) If $\mathsf{M}_j, \ldots, \mathsf{M}_k$ $(j < k)$ contain $\beta + 1$ D's, they contain at least one A.

We prove (A); (B) is symmetric. Without loss of generality, assume that $\mathsf{M}_j = \mathsf{M}_k = \mathsf{A}$. First, suppose that $\mathsf{M}_j, \ldots, \mathsf{M}_k$ contain a G. Then, between this G and the nearest A left to it there must be an D, since the symbol before any G must be a G or D. Second, suppose that $\mathsf{M}_j, \ldots, \mathsf{M}_k$ neither contain G's nor contain D's. Now, applying (I) and (II), $b_j, \ldots, b_k$ contain $\alpha + 1$ ascends but no descends, which contradicts the assumption that **b** is (α, β)-fluctuated. Therefore, no matter $\mathsf{M}_j, \ldots, \mathsf{M}_k$ contain a G or not, they contain a D.

Proof of ②: Suppose to the contrary that **c** has some elements smaller than γ. We construct a new sequence **c**′ based on **c**: define $c_i' = \max(c_i, \gamma)$ for $1 \leq i \leq n$. Since **c** is circularly-(α, β)-fluctuated, $(\mathbf{c}, \mathbf{c})$ is (α, β)-fluctuated. Then, by applying the proof of ①, we see that $(\mathbf{c}', \mathbf{c}')$ is (α, β)-fluctuated. Therefore, **c**′ is circularly-(α, β)-fluctuated. Besides, it is obvious that **c**′ is bounded by **a** and with distance less than **c**. All the above properties together contradicts the assumption of **c**. Thus, **c** has no elements smaller than γ.

Lemma 3. *Given* $\mathbf{a} = (a_1, \ldots, a_n)$, *where* a_1 *is the minimum element in* **a**. *Assume that* $(b_1, \ldots, b_{n+1})$ *is bounded-good with respect to (w.r.t.)* $(a_1, \ldots, a_n, a_1)$. *Then,* $(b_1, \ldots, b_n)$ *is C-bounded-good with respect to* **a**.

Proof. Since $(b_1, \ldots, b_{n+1})$ is bounded-good with respect to $(a_1, \ldots, a_n, a_1)$ and $\min(a_1, \ldots, a_n, a_1) = a_1$, we get $b_1 = b_{n+1} = a_1$ applying Lemma 2 (1).

First, we prove that $(b_1, \ldots, b_n)$ is circularly-(α, β)-fluctuated.

Define $\mathbf{d} = (d_1, \ldots, d_{2n}) = (b_1, \ldots, b_n, b_1, \ldots, b_n)$. We need to show that **d** is (α, β)-fluctuated. This reduces to prove the following arguments.

(i) If $1 < j < k \le 2n$ and d_j, d_k are both ascends and there are more than α ascends in $d_j, \ldots, d_k$, then there is at least one descend in $d_j, \ldots, d_k$.

(ii) If $1 < j < k \le 2n$ and d_j, d_k are both descends and there are more than β descends in $d_j, \ldots, d_k$, then there is at least one ascend in $d_j, \ldots, d_k$.

We only show the proof of (i); (ii) is symmetric. We discuss three cases.

1. $j \ge n+2$. In this case, $d_j, \ldots, d_k$ is a subset of $d_{n+1}, \ldots, d_{2n}$. If $d_j, \ldots, d_k$ contain more than α ascends but no descend, then $(d_{n+1}, \ldots, d_{2n}) = (b_1, \ldots, b_n)$ is not (α, β)-fluctuated, which contradicts the assumption of $(b_1, \ldots, b_{n+1})$.

2. $k \le n+1$. Here, $d_j, \ldots, d_k$ is a subset of $d_1, \ldots, d_{n+1}$. If $d_j, \ldots, d_k$ contain more than α ascends but no descend, then $(d_1, \ldots, d_{n+1}) = (b_1, \ldots, b_n, b_1) = (b_1, \ldots, b_{n+1})$ is not (α, β)-fluctuated, which contradicts the assumption.

3. $j \le n+1$ and $k \ge n+2$. Clearly, $d_{n+1} = b_1$ is the minimum element in $d_1, \ldots, d_{2n}$. Further since d_j is an ascend, there must be a descend between d_j and d_{n+1}. Further since $k \ge n+2$, there is a descend in $d_j, \ldots, d_k$.

Second, we prove that $(b_1, \ldots, b_n)$ is C-bounded-good w.r.t. **a**. Let **c** $= (c_1, \ldots, c_n)$ be a C-bounded-good sequence w.r.t. **a**. We should prove that $\mathbb{D}(\mathbf{b}) \le \mathbb{D}(\mathbf{c})$, where $\mathbb{D}(\mathbf{b}), \mathbb{D}(\mathbf{c})$ denote the distance of **b**, **c** w.r.t. $(a_1, \ldots, a_n)$.

For convenience, let $\bar{\mathbf{c}} = (c_1, \ldots, c_n, c_1)$ and $\bar{\mathbf{b}} = (b_1, \ldots, b_n, b_{n+1})$, and let $\mathbb{D}(\bar{\mathbf{b}}), \mathbb{D}(\bar{\mathbf{c}})$ denote the distance of $\bar{\mathbf{b}}, \bar{\mathbf{c}}$ w.r.t. $(a_1, \ldots, a_n, a_1)$. We claim: (i) $\bar{\mathbf{c}}$ is bounded by $(a_1, \ldots, a_n, a_1)$ and (ii) $\bar{\mathbf{c}}$ is (α, β)-fluctuated.

Proof of (i): Since **c** is C-bounded-good w.r.t. **a**, it is bounded by $(a_1, \ldots, a_n)$. So, $\bar{\mathbf{c}} = (c_1, \ldots, c_n, c_1)$ is bounded by $(a_1, \ldots, a_n, a_1)$.

Proof of (ii): Since **c** is C-bounded-good, it is circularly-(α, β)-fluctuated. So, $(c_1, \ldots, c_n, c_1, \ldots, c_n)$ is (α, β)-fluctuated. Therefore, $\bar{\mathbf{c}}$ is (α, β)-fluctuated.

According to (i), (ii) and the assumption which says that $(b_1, \ldots, b_{n+1})$ is bounded-good w.r.t. $(a_1, \ldots, a_n, a_1)$, we get $\mathbb{D}(\bar{\mathbf{b}}) \le \mathbb{D}(\bar{\mathbf{c}})$.

Applying Lemma 2 (1), we get $b_{n+1} = b_1 = a_1$. Therefore, $\mathbb{D}(\bar{\mathbf{b}}) = \mathbb{D}(\mathbf{b})$.

Applying Lemma 2 (2), we get $c_1 = a_1$, which follows that $\mathbb{D}(\bar{\mathbf{c}}) = \mathbb{D}(\mathbf{c})$.

Altogether, we obtain $\mathbb{D}(\mathbf{b}) \le \mathbb{D}(\mathbf{c})$.

By applying Lemma 3, computing C-bounded-good sequence reduces to computing bounded-good sequence. This is made precise in the next theorem.

Theorem 2. *Given* **a** $= (a_1, \ldots, a_n)$ *and parameters* α, β, *we can compute a C-bounded-good sequence in* $O((\alpha + \beta) \cdot n)$ *time.*

Proof. Assume $a_i = \min(a_1, \ldots, a_n)$. Computing a C-bounded-good sequence of **a** is equivalent to computing a C-bounded-good sequence of

$$(a_i, \ldots, a_n, a_1, \ldots, a_{i-1}).$$

In other words, we can assume that $a_1 = \min(a_1, \ldots, a_n)$. Then, according to Lemma 3, computing an C-bounded-good sequence of **a** reduces to computing a bounded-good sequence of $(a_1, \ldots, a_n, a_1)$. We can compute the bounded-good sequence of $(a_1, \ldots, a_n, a_1)$ in $O((\alpha + \beta) \cdot n)$ time according to Theorem 1.

5   Conclusion and Future Work

In this paper, we fit a sequence of data **a** by an (α, β)-fluctuated sequence **b**. We consider three types of constraints on **b**: 1. no constraints; 2. bounded (by **a**); 3. bounded and circularly-fluctuated. All of them are solved in $O((\alpha + \beta) \cdot n)$ time. Our algorithm is mainly based on a nontrivial observation on the best fitting sequences.

We can consider another variant where **b** is only required to be circularly-fluctuated. This is an open problem and is the subject of future work.

References

1. Aronov, B., Asano, T., Katoh, N., Mehlhorn, K., Tokuyama, T.: Polyline fitting of planar points under min-sum criteria. Int. J. Comput. Geom. Appl. **16**, 97–116 (2006)
2. Chen, D.Z., Wang, H.: Approximating points by a piecewise linear function. Algorithmica **66**(3), 682–713 (2013)
3. Chen, D.Z., Healy, M.A., Wang, C., Xu, B.: Geometric algorithms for the constrained 1-D K-means clustering problems and IMRT applications. In: Preparata, F.P., Fang, Q. (eds.) FAW 2007. LNCS, vol. 4613, pp. 1–13. Springer, Heidelberg (2007). doi:10.1007/978-3-540-73814-5_1
4. Chen, D., Wang, C., Wang, H.: Representing a functional curve by curves with fewer peaks. Discret. Comput. Geom. **46**(2), 334–360 (2011)
5. Chun, J., Sadakane, K., Tokuyama, T.: Linear time algorithm for approximating a curve by a single-peaked curve. Algorithmica **44**(2), 103–115 (2006)
6. Chun, J., Sadakane, K., Tokuyama, T., Yuki, M.: Peak-reducing fitting of a curve under the l_p metric. Interdisc. Inf. Sci. **2**, 191–197 (2005)
7. Cleju, I., Fränti, P., Wu, X.: Clustering based on principal curve. In: Kalviainen, H., Parkkinen, J., Kaarna, A. (eds.) SCIA 2005. LNCS, vol. 3540, pp. 872–881. Springer, Heidelberg (2005). doi:10.1007/11499145_88
8. Douglas, D.H., Peucker, T.K.: Algorithms for the Reduction of the Number of Points Required to Represent a Digitized Line or its Caricature. Wiley, Hoboken (2011). pp. 15–28
9. Fournier, H., Vigneron, A.: A deterministic algorithm for fitting a step function to a weighted point-set. Inf. Process. Lett. **113**(3), 51–54 (2013)
10. Fournier, H., Vigneron, A.: Fitting a step function to a point set. In: Halperin, D., Mehlhorn, K. (eds.) ESA 2008. LNCS, vol. 5193, pp. 442–453. Springer, Heidelberg (2008). doi:10.1007/978-3-540-87744-8_37
11. Goodrich, M.: Efficient piecewise-linear function approximation using the uniform metric. Discret. Comput. Geom. **14**(1), 445–462 (1995)
12. Guha, S., Shim, K.: A note on linear time algorithms for maximum error histograms. Knowl. Data Eng. **19**(7), 993–997 (2007)
13. Guha, S., Koudas, N., Shim, K.: Data-streams and histograms. In: Proceedings of 33rd Symposium on Theory of Computing, STOC 2001, pp. 471–475. ACM (2001)
14. Haiminen, N., Gionis, A., Laasonen, K.: Algorithms for unimodal segmentation with applications to unimodality detection. Knowl. Inf. Syst. **14**(1), 39–57 (2008)
15. Luebke, D.P.: A developer's survey of polygonal simplification algorithms. IEEE Comput. Graph. Appl. **21**(3), 24–35 (2001)

16. Ramesh, N., Yoo, J.H., Sethi, I.: Thresholding based on histogram approximation. IEEE Proc. Vis. Image Sig. Process. **142**(5), 271–279 (1995)
17. Stout, Q.F.: Unimodal regression via prefix isotonic regression. Comput. Stat. Data Anal. **53**(2), 289–297 (2008)
18. Stout, Q.F.: Isotonic regression via partitioning. Algorithmica **66**(1), 93–112 (2013)
19. Stout, Q.F.: Isotonic regression for multiple independent variables. Algorithmica **71**(2), 450–470 (2015)
20. Tokuyama, T.: Recent progress on geometric algorithms for approximating functions: toward applications to data analysis. Electron. Commun. Jpn. (Part III: Fundam. Electron. Sci.) **90**(3), 1–12 (2007)

Optimal Partitioning Which Maximizes the Weighted Sum of Products

Kai Jin[(✉)]

University of Hong Kong, Pokfulam, Hong Kong SAR, China
cscjjk@gmail.com

Abstract. We consider the problem of partitioning n real numbers to K nonempty groups, so that the weighted sum of products over all groups is maximized. Formally, given $S = \{r_1, \ldots, r_n\}$ and $W = (w_1, \ldots, w_K)$ where $w_i \geq 0$, we look for a partition of S into K nonempty groups $S_1, \ldots, S_K$, so that $\sum_{g=1}^{K}(w_g \cdot \prod_{r_j \in S_g} r_j)$ is maximized. Our main result is an $O(n^2)$ time algorithm for finding an optimal partition.

Keywords: Partitioning with additive objective · K-partition · Sum of products · Greedy algorithms · Rearrangement inequality

1 Introduction

Let $S = \{r_1, \ldots, r_n\}$ be a set of n real numbers. We consider the problem of partitioning S into K nonempty groups, so as the weighted sum of the products over all groups is maximized, where *the product of a group* refers to the product of all elements in this group. Thus, we look for a partition which maximizes

$$\sum_{g=1}^{K} w_g \cdot \prod_{r_j \in S_g} r_j, \tag{1}$$

subject to $S_g \neq \varnothing$ for any g, and $\bigcup_{g=1}^{K} S_g = S$, and $S_g \cap S_{g'} = \varnothing$ for $g \neq g'$.

We assume that the number of groups K is fixed (and $1 \leq K \leq n$), and that all the weights $w_1, \ldots, w_K$ are nonnegative. For convenience, assume that $0 \leq w_1 \leq \ldots \leq w_K$ and $r_1 \leq \ldots \leq r_n$.

A partition of S into K mutually exclusive and nonempty groups is called a *K-partition*. Throughout this paper, K will be fixed. So, unless otherwise stated, a partition means a K-partition. We call (1) the *score* of the partition, and we call $w_g \cdot \prod_{r_j \in S_g} r_j$ the *score* of group S_g. A partition is *optimal* if it has maximum score. We are interested in finding an optimal partition.

Our problem belongs to the class of optimal partitioning problem with additive objective function, in which the score is separable in the groups. For a general choice of the group score function this problem is NP-hard even for $K = 2$. For example, let the score of group S_g be $-(\sum_{r_j \in S_g} r_j)^2$. The well-known NP-complete problem Number Partitioning [13] can be formulated to check whether

© Springer International Publishing AG 2017
M. Xiao and F. Rosamond (Eds.): FAW 2017, LNCS 10336, pp. 127–138, 2017.
DOI: 10.1007/978-3-319-59605-1_12

the optimum 2-partition has score $-((R/2)^2 + (R/2)^2)$, where R is the sum of all elements in S. Moreover, the problem of minimize (1) is also NP-hard for $K = 2$. (We may discuss it in the full version of this paper.[1])

The optimal partitioning problems arise in many fields including system reliability, circuit design, vehicle routing, inventory grouping, location problems, scheduling, grouping testing, clustering, and bin-packing. (See [1,9] and the references within; and also see Subsect. 1.2 for some examples.)

For some partitioning problems, there exists an optimal partition which admits some special properties, such as "consecutive" (i.e. the indices of the elements in all groups are consecutive), or "ordered" (i.e. the indices of the elements in S_{g+1} are larger than those in S_g). If there exists an ordered optimal partition, the problem can be solved in polynomial time. If there exists a consecutive optimal partition, the problem can be solved in $O(n^{O(K)})$. Researches have proposed several sufficient conditions on the objective function under which an optimal partition with a special property exists, see [1,4,5,8–10]. Examples satisfying these conditions are discussed in these literatures.

Our objective function does not satisfy any sufficient conditions given in the aforementioned literatures. Consequently, there are cases where all optimal partitions are not consecutive. For example, for $S = \{r_1, r_2, r_3, r_4\} = \{-2, -2, 1, 2\}$ and $K = 2$ and unit weights, the unique optimal partition assign $\{-2, -2, 2\}$ to one group and $\{1\}$ to the other, which is not consecutive because the indices in the first group are $\{1, 3, 4\}$. Nevertheless, we are able to derive an efficient algorithm for our problem. It runs in $O(n^2)$ time, which is near-optimal, because computing the score of a given partition requires $O(n^2)$ time. (In realistic computational model, computing the product of n reals would take $O(n^2)$ time.)

1.1 Technique Overview and Organization of This Paper

Definition 1. *A partition is **ordered** if for any two groups S_g, S_{g+1}, the indices of elements in S_g are smaller than those in S_{g+1}. A partition is **extremal**, if beside the first and last group, each group contains exactly one element.*

We start from solving two special cases: all the elements are nonnegative; or there is exactly one negative element. We prove that in these cases, there exists an optimal partition which is ordered and extremal.

We then consider the general case. Our basic idea is to reduce the general case to the preceding special cases. To this end, we design a suit of four policies (see Lemma 5) and prove that there always exists an optimal partition obeying these policies. The design of the policies applies greedy strategies. For example, one policy roughly states that most of the negative elements must be paired in a monotone order (e.g. r_1, r_2 are paired and assigned to the same group; r_3, r_4 are paired and assigned to the same group).

We then presents an algorithm framework for computing an optimal partition. It generates several candidate partitions $\chi_0, \ldots, \chi_{K-1}$ and select the best

---

[1] According to one of the reviewers, this can be proved from "exact cover by 3-sets" problem, similar to Yao's NP-hardness proof for the "subset product" problem.

of them. We prove that $\chi_0, \ldots, \chi_{K-1}$ dominate all the partitions that obey the aforementioned four policies. So, the best among $\chi_0, \ldots, \chi_{K-1}$ is optimal. (Note that we do not enumerate out all the partitions that obey the four policies.)

Finally, we present the detailed algorithm. Note that computing the score of a candidate partition requires $O(n^2)$ time. So, a direct implementation of the above framework would take $O(n^3)$ time. However, in $O(n)$ time we can compute the score of a candidate partition based on the score of the previous one. Therefore, an improved algorithm only runs in $O(n^2)$ time.

Theorem 1. *An optimal K-partition of set S can be found in $O(n^2)$ time.*

Alternatively, Sect. 5 provides a different approach that uses dynamic programming. It exploits the characterization of the optimal partition that the last group S_K always consists of a union of a prefix and a suffix of $r_1, \ldots, r_n$.

1.2 Related Work - Typical Partitioning Problems

Example 1 (System reliability [10]*).* Assume $r_1, \ldots, r_n$ are reals in $[0, 1]$. Given $n_1, \ldots, n_K$ so that $\sum n_i = n$. Minimize $\sum_{g=1}^{K} \log(1 - \prod_{r_j \in S_g} r_j)$, in which S_g contains n_g elements.

Example 2 (Grouping testing [8]*).* Assume $r_1, \ldots, r_n$ are reals in $[0, 1]$. Minimize $\sum_{g=1}^{K} \{1 + (1 - \prod_{r_j \in S_g} r_j) n_g\}$, where n_g is the cardinality of S_g.

Example 3 (Clustering [6]*).* Assume object i in S is characterized by two attributes w_i and r_i. Minimize $\sum_{g=1}^{K} w_i \times (r_i - \bar{r}_i)$, where $\bar{r}_i$ is the weighted arithmetic mean of those r's that are assigned to the subset containing i.

Example 4 (Scheduling [7]*).* Assume each object i in S is characterized by two attributes r_i and p_i. Maximize $\sum_{g=1}^{K} (\sum_{i \in S_g} p_i r_i / \sum_{i \in S_g} p_i)$.

Example 5 (Bin-packing [11,12]*).* Given a concave function F, minimize $\sum_{g=1}^{K} F(\sum_{r_j \in S_g} r_j)$, so that the sum of items in each subset does not exceeds the size of bin.

In addition, [2] considered a generalization where each object in S is characterized by a point in p-dimensional space. [3] considered a constrained partition problem where the groups in the partition must be basics of given matroids.

2   The Structural Properties of the Optimal Partition

Recall that the product of S_g is defined to be $\prod_{r_j \in S_g} r_j$.

Proposition 1. *Let χ be an arbitrary partition. Consider two groups S_i and S_j of χ, where $i < j$. If the product of S_i is no less than the product of S_j, by interchanging the elements in these two groups, the score of χ does not decrease.*

Lemma 1. *Consider the special case where $K = n$. Notice that in this case there is a unique ordered partition, which assigns r_i to S_i for $1 \le i \le K$. We claim that this partition is optimal.*

The above proposition and lemma follow from the Rearrangement Inequality.

Lemma 2. *Two claims on two restricted cases are given below.*

1. *Consider a restricted case where all elements in S are in range $[1, +\infty)$. Notice that there is a unique ordered partition in which each of the first $K - 1$ groups contains a single element. It is optimal in this restricted case.*
2. *Consider a restricted case where all elements are in range $[0, 1]$. Notice that there is a unique ordered partition in which each of the last $K - 1$ groups contains a single element. It is optimal in this restricted case.*

To prove the first claim, we will show that *for $K > 1$, every partition can be modified without loss of score so that r_1 is assigned solely to the first group*. Then, the first claim follows by induction. The proof of the second claim is symmetric. The trivial proofs can only be given in the full version due to space limit.

Remark 1. To find an optimal partition, one may consider the following greedy strategy: always make the last group as large as possible. This works for the two restricted cases considered in Lemma 2. But it fails in general case. For example, let $S = \{0.9, 0.9, 1.01, 1.01\}, K = 3$ and let the weights be unit. The greedy strategy outputs $\{0.9\}, \{0.9\}, \{1.01, 1.01\}$, which has a score of 2.8201; but the optimal one is $\{0.9, 0.9\}, \{1.01\}, \{1.01\}$, which has a score of 2.83.

2.1   The Special Case: Zero or One Negative Element

The following lemmas show that in the case where there is at most one negative in $r_1, \ldots, r_n$, there exists an optimal partition which is ordered and extremal.

Lemma 3. *When $r_1, \ldots, r_n$ are nonnegative, there exists an optimal partition which is ordered and extremal and satisfies the following additional property:*

If $K > 1$, either the first group contains only one element, or all the elements contained in the first group are at most 1.

Lemma 4. *When there is exactly one negative in $r_1, \ldots, r_n$, there exists an optimal partition which is ordered and extremal.*

To prove Lemmas 3 and 4, we apply the following proposition.

Proposition 2. *Assume the following hold.*

(1) $r_1, \ldots, r_n$ are nonnegative.
(2) $K > 1$.
(3) Not all weights are zeros (i.e. $w_k > 0$).

Then, in any optimal partition, an element larger than 1 is not assigned together with an element smaller than 1 into the same group.

We use indirect method to prove this proposition. When an element larger than 1 is assigned with an element smaller than 1, we can modify the partition to increase the total value. The trivial proof is given in the full version.

Proof (of Lemma 3). Assume that $K > 1$ and not all weights are zeros. Otherwise the result is obvious. Further assume that there exists an element larger than 1 and an element smaller than 1. Otherwise the elements are in $[1, +\infty)$ or $[0, 1]$ and the result follows from Lemma 2.

Suppose χ is an optimal partition. Some groups in χ contain elements smaller than 1 and are called *light groups*. Others do not contain elements smaller than 1 and are called *heavy groups*. By Proposition 2, the light groups does not contain elements larger than 1. So, their respective products are smaller than 1. On the other hand, since the heavy groups do not contain elements smaller than 1, their respective products are at least 1. Therefore, by Proposition 1, we can modify χ without loss of score so that the light groups are in front of the heavy groups.

Now, assume that in χ the light groups are in front of the heavy groups. Let l denote the number of elements in the light groups. By the definition of light and heavy groups, *the elements in the light groups are the smallest l elements in S.* So, we can assume that $r_1, \ldots, r_l$ are in the light groups while the remaining elements $r_{l+1}, \ldots, r_n$ are in the heavy groups. Next, we repartition the elements in the light groups using the method given in Lemma 2.2, and we repartition the elements in the heavy groups using the method given in Lemma 2.1. This would not cut the total score. As a consequence, the final partition is ordered and extremal. The additional property is also satisfied since the first group is light and all elements in the light groups are at most 1.   □

Proof (of Lemma 4). Notice that the only negative element is r_1.

Let χ be an optimal partition. First, we modify χ so that r_1 is assigned to S_1. On the opposite case, we interchange S_1 with the group containing r_1. Notice that group containing r_1 has nonpositive product while any other group has nonnegative product, this interchange would not cut the score due to Proposition 1.

Then, we modify χ so that the elements in S_1 are consecutive elements $r_1, \ldots, r_i$. On the opposite case, there is a pair of elements r_i, r_j where $i < j$ such that r_j is assigned to S_1 but r_i is assigned to another group. We swap r_i, r_j to each other's group. This would not cut the total score since the individual scores of the associated groups are both non-decreased.

If $K = 2$, we are done since partition χ is already ordered and extremal. In the following we assume that $K \geq 3$.

Consider all the elements in groups $S_2, \ldots, S_K$. Since all of them are nonnegative, by Lemma 3, there is an optimal partition of them into $S_2, \ldots, S_K$, which satisfies the three properties stated in Lemma 3. We reassign the elements in $S_2, \ldots, S_K$ by adopting the preceding partition. Afterwards, there are two cases: (1) S_2 contains exactly one element, or (2) all elements in S_2 are in range $[0, 1]$. Under the first case, χ is already ordered and extremal. Under the second case, we further move all elements in S_2 except for the largest one to S_1. This would not cut the total score because the elements being moved are in range $[0, 1]$. Finally, χ becomes ordered and extremal.   □

2.2   The General Case

For convenience, let n_- denote the number of negative elements.

A group is called a *Neg-group*, if it has a negative product; is called a *Pos/Zero-group* (abbreviated as P/0-group), if it has a nonnegative product.

Lemma 5. *When* $1 < K < n$, *there exists an optimal partition obeying the following four policies.*

P_1 *It has at least one P/0-group.*

P_2 *There exists an integer number* $x(0 \leq x \leq n_-)$ *such that the negative elements* $r_1, \ldots, r_x$ *are assigned to P/0-groups while the negative elements* $r_{x+1}, \ldots, r_{n_-}$ *are assigned to Neg-groups.*
Moreover, this number x *is even when there exist Neg-groups.*

P_3 *For each* i $(1 \leq i \leq \lceil x/2 \rceil)$, r_{2i-1} *and* r_{2i} *are assigned to the same P/0-group.*

P_4 *If the number of Neg-groups is* m *and* $m > 1$, *two additional policies* $P_{4.1}$ *and* $P_{4.2}$ *must be obeyed. Here, we define the "components". (1) Each nonnegative element constitutes a component. (2) For each* i $(1 \leq i \leq \lceil x/2 \rceil)$, *elements* r_{2i-1} *and* r_{2i} *constitute a component. Notice that* x *is even according to* P_2. *So, each element in the P/0-groups belongs to a unique component. The price of a component is defined to be the product of the elements contained in this component. For convenience, let* $c_1, \ldots, c_{n'}$ *be an enumeration of all components from smallest price to largest price.*

$P_{4.1}$ *Each Neg-group contains exactly one element. More specifically,*

$$S_1 = \{r_{x+1}\}, S_2 = \{r_{x+2}\}, etc., S_m = \{r_{x+m}\} = \{r_{n_-}\}.$$

$P_{4.2}$ *Except for the last P/0-group, each P/0-group contains exactly one component. Specifically,* $S_{m+1} = \{c_1\}$, $S_{m+2} = \{c_2\}$, *etc.,* $S_{K-1} = \{c_{K-m-1}\}$; *and other components are assigned to* S_K.

Remark 2. Our basic idea is to reduce the general case to the preceding special cases. We will see this more clearly when we design algorithms in the next section. (See Remark 3.)

In designing the suite of policies, we mainly apply greedy strategies. For example, P_3 roughly states that most of the negative elements must be paired in a monotone order (e.g. r_1, r_2 are paired and assigned to the same group).

The proof of Lemma 5 is complicated and can only be given in the full version. We state the following facts of the partitions that obey $P_1, \ldots, P_4$.

Proposition 3. *Let* χ *be any partition that obeys* $P_1, \ldots, P_4$ *and has zero Neg-groups. For each* i $(1 \leq i \leq \lceil \frac{n_-}{2} \rceil)$, *elements* r_{2i-1} *and* r_{2i} *are assigned to the same group under* χ.

Proposition 4. *Let* χ *be any partition that obeys* $P_1, \ldots, P_4$ *and has one Neg-group. For each* i $(1 \leq i \leq \lfloor \frac{n_-}{2} \rfloor)$, *elements* r_{2i-1} *and* r_{2i} *are assigned to the same group under* χ.

Proof (of Proposition 3). Recall that x denotes the number of negative elements assigned to the P/0-groups. When there are no Neg-groups, we have $x = n_-$. Proposition 3 follows from P_3. □

Proof (of Proposition 4). Recall the definition of x as above. Since there is one Neg-group, x must be even according to P_2. According to P_3, for each i ($1 \leq i \leq \frac{x}{2}$), elements r_{2i-1} and r_{2i} are assigned to the same group. Moreover, for each i ($\frac{x}{2} < i \leq \lfloor \frac{n_-}{2} \rfloor$), elements r_{2i-1} and r_{2i} are both assigned to the unique Neg-group. Together, we obtain Proposition 4. □

3 The Framework for Computing an Optimal Partition

The following proposition is trivial (proof omitted) and it briefly states that there are only $O(n)$ partitions which are ordered and extremal.

Proposition 5. *If a partition is ordered and extremal, there exists $x(1 \leq x \leq n - (K - 1))$, such that $S_1 = \{r_1, \ldots, r_{n-K+2-x}\}$ and $S_K = \{r_{n-x+1}, \ldots, r_n\}$ and the remaining $K - 2$ elements are assigned to the middle $K - 2$ groups in the monotone order.*

We now present an algorithm framework for finding an optimal partition.

Definition 2. *Recall policies $P_1, \ldots, P_4$ in Lemma 5. When $1 < K < n$, for any integer m ($0 \leq m < K$), we denote by $\mathcal{A}_m$ the set of partitions which obey $P_1, \ldots, P_4$ and contain m Neg-groups.*

Our idea for finding the optimal partition is as follows. For each integer m ($0 \leq m < K$), we find a partition χ_m which has score larger than or equal to any partition in $\mathcal{A}_m$. (Note that χ_m may not belong to $\mathcal{A}_m$.) Then, we select the partition χ^* with maximum score among $\chi_0, \ldots, \chi_{K-1}$. Clearly, χ^* must be an optimal partition and the reason is the following. First, χ^* has score larger than or equal to any partition in $\mathcal{A}_0 \cup \ldots \cup \mathcal{A}_{K-1}$. On the other hand, according to Lemma 5, there must an optimal partition in $\mathcal{A}_0 \cup \ldots \cup \mathcal{A}_{K-1}$. Our algorithm is given below.

procedure COMPUTE_0.   ▷ This procedure computes χ_0.
    Let $\mathcal{T}_0$ be an empty set.
    For each i ($0 \leq i \leq \lceil \frac{n_-}{2} \rceil$), put the number ($r_{2i-1} \times r_{2i}$) into set $\mathcal{T}_0$.
    For each j ($j > 2 \cdot \lceil \frac{n_-}{2} \rceil$), put the number r_j into set $\mathcal{T}_0$.
    Compute the optimal partition of $\mathcal{T}_0$.   ▷ Reduce to special case.
    Recover this partition of $\mathcal{T}_0$ to be a partition of S and store it to χ_0.
    To "recover", we split the number ($r_{2i-1} \times r_{2i}$) to r_{2i-1} and r_{2i}.
end procedure

procedure COMPUTE_1.   ▷ This procedure computes χ_1.
    Let $\mathcal{T}_1$ be an empty set.
    For each i ($0 \leq i \leq \lfloor \frac{n_-}{2} \rfloor$), put the number ($r_{2i-1} \times r_{2i}$) into set $\mathcal{T}_1$.
    For each j ($j > 2 \cdot \lfloor \frac{n_-}{2} \rfloor$), put the number r_j into set $\mathcal{T}_1$.

Compute the optimal partition of $\mathcal{T}_1$.   ▷ Reduce to special case.
Recover this partition of $\mathcal{T}_1$ to be a partition of S and store it to χ_1.
To "recover", we split the number $(r_{2i-1} \times r_{2i})$ to r_{2i-1} and r_{2i}.
end procedure

procedure COMPUTE_M(m) ▷ This procedure computes χ_m for $m > 1$.
  if $(n_- - m)$ is even and $(n_- - 2(n - K) \le m \le n_-)$ **then**
    Let $\mathcal{T}_m$ be an empty set.
    For each i $(0 \le i \le \frac{n_- - m}{2})$, put the number $(r_{2i-1} \times r_{2i})$ into set $\mathcal{T}_m$.
    For each j $(j > n_-)$, put the number r_j into set $\mathcal{T}_m$.
    Let $c_1, \ldots, c_{n'}$ be an enumeration of $\mathcal{T}_m$ from small to large.
    Let χ_m be the following partition: Assign negatives $r_{n_- - m+1}, \ldots, r_{n_-}$
  to the first m groups, from small to large. Assign c_1 to S_{m+1}, c_2 to S_{m+2}, etc.,
  c_{K-m-1} to S_{K-1}. Assign $c_{K-m}, \ldots, c_{n'}$ to S_K.
    else
      Let χ_m be an arbitrary partition.
    end if
end procedure

Algorithm 1. Algorithm for computing the optimal partition.

1: **if** $(K = 1) \vee (K = n)$ **then**
2:     Let χ^* be the unique ordered partition.
3: **else**
4:     Call COMPUTE_0.
5:     Call COMPUTE_1.
6:     **for** $m : 1 < m < K$ **do**
7:         Call COMPUTE_M(m).
8:     **end for**
9:     Let χ^* be the partition with maximum score among $\chi_0, \chi_1, \ldots, \chi_{K-1}$.
10: **end if**
11: Output χ^*.

Lemma 6. *1. In Procedure* COMPUTE_0,
  (a) Set $\mathcal{T}_0$ contains zero of one negative number.
  (b) Partition χ_0 has score larger than or equal to any partition in $\mathcal{A}_0$.
2. In Procedure COMPUTE_1,
  (a) Set $\mathcal{T}_1$ contains zero of one negative number.
  (b) Partition χ_1 has score larger than or equal to any partition in $\mathcal{A}_1$.
3. For $1 < m < K$,
  (a) Set $\mathcal{A}_m$ is nonempty if and only if

$$n_- - m \text{ is even}; m \le n_-; \text{ and } m \ge n_- - 2(n - K). \tag{2}$$

  (b) Partition χ_m has score larger than or equal to any partition in $\mathcal{A}_m$.
4. The partition χ^* outputted by Algorithm 1 is an optimal partition.

Proof

1-(a). When $i < \lceil \frac{n_-}{2} \rceil$, we have $2i < n_-$ and so $r_{2i-1} \times r_{2i}$ is positive. When $j > 2 \cdot \lceil \frac{n_-}{2} \rceil$, we have $j > n_-$ and so r_j is nonnegative. Therefore, the only possible negative element that may be added to $\mathcal{T}_0$ is $(r_{2i-1} \times r_{2i})$ for $i = \lceil \frac{n_-}{2} \rceil$.

1-(b). By Proposition 3, any partition in $\mathcal{A}_0$ obeys the policy that it assigns r_{2i-1}, r_{2i} to the same group for $1 \le i \le \lceil \frac{n_-}{2} \rceil$. On the other hand, χ_0 is the optimal partition that obeys this policy. Therefore, χ_0 has score larger than or equal to any partition in $\mathcal{A}_0$. (Note: if the size of $\mathcal{T}_0$ is smaller than K, set $\mathcal{A}_0$ must be empty; in this case we let χ_0 be an arbitrary partition.)

2-(a). When $i \le \lfloor \frac{n_-}{2} \rfloor$, we have $2i \le n_-$ and so $r_{2i-1} \times r_{2i}$ is positive. When $j > 2 \cdot \lfloor \frac{n_-}{2} \rfloor + 1$, we have $j > n_-$ and so r_j is nonnegative. Therefore, the only possible negative element that may be added to $\mathcal{T}_1$ is r_j for $j = 2 \cdot \lfloor \frac{n_-}{2} \rfloor + 1$.

2-(b). By Proposition 4, any partition in $\mathcal{A}_1$ obeys the policy that it assigns r_{2i-1}, r_{2i} to the same group for $1 \le i \le \lfloor \frac{n_-}{2} \rfloor$. On the other hand, χ_1 is the optimal partition that obeys this policy. Therefore, χ_1 has score larger than or equal to any partition in $\mathcal{A}_1$. (Note: if the size of $\mathcal{T}_1$ is smaller than K, set $\mathcal{A}_1$ must be empty; in this case we let χ_1 be an arbitrary partition.)

3-(a). First, assume that A_m is nonempty. Then, there is a partition χ which has m Neg-groups and obeys P_1, P_2, P_3, P_4. Since it obeys P_2, the number of negatives in the P/0-groups should be even, i.e., $n_- - m$ is even. Since the number of Neg-groups must be at most the number of negative elements, we get $m \le n_-$. Since the number of components must be at least the number of P/0-groups, we get $n - n_- + (n_- - m)/2 \ge K - m$, i.e. $m \ge n_- - 2(n - K)$. Together, we get (2).

Now, suppose (2) holds. Then, we can find a partition χ_m (by Procedure COMPUTE_M) which has m Neg-groups and obeys P_1, P_2, P_3, P_4. This implies that A_m is nonempty.

3-(b). Assume (2) holds. Otherwise A_m is empty and the claim is trivial. The elements in $\mathcal{T}_m$ are exactly the "components" defined in Lemma 5. According to P_4, there could be only one "essentially distinct" partition in A_m, which is exactly χ_m. This implies 3-(b). (Note: there might be distinct partitions χ, χ' in A_m; however, the corresponding groups in χ and χ' have the same product (and score); in this manner, χ, χ' are essentially the same.)

4. When $K = 1$, this claim is trivial. When $K = n$, this claim follows from Lemma 1. When $1 < K < n$, we have: (i) The best partition in $\mathcal{A}_0, \ldots, \mathcal{A}_{K-1}$ is an optimal partition (due to Lemma 5). (ii) for $0 \le i < K$, χ_i has score larger than or equal to any partition in $\mathcal{A}_i$ (due to 1-(b), 2-(b), and 3-(b)). (iii) χ^* is the best partition among $\chi_0, \ldots, \chi_{K-1}$. Altogether, we get Claim 4. $\square$

Since sets $\mathcal{T}_0, \mathcal{T}_1$ both contain at most one negative, their optimal partition can be computed efficiently. From what we have shown in Subsect. 2.1, there is an optimal partition of $\mathcal{T}_0$ (respectively, $\mathcal{T}_1$) which is ordered and extremal. On the other hand, there are only $O(n)$ such partitions due to Proposition 5. We can enumerate all of these partitions and find the best one among them.

Therefore, Procedure COMPUTE_0 and COMPUTE_1 and thus Algorithm 1 can be implemented in polynomial time. We show a more delicate implementation of Algorithm 1 in the next section.

Remark 3. According to the above framework, solving an instance of the general case reduces to solving two instances of the aforementioned special cases.

4   An $O(n^2)$ Time Implementation of Algorithm 1

To compute χ_0, χ_1, we have to consider $O(n)$ candidate partitions. Moreover, Procedure COMPUTE_M generates a partition for each m so that $1 < m < K$. Therefore, Algorithm 1 generates in total $O(n)$ candidate partitions. For each partition, we can use $O(n^2)$ time to compute its score. (In the realistic computational model, the computation of the product of n elements would take $O(n^2)$ time, not $O(n)$ time. This is because we need to do $O(n)$ multiplication between a large $O(n)$ bytes number with a small number.) Therefore, Algorithm 1 can easily be implemented in $O(n^3)$ time.

In this section, we show an improved implementation of Algorithm 1, which only takes $O(n^2)$ time.

The basic idea for the improvement is simple. We arrange the candidate partitions in a queue; and in computing the score of the current partition, we always borrow the partial results for computing the previous one. In this way, computing the new score only takes $O(n)$ time.

First, consider the procedure for computing χ_0. (The same analysis applies for χ_1.) Assume that $T_0 = (a_1, \ldots, a_{n'})$ are sorted. There are $n' - (K - 1)$ candidate partitions, which are those partitions that are ordered and extremal (see Proposition 5), and they can be arranged as follows.

Index	S_1	S_2	...	S_{K-1}	S_K
1	a_1	a_2	...	a_{K-1}	$a_K, \ldots, a_{n'}$
2	a_1, a_2	a_3	...	a_K	$a_{K+1}, \ldots, a_{n'}$
...	...	...	...	...	...
$n' - (K-1)$	$a_1, \ldots, a_{n'-K-1}$	$a_{n'-K+2}$	...	$a_{n'-1}$	$a_{n'}$

To compute the score of the i-th $(i > 1)$ candidate partition, we need to sum up the score of its K groups. It is easy to compute the summation of the middle $K-2$ groups in $O(n)$ time. To compute the score of group S_1 (i.e. $a_1 \times \ldots \times a_i$), we borrow the score of group S_1 in the previous candidate, which is $(a_1 \times \ldots \times a_{i-1})$; so, we compute the product of $(a_1 \times \ldots \times a_{i-1})$ and a_i; this only takes $O(n)$ time. Similarly, we can compute the score of group S_K in $O(n)$ time.

The same idea can be applied in computing the scores of $\chi_2, \ldots, \chi_{K-1}$.

We first illustrate it by the following example. Assume $n = 13, K = 7$, $n_- = 8$. We have three candidate partitions, which are arranged in a queue:

$$\chi_2 = r_7 \mid r_8 \parallel r_9 \mid r_{10} \mid (r_5, r_6) \mid (r_3, r_4) \mid \underline{r_{11}, r_{12}, r_{13}, (r_1, r_2)}$$
$$\chi_4 = r_5 \mid r_6 \mid r_7 \mid r_8 \parallel r_9 \mid r_{10} \mid \underline{r_{11}, r_{12}, r_{13}, (r_3, r_4), (r_1, r_2)}$$
$$\chi_6 = r_3 \mid r_4 \mid r_5 \mid r_6 \mid r_7 \mid r_8 \parallel \underline{r_9, r_{10}, r_{11}, r_{12}, r_{13}, (r_1, r_2)}$$

In these partitions, the first $K-1$ groups always contain one or two elements; thus the total score of the first $K-1$ groups can be computed in $O(n)$ time. The challenge lies on computing the score (or product) of the last group - we should compute it in $O(n)$ time.

To compute the product of S_K of χ_m, we use the product of S_K in χ_{m-2}. In the above example,

- for $m = 4$, we should multiply $(r_3 \times r_4)$ with the previous product;
- for $m = 6$, we should multiply r_9 as well as r_{10}, and then divide $(r_3 \times r_4)$.

To make the general method more clear and to make the following claim (namely, Proposition 6) rigorous, here we have to give a more specific definition for the last group of χ_m. (Recall that in Policy P₄ or Procedure COMPUTE_M, the assignment of components was not uniquely specified.)

Definition 3 (Specific definition of the last group of χ_m). *Let $\alpha :=$ $(r_{n_-}, \ldots, r_n)$. Let $\beta = (\beta_1, \ldots, \beta_b) := (r_{2\lfloor n_-/2\rfloor - 1} \times r_{2\lfloor n_-/2\rfloor}, \ldots, r_1 \times r_2)$. Recall that T_m denote all components in χ_m. Notice that the elements in α, β include T_m for any $m > 1$.*

Suppose that group S_K in χ_m should contain t_m components. Clearly, $t_m = |T_m| - (K - M - 1)$.

We define $S_k := \{r_{p_m}, \ldots, r_n, \beta_{q_m}, \ldots, \beta_b\}$, where (p_m, q_m) are the unique pair of numbers (p, q) such that (i) $\{r_p, \ldots, r_n, \beta_q, \ldots, \beta_b\}$ are the largest t_m components in T_m and (ii) p is minimized.

Proposition 6. *S_K may add 1 or 2 and remove 0 or 1 component, comparing χ_m to χ_{m-2}.*

Proof (Sketch). If $\beta_{q_{m-2}} \notin T_m$, we must have $p_m = p_{m-2} - 2$ and $q_m = q_{m-2} + 1$. In this case, we add two components to S_K and remove one. If $\beta_{q_{m-2}} \in T_m$, there are two subcases. If $\{\alpha_{p_{m-2}-1}, \ldots \alpha_n, \beta_{q_{m-2}}, \ldots, \beta_b\}$ are the largest t_m components in T_m. Then, $p_m = p_{m-2} - 1, q_m = q_{m-2}$. Otherwise, $p_m = p_{m-2}, q_m = q_{m-2} - 1$. In each subcase, we add one component. $\square$

When computing χ_m, we can store the change of S_K compared with χ_{m-2}. Based on this change, we can compute the product of S_K in χ_m from the product of S_K in χ_{m-2} in $O(n)$ time.

5   An Alternative Approach - Dynamic Programming

Lemma 7. *There is an optimal partition, in which the last group is assigned with a union of a suffix and a prefix of $r_1, \ldots, r_n$, namely, $S_K = \{r_1, \ldots, r_x\} \cup \{r_y, \ldots, r_n\}$ for some x, y.*

Proof. Due to the definition of $\chi_0, \ldots, \chi_{K-1}$, the last group in any of these partitions is assigned with a suffix and a prefix of $r_1, \ldots, r_n$. Moreover, there is an optimal partition among $\chi_0, \ldots, \chi_{K-1}$. Together, we obtain the lemma. $\square$

According to this lemma and by using Dynamic-Programming, an alternative algorithm can be designed for finding an optimal partition. The transition equation is given below. Let $f_{i,j,K}$ denote the maximum score we can get by partitioning $r_i, \ldots, r_j$ into the first K groups.

$$f_{i,j,K} = \begin{cases} w_1 \prod_{x=i}^{j} r_x, & K = 1; \\ max\{g_{i,j,k,i',j'} \mid i \le i' \le j' \le j, j' - i' < j - i\}, & K > 1. \end{cases} \quad (3)$$

where

$$g_{i,j,k,i',j'} = f_{i',j',K-1} + w_K \cdot \prod_{x=i}^{i'-1} r_x \prod_{x=j'+1}^{j} r_x.$$

References

1. Anily, S., Federgruen, A.: Structured partitioning problems. Oper. Res. **39**(1), 130–149 (1991)
2. Barnes, E.R., Hoffman, A.J., Rothblum, U.G.: Optimal partitions having disjoint convex and conic hulls. Math. Program. **54**, 69–86 (1992)
3. Burkard, R.E., Yao, E.: Constrained partitioning problems. Discret. Appl. Math. **28**(1), 21–34 (1990)
4. Chakravarty, A.K., Orlin, J.B., Rothblum, U.G.: Technical note - a partitioning problem with additive objective with an application to optimal inventory groupings for joint replenishment. Oper. Res. **30**(5), 1018–1022 (1982)
5. Chakravarty, A.K., Orlin, J.B., Rothblum, U.G.: Consecutive optimizers for a partitioning problem with applications to optimal inventory groupings for joint replenishment. Oper. Res. **33**(4), 820–834 (1985)
6. Fisher, W.D.: On grouping for maximum homogeneity. J. Am. Stat. Assoc. **53**(284), 789–798 (1958)
7. Gal, S., Klots, B.: Optimal partitioning which maximizes the sum of the weighted averages. Oper. Res. **43**(3), 500–508 (1995)
8. Hwang, F.K.: Optimal partitions. J. Optim. Theory Appl. **34**(1), 1–10 (1981)
9. Hwang, F.K., Rothblum, U.G., Yao, Y.: Localizing combinatorial properties of partitions. Discret. Math. **160**(1–3), 1–23 (1996)
10. Hwang, F.K., Sun, J., Yao, E.Y.: Optimal set partitioning. SIAM J. Algebraic Discret. Methods **6**(1), 163–170 (1985)
11. Leung, J.Y., Li, C.: An asymptotic approximation scheme for the concave cost bin packing problem. Eur. J. Oper. Res. **191**(2), 582–586 (2008)
12. Li, C., Chen, Z.: Bin-packing problem with concave costs of bin utilization. Nav. Res. Logist. **53**(4), 298–308 (2006)
13. Wikipedia: Partition problem. Technical report, Wikipedia (2016). https://en.wikipedia.org/wiki/Partition

Modular-Width: An Auxiliary Parameter for Parameterized Parallel Complexity

Faisal N. Abu-Khzam[1,4], Shouwei Li[2(✉)], Christine Markarian[3],
Friedhelm Meyer auf der Heide[2], and Pavel Podlipyan[2]

[1] Department of Computer Science and Mathematics,
Lebanese American University, Beirut, Lebanon
[2] Heinz Nixdorf Institute and Department of Computer Science,
Paderborn University, Paderborn, Germany
sli@mail.uni-paderborn.de
[3] Department of Mathematical Sciences, Haigazian University, Beirut, Lebanon
[4] School of Engineering and Information Technology,
Charles Darwin University, Darwin, Australia

Abstract. Many graph problems such as maximum cut, chromatic number, hamiltonian cycle, and edge dominating set are known to be *fixed-parameter tractable* (FPT) when parameterized by the treewidth of the input graphs, but become W-hard with respect to the clique-width parameter. Recently, Gajarský *et al.* proposed a new parameter called modular-width using the notion of modular decomposition of graphs. They showed that the chromatic number problem and the partitioning into paths problem, and hence hamiltonian path and hamiltonian cycle, are FPT when parameterized by this parameter. In this paper, we study modular-width in parameterized parallel complexity and show that the weighted maximum clique problem and the maximum matching problem are *fixed-parameter parallel-tractable* (FPPT) when parameterized by this parameter.

1 Introduction

Parameterized complexity has become a mainstream framework of theoretical computer science in the last two decades. The central idea of this theory is to study the complexity of NP or even PSPACE-hard problems with respect to one (or more) parameter(s) rather than restricting the analysis to the input size. This has led to the development of the class of *fixed-parameter tractable* (FPT) problems. In short, a parameterized problem is FPT if it has a deterministic algorithm with running time $f(k) \cdot n^{\mathcal{O}(1)}$, where n is the input size, k is the parameter, and f is an arbitrary computable function.

To further reduce the running time of fixed-parameter algorithms, parallel computing is employed, and this is broadly known as *parameterized parallel complexity*. To the best of our knowledge, there are three complexity classes

This work was partially supported by the German Research Foundation (DFG) within the Collaborative Research Center "On-The-Fly Computing" (SFB 901) and the International Graduate School "Dynamic Intelligent Systems".

M. Xiao and F. Rosamond (Eds.): FAW 2017, LNCS 10336, pp. 139–150, 2017.
DOI: 10.1007/978-3-319-59605-1_13

defined in the literature with the aim of capturing parametrized analogues of the class NC, known as PNC (parameterized analog of NC), introduced in [2], FPP (fixed-parameter parallelizable), introduced in [6], and FPPT (fixed-parameter parallel-tractable), introduced in [1], respectively. Let $f(k)$, $g(k)$, and $h(k)$ be arbitrary functions of the parameter k, and α, β be positive constants. (1) PNC is the class of all parametrized problems solvable in time $f(k) \cdot (\log n)^{h(k)}$ using $g(k) \cdot n^\beta$ parallel processors. (2) FPP is the class of all parametrized problems solvable in time $f(k) \cdot (\log n)^\alpha$ using $g(k) \cdot n^\beta$ parallel processors. (3) FPPT is the class of all parametrized problems solvable in time $f(k) \cdot (\log n)^\alpha$ using n^β parallel processors. It is easy to observe that FPPT $\subseteq$ FPP $\subset$ PNC according to the definition.

We also note here that FPP $\subseteq$ FPPT, because a parallel algorithm with runtime $f(k) \cdot (\log n)^\alpha$ using $g(k) \cdot n^\beta$ parallel processors can be simulated by one with running time $g(k)f(k) \cdot (\log n)^\alpha$ using n^β parallel processors. The definition of FPPT is, therefore, a simplified version of FPP, and it was introduced in [1] to emphasize the need to have a polynomial number of processors. In fact, the parameter k may be treated like another input variable (rather than a constant), so a number of processors that varies as an arbitrary (super-polynomial) function of k is not desired. Yet, from a theoretical standpoint, an FPP-algorithm may give more information about the parametrized complexity of a problem than FPPT, if we are interested in the maximum possible size of k (as a function $k(n)$ of n) so that the problem is in NC as long as $k \leq k(n)$. For example, an algorithm with running time $\mathcal{O}(k \cdot \log^\alpha n)$ using $\mathcal{O}\left(2^k \cdot n^\beta\right)$ parallel processors is an NC-algorithm for $k = \mathcal{O}(\log n)$. Expressing its performance in the FPPT model $\mathcal{O}\left(k \cdot 2^k \cdot \log^\alpha n\right)$ using $\mathcal{O}(n^\beta)$ processors would only yield the bound $k = \mathcal{O}(\log \log n)$. The results in this paper can be interpreted in either of these models.

One of the most ubiquitous parameters studied in both sequential and parallel classes is *treewidth*, which roughly measures how tree-like a graph is. Algorithms for problems on graphs of bounded treewidth are much more efficient than their counterparts in general graphs (e.g. see [3,4]). A celebrated meta-theorem of Courcelle and Di Ianni [8] states that any problem expressible in monadic second-order logic is FPT when parameterized by the treewidth of the input graphs. Similarly, Cesati *et al.* in [6] showed that all problems involving *MS* (definable in monadic second-order logic with quantifications over vertex and edge sets) or *EMS properties* (that involve counting or summing evaluations over sets definable in monadic second-order logic) are FPP when restricted to graphs of bounded treewidth. Moreover, Lagergren [17] presented an efficient parallel algorithm for the tree decomposition problem for fixed treewidth. Hagerup *et al.* [16] showed that the maximum network flow problem is in NC when parameterized by treewidth. Despite the fact that these results are noteworthy, one of the main drawbacks of treewidth is that a large number of interesting instances are excluded since graphs of small treewidth are necessarily sparse. The notion of clique-width (as well as rank-width [20], boolean-width [5] and shrub-depth [14]), which is stronger than treewidth, tries to address this problem by covering a larger family of graphs, including many dense graphs. However, the price for this generality is exorbitant. Several natural problems such as maximum cut, chromatic number, hamiltonian cycle, and edge dominating set, which were shown to be FPT for bounded treewidth, become W[1]-hard when parameterized by these measures [10–12].

Consequently, a parameter called *modular-width*, that covers a significantly larger class of graphs has been introduced by Gajarský *et al.* in [13] where it was shown that several problems, such as the chromatic number and the partitioning into paths, and hence hamiltonian path and hamiltonian cycle, which are W[1]-hard for both clique-width and shrub-depth, are FPT when parameterized by modular-width.

In this paper, we extend the study of modular-width to parameterized parallel complexity and show that the weighted maximum clique problem and the maximum matching problem are FPPT for bounded modular-width. Our algorithms are based on the following ideas/techniques: we use an algebraic expression to represent the input graph. Then, we construct the modular decomposition tree of the input graph and a computation tree that corresponds to the maximal strong modules in the maximal decomposition tree. For each node of the computation tree, we compute an optimal solution by giving the optimal solutions for the children of this node in the modular decomposition tree. The optimal solution for each node can be obtained by integer linear programming. Then, we use a bottom-up dynamic programming approach along with the modular decomposition tree to obtain the global solution. In order to explore and evaluate the tree efficiently, we use a parallel tree contraction technique due to Miller and Reif [18].

2 Preliminaries

All graphs considered in this paper are simple, undirected and loopless. We use the classical graph theoretic notations and definitions (e.g. see [15]). The neighborhood of a vertex x in a graph $G = (V, E)$ is denoted by $N(x)$. Given a subset of vertices $X \subseteq V$, $G[X]$ denotes the subgraph induced by X.

Let M be a subset of vertices of a graph G, and x be a vertex of $V \setminus M$. We say that vertex x *splits* M (or is a *splitter* of M) if x has both a neighbor and a non-neighbor in M. If x does not split M, then M is *homogeneous* with respect to x.

Definition 1. Given a graph $G = (V, E)$, $M \subseteq V$ is called a *module* if M is homogeneous with respect to any vertex $x \in V \setminus M$ (i.e. $M \subseteq N(x)$ or $M \cap N(x) = \emptyset$).

Let M and M' be disjoint sets. We say that M and M' are adjacent if any vertex of M is adjacent to all the vertices in M' and non-adjacent if the vertices of M are non-adjacent to the vertices of M'. Thus, it is not hard to observe that two disjoint modules are either adjacent or non-adjacent.

A module M is *maximal* with respect to a set S of vertices if $M \subset S$ and there is no module M' such that $M \subset M' \subset S$. If the set S is not specified, we assume that $S = V$. A module M is a *strong module* if it does not overlap with any other module. Note that, one-vertex subsets and the empty set are modules and are known as the *trivial* modules. A graph is called *prime* if all of its modules are trivial.

Definition 2. Let $P = \{M_1, \ldots, M_k\}$ be a partition of the vertex set of a graph $G = (V, E)$. If for all i, $1 \leq i \leq k$, M_i is a non-trivial module of G, then P is a *modular partition* of G.

A non-trivial modular partition $P = \{M_1, \ldots, M_k\}$ which only contains maximal strong modules is a *maximal modular partition*. Note that each undirected graph has a unique maximal modular partition. If G (resp. $\overline{G}$) is not connected then its connected (resp. co-connected) components are the elements of the maximal modular partition.

Definition 3. For a modular partition $P = \{M_1, \ldots, M_k\}$ of a graph $G = (V, E)$, we associate a *quotient graph* $G_{/p}$, whose vertices are in one-to-one correspondence with the parts of P. Two vertices v_i and v_j of $G_{/p}$ are adjacent if and only if the corresponding modules M_i and M_j are adjacent in G.

The inclusion tree of the strong modules of G, called the *modular decomposition tree*, entirely represents the graph if the representative graph of each strong module is attached to each of its nodes (see Fig. 1). It is easy to observe that there are only three relations, $M \subseteq M'$, $M' \subseteq M$, or $M \cap M' = \emptyset$, for any

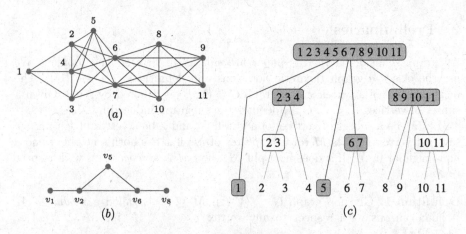

Fig. 1. (*a*) Shows the graph G; $\{\{1\}, \{2, 3\}, \{4\}, \{5\}, \{6, 7\}, \{9\}, \{8, 10, 11\}\}$ is a modular partition of G. The maximal modular partition of G is $P = \{\{1\}, \{2, 3, 4\}, \{5\}, \{6, 7\}, \{8, 9, 10, 11\}\}$ and (*b*) represents its quotient graph. (*c*) is the modular decomposition tree of G. The maximal strong modules are in blue. The green edges indicate that the root node is parallel, the red edges indicate that the root is series, and the black edges indicate that the root is a prime graph. (Color figure online)

two nodes M and M' in the modular decomposition tree. The *modular-width* is the maximum degree of the modular decomposition tree. An excellent feature of modular decomposition is that it can be computed in $\mathcal{O}(\log^2 n)$ time with $\mathcal{O}(n + m)$ parallel processors [9], thus the modular-width stays also within the same resource bounds.

Theorem 1 (Modular decomposition theorem [7]**).** *For any graph $G = (V, E)$, one of the following three conditions is satisfied:*

1. *G is not connected;*
2. *$\overline{G}$ is not connected;*
3. *G and $\overline{G}$ are connected and the quotient graph $G_{/P}$, where P is the maximal modular partition of G, is a prime graph.*

Theorem 1 indicates that, the quotient graphs associated with the nodes of the modular decomposition tree of the strong modules are of three types: an *independent set* if G is not connected (the node is labeled *parallel*); a clique if $\overline{G}$ is not connected (the node is labeled *series*); a prime graph otherwise.

Parallel tree contraction is a "bottom-up" technique for constructing parallel algorithms on trees. There are two basic operations called *rake* and *compress*. During each contraction, processors are assigned to leaves of the tree and perform local modifications by removing these leaves, hence creating new leaves that are processed at the next round. This operation is called rake. Clearly, removing leaves is not sufficient for a tree that is thin and tall, like a linked list, which would take a linear number of rounds to reduce the tree to a point. Thus, a complementary operation called compress that reduces a chain of vertices, each with a single child to a chain of half the length is introduced. Ideally, rake and compress work on different parts of the tree simultaneously. During the run of the algorithm, the rake operation tends to produce chains that are then reduced by the compress operation. Thus, the whole tree can be evaluated in $\mathcal{O}(\log n)$ time using $\mathcal{O}(n)$ parallel processors.

3 Parallel Algorithms on Modular Decomposition

In this section, we show how modular-width can be used to derive efficient parallel algorithms for the weighted maximum clique problem and the maximum matching problem. Our results imply that these two problems are FPPT.

The input to our algorithms is assumed to be a graph of modular-width at most k, and we shall represent the input graph as an algebraic expression consisting of the following operations:

1. G has only one vertex. This corresponds to a leaf node in the modular decomposition tree.
2. G is a *disjoint union* of two graphs G_1 and G_2 of modular-width at most k. The disjoint union of G_1 and G_2 defined as a graph with vertex set $V_1 \cup V_2$ and edge set $E_1 \cup E_2$. This corresponds to a parallel node in the modular decomposition tree.

3. G is a *complete join* of two graphs G_1 and G_2 of modular-width at most k. The complete join of G_1 and G_2 is defined as a graph with vertex set $V_1 \cup V_2$ and edge set $E_1 \cup E_2 \cup \{\{u,v\} : u \in V_1 \text{ and } v \in V_2\}$. This corresponds to a series node in the modular decomposition tree.
4. The *substitution* operation with respect to a graph G is the reverse of the quotient operation and defined as replacing a vertex of G by $G_i = (V_i, E_i)$ of modular-width at most k while preserving the neighborhood,

$$G_{x \to G_i} = (V \setminus \{x\} \cup V_i,$$
$$(E \setminus \{(x,y) \in E\} \cup E_i \cup \{(y,z) : (x,y) \in E, z \in V_i\})).$$

This is corresponding to the maximal modular partition of G, and G_i is one of the maximal strong module of G.

Throughout the rest of this paper, we may assume that a graph $G = (V,E)$ and the modular decomposition of G, of modular-width at most k, are already given. Otherwise, we can apply the algorithm presented in [9] to obtain one. Under this assumption, it is easy to note that the modular decomposition tree of G can be constructed in constant time using $\mathcal{O}(n)$ parallel processors. Moreover, the number of maximal strong modules in the decomposition tree of G is at most n, which is equal to the cardinality of the maximal modular partition of G.

The central idea of our algorithm is a bottom-up dynamic programming approach along the modular decomposition tree of the algebraic expression as defined above. For each node of the modular decomposition tree, we compute a record for the graph represented by the subtree of the modular decomposition below that node. That is, given the optimal solutions for the children of each node in the modular decomposition tree, we can compute an optimal solution for the node itself. In order to explore the decomposition tree efficiently, we take the parallel tree contraction technique due to Miller and Reif [18].

3.1 The Weighted Maximum Clique Problem

Let us consider the weighted maximum clique problem which is known to be NP-complete for general graphs. Given a graph $G = (V,E)$ and weights on each vertex, is there a clique with maximum weight ω?

Theorem 2. *The weighted maximum clique problem parameterized by the modular-width k can be solved in $\mathcal{O}(2^k \cdot \log n)$ time using $\mathcal{O}(n)$ parallel processors. Thus it is FPPT.*

Proof. Clearly, graph G with bounded modular-width k can be represented by the four operations mentioned above according to Theorem 1. We only need to show that each operation can be done efficiently, and the whole decomposition tree can be evaluated by the parallel tree contraction technique.

First, each leaf node in the modular decomposition tree is an isolated vertex, which can be represented by the first operation of the algebraic expression. Thus, the maximum clique weight of each vertex is trivially its own weight. Obviously, this can be done in constant time with a linear number of parallel processors.

Next, we consider other operations on combining two modules to form a larger module:

- If G is the disjoint union of G_1 and G_2, then the maximum clique weight of G would be:

$$\omega(G) = \max\{\omega(G_1), \omega(G_2)\},$$

since the disjoint union operation corresponds to a parallel node, which implies two modules are non-adjacent.

- If G is the complete join of G_1 and G_2, then the maximum clique weight of G would be:

$$\omega(G) = \omega(G_1) + \omega(G_2),$$

since complete join corresponding to a series node, which implies any vertex in G_1 is adjacent to all the vertices in G_2.

The last case is for G is a substitution of G_i for $1 \leq i \leq k$, which means G is neither obtained by disjoint union operation nor complete join operation. In other words, the quotient graph of G is prime, and the vertices are in one-to-one correspondence with G_i for $1 \leq i \leq k$. In this scenario, graph G can be treated as a graph with at most k vertices, and the weight of each vertex is equal to the maximum clique weight of the corresponding module G_i for $1 \leq i \leq k$. Since each G_i looks like a black box to the other G_j in the maximal modular decomposition of G for $1 \leq i, j \leq k$ and $i \neq j$, and there has no efficient algorithm for the maximum weighted clique problem, we have no choice but take a brute-force strategy to evaluate G if the maximum clique weight of each G_i for $1 \leq i \leq k$ are given, and this can be done in $\mathcal{O}(2^k)$ time.

Now we show how to parallelize the algorithm by the parallel tree contraction technique. We construct a computation tree corresponding to the modular decomposition tree of G, such that each tree node corresponds to a maximal strong module of size at most k and has at most k children. Suppose v is an internal node in the computation tree, we call v is *half-evaluated* when all but one of its children has been evaluated. With the parallel tree contraction technique, it can be compressed later. Suppose the unevaluated child is v_1 and its maximum clique weight is ω', the maximum clique weight of v without v_1 evaluated to a, and the maximum clique weight among evaluated children $v_2, \ldots, v_k$ of v is b, a and b are known values. Then the maximum clique weight of v is

$$\omega = \max\{a, \omega' + b\}.$$

During each contraction progress, we can take the above function recursively and have

$$\omega'' = \max\{c, \omega + d\},$$

where c and d are two known values for next round, then

$$\omega'' = \max\{\max\{b + c, d\}, \omega' + (a + c)\}.$$

Thus, the running time is $\mathcal{O}(2^k \cdot \log n)$ using $\mathcal{O}(n)$ parallel processors, because $\mathcal{O}(2^k)$ time is required to compute a maximum weight clique for a prime graph with at most k vertices, the parallel tree contraction takes $\mathcal{O}(\log n)$ time using a linear number of parallel processors, and half-evaluating a node requires $\mathcal{O}(\log k)$ time. □

3.2   The Maximum Matching Problem

A matching in a graph is a set of edges such that no two edges share a common vertex. We now consider the maximum matching problem which seeks a matching of maximum size (i.e., the largest number of edges). The existence of an NC algorithm for this problem has been open for several decades, even if the graph is planar. By considering the modular-width as the parameter, we prove the following:

Theorem 3. *The maximum matching problem parameterized by the modular-width k can be solved in $\mathcal{O}(2^k \cdot \log n)$ time using $\mathcal{O}(n)$ parallel processors. Therefore the problem is FPPT.*

Proof. We follow the same strategy as Theorem 2 and evaluate different operations on combining modules. Let n_1, n_2 denote the number of vertices and u_1, u_2 denote the number of unmatched vertices of graphs G_1 and G_2. We use the pair $<n_i, u_i>$ to track the maximum matching of graph G_i.

First, each leaf node i in the modular decomposition tree is an isolated vertex, thus

$$n_i = 1 \text{ and } u_i = 1.$$

Next, we consider various operations on combining two modules to form a larger module:

– If G is the disjoint union of G_1 and G_2, the values of G would be:

$$n = n_1 + n_2 \text{ and } u = u_1 + u_2,$$

since disjoint union corresponds to a parallel node, which implies there is no edge between G_1 and G_2.
– If G is the complete join of G_1 and G_2, the values of G would depend on the values of n_1, n_2, u_1 and u_2. We have to consider different cases for this scenario. In fact, no matter for which cases,

$$n = n_1 + n_2$$

always valid, we only need to consider u.
  1. If $u_1 > n_2$, then the unmatched vertices in G_1 are more than the vertices of G_2, then the values of G would be:

$$u = u_1 - n_2,$$

because we could match all vertices of G_2 through the edges between unmatched vertices in G_1 and all vertices of G_2.

2. Symmetrically, if $u_2 > n_1$, the values of G would be:

$$u = u_2 - n_1.$$

3. The last case comes to $u_1 < n_2$ and $u_2 < n_1$. In this circumstances, we are able to match almost all vertices in G_1 and G_2, and only have one unmatched vertex left over if there is an odd number of vertices in G, the values of G would be

$$u = n_1 + n_2 \pmod 2.$$

Without loss of generality, suppose $u_1 - u_2 \geq 2$, we can further match the u_2 unmatched vertices in G_2 by the additional edges of complete join operation. After that, all vertices in G_2 are matched, and only $u_1 - u_2 \geq 2$ vertices left in G_1 still unmatched. Suppose x, y are two unmatched vertices in G_1; we know that x, y are also adjacent to all vertices in G_2 because of the complete join operation, and all vertices of G_2 are matched. Then there must be at least one edge (a, b) in the matching of G_2, such that $x - a - b - y$ is an augmenting path and the matching can be extended by 1. Thus the number of unmatched vertices in G only depends on the parity of $u_1 + u_2$, which is equal to the parity of $n_1 + n_2$.

Thus for the complete join of G_1 and G_2, we have

$$n = n_1 + n_2,$$

and

$$u = \max\{u_1 - n_2, u_2 - n_1, n_1 + n_2 \pmod 2\}.$$

Obviously, this can be done in a constant time given the values of G_1 and G_2.

– Finally, we consider the case where G is a prime graph, which is obtained by the substitution operation on modules $G_1, \ldots, G_k$. As claimed in the proof of Theorem 2, G can be treated as a prime graph with at most k vertices, and each vertex corresponding to a module G_i for $1 \leq i \leq k$ in this case.

It is well-known that the maximum matching problem can be formulated as integer linear programming. Once again, let u_i denote the number of unmatched vertices in G_i, E denotes the edge set among $G_1, \ldots, G_k$, and $e_{i,j}$ denote the number of matched edges between G_i and G_j for $1 \leq i, j \leq k$. Then finding a maximum matching in G is equivalent to solving the following problem:

$$\text{Maximize} \left\{ \sum_{(i,j) \in E} e_{i,j} + \sum_i e_{i,i} \right\} \text{ subject to}$$

$$2e_{i,i} + \sum_{(i,j) \in E} e_{i,j} \leq n_i \text{ for } i = 1, \ldots, k$$

$$e_{i,i} \leq \frac{(n_i - u_i)}{2} \text{ for } i = 1, \ldots, k$$

$$e_{i,j} \in [1, k] \text{ for } 1 \leq i, j \leq k.$$

For each prime graph, we can compute the matching by taking the maximum of our objective function at every feasible solution. This can be done in $\mathcal{O}(2^k)$ time since the graph has a bounded modular-width k.

Now, we show how to parallelize the algorithm using the parallel tree contraction technique.

Let $n_1, \ldots, n_k$ denote the number of vertices and $u_1, \ldots, u_k$ denote the number of unmatched vertices in $G_1, \ldots, G_k$. Suppose $n_1, \ldots, n_k$ and $u_2, \ldots, u_k$ are known, but u_1 is not. Then the number of unmatched vertices of the graph G can be represented as a function u of u_1 of the form $\max\{p, u_1 - q\}$ for a proper choice of constants p and q, such that $p = u(n_1 \pmod 2)$ and $q = n_1 - u(n_1)$.

As argued in the complete join operation of two graphs, u_1 must have the same parity as n_1. For any x of the same parity as n_1 between 2 and n_1, it is clear that $u(x-2) \leq u(x) \leq u(x-2)+2$. We will show that $u(u_1)$ is a piecewise linear function, consisting of a constant portion for low values of u_1 followed by a portion with slope 1 for high values of u_1 in Lemma 1. Thus $u(u_1)$ has the form $\max\{p, u_1 - q\}$. We choose $p = u(n_1 \pmod 2)$ so the formula is correct at the low end. For the high end, we choose $q = n_1 - u(n_1)$.

We now use the parallel tree contraction technique. Composing functions of the form $u(x) = \max\{p, x - q\}$ leaves another function of the same form which can be computed in constant time. Therefore, the tree contraction can be done in $\mathcal{O}(\log n)$ time with $\mathcal{O}(n)$ parallel processors. $\square$

Lemma 1. *If x has the same parity as n_1 and also $4 \leq x \leq n_1$, then it cannot satisfy:* $u(x) = u(x-2) = u(x-4)+2$.

Proof. Let M' be the matching used to calculate $u(x-4)$ and M be the matching used to calculate $u(x)$. Then we have

$$n_1 = 2 * |M'| + (x - 4) = 2 * |M| + x;$$

thus,

$$|M'| = |M| + 2;$$

It follows that M' contains at least two edges between vertices in G_1.

Let M'' be the matching M' without an edge e between two of the vertices in G_1, then M'' will be a matching used to calculate $u(x-2)$. If we choose $u_1 = x - 2$, both matchings M and M'' have the same cardinality and both are also maximum. Let G' be the resultant graph from taking the symmetric difference of M and M''; i.e. $(M - M'') \cup (M'' - M)$. Every connected component of G' must be either an even cycle whose edges alternate between M and M'' or an even length path whose edges alternate between M and M'' with distinct endpoints. Now add the edge e that is in $M' \setminus M''$, its connected component must be an odd path. If there is another matched edge e' between two of the vertices in G_1, and e' is not in the same connected component, then we can take the edges from M in the component of e', add them to the rest of M', and have a larger matching for the case $u_1 = x - 2$. Alternatively, if there does not exist a matching edge between vertices in G_1, that is also in a different component from e, we can still modify the matching to obtain a larger one when $u_1 = x - 2$. Let e' be another edge from M' that is contained in G_1. Any vertex outside of G_1 that is adjacent to the vertices of e is also adjacent to the vertices of e'. We can add an edge between one of these vertices and a vertex of e' so that an even cycle is created. We use this even cycle to take a different set of edges. This new set has the property that it no longer includes edge e'. We have found a larger matching for the case $u_1 = x - 2$, contradicting our assumption. In both cases, $u(x - 2) = u(x) + 2$ follows from our premises. $\qquad\square$

4 Concluding Remarks and Future Work

In this paper, we showed that the weighted maximum clique problem and the maximum matching problem are FPPT when parameterized by modular-width. It would be interesting to find out whether other problems are FPPT when parameterized by this parameter. It was shown that the maximum network flow problem is FPPT with respect to treewidth as parameter [16]. We know that the maximum network flow problem is not easier than the maximum matching problem from a parallel complexity standpoint, being P-Complete. However, we believe that the maximum network flow problem would also fall in FPPT when parameterized by modular-width. Moreover, it was shown that the chromatic number, hamiltonian cycle, maximum cut, and edge dominating set problems are FPT when parameterized by treewidth but become W[1]-hard when parameterized by clique-width. Also, when parameterized by modular-width, chromatic number and hamiltonian cycle are FPT, while the other two are still open. We conjecture that the chromatic number problem parameterized by modular-width is not FPPT, mainly because Miyano [19] showed that most of the lexicographically first maximal subgraph problems are still P-complete even if the instances are restricted to graphs with bounded degree three.

References

1. Abu-Khzam, F.N., Li, S., Markarian, C., Meyer auf der Heide, F., Podlipyan, P.: On the parameterized parallel complexity and the vertex cover problem. In: Chan, T.-H.H., Li, M., Wang, L. (eds.) COCOA 2016. LNCS, vol. 10043, pp. 477–488. Springer, Cham (2016). doi:10.1007/978-3-319-48749-6_35
2. Bodlaender, H., Downey, R., Fellows, M.: Applications of parameterized complexity to problems of parallel and distributed computation (1994, unpublished extended abstract)
3. Bodlaender, H.L.: Treewidth: characterizations, applications, and computations. In: Fomin, F.V. (ed.) WG 2006. LNCS, vol. 4271, pp. 1–14. Springer, Heidelberg (2006). doi:10.1007/11917496_1
4. Bodlaender, H.L., Koster, A.M.: Combinatorial optimization on graphs of bounded treewidth. Comput. J. 51(3), 255–269 (2008)
5. Bui-Xuan, B.-M., Telle, J.A., Vatshelle, M.: Boolean-width of graphs. Theoret. Comput. Sci. 412(39), 5187–5204 (2011)

6. Cesati, M., Di Ianni, M.: Parameterized parallel complexity. In: Pritchard, D., Reeve, J. (eds.) Euro-Par 1998. LNCS, vol. 1470, pp. 892–896. Springer, Heidelberg (1998). doi:10.1007/BFb0057945

7. Chein, M., Habib, M., Maurer, M.-C.: Partitive hypergraphs. Discret. Math. **37**(1), 35–50 (1981)

8. Courcelle, B.: The monadic second-order logic of graphs. I. Recognizable sets of finite graphs. Inf. Comput. **85**(1), 12–75 (1990)

9. Dahlhaus, E.: Efficient parallel modular decomposition (extended abstract). In: Nagl, M. (ed.) WG 1995. LNCS, vol. 1017, pp. 290–302. Springer, Heidelberg (1995). doi:10.1007/3-540-60618-1_83

10. Fomin, F.V., Golovach, P.A., Lokshtanov, D., Saurabh, S.: Clique-width: on the price of generality. In: Proceedings of the Twentieth Annual ACM-SIAM Symposium on Discrete Algorithms, pp. 825–834. Society for Industrial and Applied Mathematics (2009)

11. Fomin, F.V., Golovach, P.A., Lokshtanov, D., Saurabh, S.: Algorithmic lower bounds for problems parameterized by clique-width. In: Proceedings of the Twenty-First Annual ACM-SIAM Symposium on Discrete Algorithms, pp. 493–502. Society for Industrial and Applied Mathematics (2010)

12. Fomin, F.V., Golovach, P.A., Lokshtanov, D., Saurabh, S.: Intractability of clique-width parameterizations. SIAM J. Comput. **39**(5), 1941–1956 (2010)

13. Gajarský, J., Lampis, M., Ordyniak, S.: Parameterized algorithms for modular-width. In: Gutin, G., Szeider, S. (eds.) IPEC 2013. LNCS, vol. 8246, pp. 163–176. Springer, Cham (2013). doi:10.1007/978-3-319-03898-8_15

14. Ganian, R., Hliněný, P., Nešetřil, J., Obdržálek, J., Ossona de Mendez, P., Ramadurai, R.: When trees grow low: shrubs and fast MSO_1. In: Rovan, B., Sassone, V., Widmayer, P. (eds.) MFCS 2012. LNCS, vol. 7464, pp. 419–430. Springer, Heidelberg (2012). doi:10.1007/978-3-642-32589-2_38

15. Habib, M., Paul, C.: A survey of the algorithmic aspects of modular decomposition. Comput. Sci. Rev. **4**(1), 41–59 (2010)

16. Hagerup, T., Katajainen, J., Nishimura, N., Ragde, P.: Characterizing multiterminal flow networks and computing flows in networks of small treewidth. J. Comput. Syst. Sci. **57**(3), 366–375 (1998)

17. Lagergren, J.: Efficient parallel algorithms for graphs of bounded tree-width. J. Algorithms **20**(1), 20–44 (1996)

18. Miller, G.L., Reif, J.H.: Parallel tree contraction-part I: fundamentals (1989)

19. Miyano, S.: The lexicographically first maximal subgraph problems: P-completeness and NC algorithms. Math. Syst. Theory **22**(1), 47–73 (1989)

20. Oum, S.-I.: Rank-width and vertex-minors. J. Comb. Theory Ser. B **95**(1), 79–100 (2005)

Online Strategies for Evacuating from a Convex Region in the Plane

Songhua Li[1,2(✉)] and Yinfeng Xu[1,2]

[1] School of Management, Xi'an Jiaotong University, Xi'an 710049, China
lisonghua@stu.xjtu.edu.cn, yfxu@mail.xjtu.edu.cn
[2] The State Key Lab for Manufacturing Systems Engineering,
Xi'an 710049, China

Abstract. This paper studies an evacuation problem that evacuees inside an affected convex region in the plane try to escape to a boundary of the region as quickly as possible. The boundary information of the region is usually unknown to the evacuees at the beginning during an emergency. But with the help of helicopters or even satellite remote sensing technology, outside rescuers can easily get complete boundary information, and rescuers can share the information with evacuees once getting in touch with the evacuee who firstly reaches a boundary. For the scenario that people evacuate from several different positions, we first show that 3 is a lower bound on the competitive ratio, and present an online strategy with its competitive ratio proved to be no more than $2 + \sqrt{5}$. For the scenario that people evacuate from a single initial position, we present a strategy with its competitive ratio very close to the lower bound.

Keywords: Evacuation strategy · Competitive analysis · Convex region

1 Introduction

Emergencies have occurred frequently in recent years, such as the Tianjin Tanggu blast happened in 2015, and the Haiti earthquake happened in 2010. Thus, much attention has recently been devoted to the evacuation problem, which becomes increasingly important for public crisis management and social safety. In this paper, suppose evacuees lack information about the boundary, but they can share information with each other during the evacuation. We seek strategies for evacuating from an affected region as quickly as possible when an emergency happens.

Previous Work. Previous research has focused on two main problems, search problem and evacuation problem. Deng [1] proved that there exists a competitive algorithm for exploring interiors of general polygons which has a bounded number of polygonal obstacles. Papadimitriou [2] tried to seek dynamic decision rules that optimize the worst-case ratio of the distance covered to the length of the optimal path. Berman [3] surveyed the problems of online searching and navigation. In recent works, Xu et al. [4] and Wei et al. [5, 6] considered the evacuation problem that evacuees don't know any boundary information about the affected region, this really occurs in an emergency, as the affected region is usually unknown to the evacuees. Xu et al. [4] presented the

© Springer International Publishing AG 2017
M. Xiao and F. Rosamond (Eds.): FAW 2017, LNCS 10336, pp. 151–162, 2017.
DOI: 10.1007/978-3-319-59605-1_14

EDES evacuation strategy for k (k > 3) groups of evacuees with a competitive ratio of $3/\cos(\pi/k)$, and they proved that the evacuate ratio of any groups strategy in the plane is no less than 3. For the case k = 1, Wei et al. [5] gave a 19.64-competitive evacuation strategy SEP. Qin et al. [8] improved the result by presenting a 19.5-competitive strategy FSS based on the Fibonacci sequence. For k = 3, Wei et al. [6] gave the EET strategy with an improved competitive ratio of $2 + 2\sqrt{3}$.

Our Work. The previous studies mainly focus on the evacuation problem that people evacuate from a single initial position. But in reality, the evacuees may be located at several different initial positions. In this paper, we further consider a more practical scenario that people evacuate from several different initial positions.

For the problem that people evacuate from a single initial position, recall the best previous strategy EET proposed by Wei et al. [6], the evacuees still in the affected region are just required to move towards the boundary point that evacuees firstly reach. We present the SES strategy by improving the last step in strategy EET [6], thus SES can be proved to perform better.

For the problem that people evacuate from several different initial positions, we first show that 3 is a lower bound on the competitive ratio for this problem, and then present the strategy MES derived by some observations about a convex region. The competitive ratio of MES is proved to be no more than $2 + \sqrt{5} \approx 4.237$, this is better than the best previous result of $4\sqrt{2} \approx 5.656$ proposed by Liu et al. [7].

The rest of this paper is organized as follows. Section 2 gives some basic definitions. The lower bound on the competitive ratio for the problem is given in Sect. 3. In Sect. 4, we present the algorithm SES for the evacuation problem that evacuates from a single initial position, with its competitive analysis. Section 5 studies a more practical scenario that evacuates from several different initial positions, we present the strategy MES with its competitive ratio very close to the lower bound. In Sect. 6, we analyze the performances of our presented algorithms. Section 7 concludes this paper with a discussion of further research.

2 Preliminaries

2.1 Problem Statement

In this paper, we define the affected region P as a convex region in the plane and define the evacuees as the points inside P. The evacuees can't get any boundary information of P until someone of them reaches an arbitrary boundary line, they try to evacuate from P to its boundary as quickly as possible. A successful evacuation requires all evacuees reach a boundary of P. The goal is to design an online strategy for the evacuees that minimizes the competitive ratio, which is the ratio of the path length (or evacuation time, used interchangeably) in the online strategy without boundary information to that in the optimal strategy with the full information. We call the problem *k-source evacuation problem* when people evacuates from k different initial positions. Particularly, we call the problem *single-source evacuation problem* when

$k = 1$ and ***multi-source evacuation problem*** when $k \geq 2$. Our discussions are based on the following assumptions.

Assumptions

(a) The evacuees in P can't get any boundary information until someone of them reaches an arbitrary point on the boundary.

(b) The evacuees can share real-time information during the evacuation.

(c) The evacuees move at unit speed.

Assumption (a) is derived by the fact that, the boundary information of the affected region is usually unknown to the evacuees during an emergency, but with the help of helicopters or even satellite remote sensing technology, outside rescuers can easily get complete boundary information of the region, and share it with other evacuees once getting in touch with the evacuee who firstly reaches a boundary. According to Assumption (c), we use path length and evacuation time interchangeably in this paper.

2.2   Competitive Ratio

For *k-source evacuation problem*, the competitive ratio is defined as follows. Let P denote an affected area restricted to a convex region in the plane. The initial positions of evacuees in P are modeled as position points $I = \{I_1, I_2, ..., I_k\}$ in P. For an arbitrary online algorithm S, we denote $l_S^{I_j}(P)$ as the path length (or evacuation time) of evacuees to evacuate from position I_j to a boundary of P by algorithm S, let $l_{OPT}^{I_j}(P)$ denote the optimal (offline) evacuation path length of evacuees to evacuate from position I_j to a boundary of P. Thus, $l_S(P, I) = \max\limits_{j \in \{1,2,...k\}} \{l_S^{I_j}(P)\}$ can denote the path length of all evacuees by algorithm S, and $l_{OPT}(P, I) = \max\limits_{j \in \{1,2,....,k\}} \{l_{OPT}^{I_j}(P)\}$ can denote the optimal path length that all evacuees evacuate with complete boundary information of P at the beginning. The competitive ratio of algorithm S can be defined as

$$c_S = \sup_{P,I} \frac{l_S(P,I)}{l_{OPT}(P,I)}$$

The quantity $\alpha = \inf\limits_{S} c_S$ is the limit of the competitive ratios for *k-source evacuation problem* that can ever be achieved by online algorithms. Note that $c_S \geq 1$ for any online algorithm S, thus 1 is a trivial lower bound on α.

3   A Lower Bound

In this section, we prove a lower bound on the competitive ratio for the evacuation problem in the plane.

Lemma 1 Xu et al. [4]. *For single-source evacuation problem, no online algorithm can achieve a competitive ratio less than 3.*

Theorem 1. *For multi-source evacuation problem, no online algorithm can achieve a competitive ratio less than 3.*

Proof. We prove the lower bound by the following instance, which is similar to the proof in Lemma 1, see Fig. 1. All evacuees are located at the same segment in P initially, with their positions denoted by $\{O_1, O_2, \ldots, O_n\}$ respectively, the closest boundary line to evacuees is supposed to be parallel to segment O_1O_n, denoted by l_0, other boundary lines are very far from evacuees. Let G_i denote the group of evacuees that evacuates from O_i. Suppose G_1 to be the first one that reaches l_0, with the shortest distance denoted by l_{OPT}. And suppose group G_k to be the one that moves towards the opposite direction. Clearly, the distance covered by group G_k is no less than $3l_{OPT}$. Therefore, the path length covered by any strategy is at least 3 times as the optimal path length.

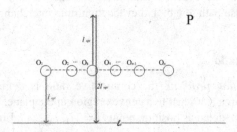

Fig. 1. The proof of a lower bound

4   Single-Source Evacuation Problem

In this section, we study the single-source evacuation problem that evacuates from a single ($k = 1$) initial position O in the affected region P. Recall that with the help of helicopters or even satellite remote sensing technology, outside rescuers can easily get complete boundary information of the region, and share it with other evacuees once getting in touch with the evacuee who firstly reaches a boundary.

Algorithm 1. Single-source Evacuation Strategy (SES)

1. Divide the evacuees into $m \geq 3$ groups, denoted by $G = \{G_1, G_2, \ldots, G_m\}$;

2. Let G_1 move towards an arbitrary direction denoted by R_1 , let G_k move towards direction R_k with the angle from R_{k-1} to R_k equal to $\dfrac{2\pi}{m}$ in clockwise direction, for

$\forall k \in \{2, 3, \ldots, m\}$;

3. Once some evacuee arrives at a boundary point of P, the others in P stop to move towards the corresponding nearest boundary to their current positions.

Theorem 2. *The competitive ratio of algorithm SES is no more than 5.*
    The proof of Theorem 2 is straightforward.

5 Multi-source Evacuation Problem

In this section, we study the multi-source evacuation problem, that evacuees are located at $k \geq 2$ different initial evacuation positions in the affected region P. The set of initial evacuation positions can be denoted as $E = \{e_1, e_2, ..., e_k\}$, let $d(e_i, e_j)$ denote the distance between any two evacuation positions e_i and e_j, and $d(e_i, vP)$ denote the shortest distance from position e_i to a boundary of the affected area P. Derived by the fact that, initial positions of evacuees in one group is usually visible by each other during an emergency, while boundaries are usually invisible by the evacuees. Thus, the maximum distance among initial evacuation positions is supposed to be no more than $l_{OPT}(P) = \max_{\forall e_i \in E}\{d(e_i, P)\}$ meaning that $\max_{\forall e_i, e_j \in E}\{d(e_i, e_j)\} \leq \max_{\forall e_i \in E}\{d(e_i, P)\}$.

Derived by the following three facts about a convex polygon, we design an online evacuation strategy MES. What's more, the evacuees can also take the strategy respectively by being divided into several groups when the maximum distance among initial positions is more than $l_{OPT}(P)$.

Fact 1. For any rectangular region R, if its 4 vertices are in a convex region P or on the boundary of P, then any interior point of R must be an interior point of P.

Fact 2. For any boundary of a convex region P, it can't intersect with the interior of P.

Fact 3. For an arbitrary interior point M of a convex region P, the boundary point N with the shortest distance to M can't be a vertex of P. Besides, line MN must be perpendicular to the boundary line that N lies on.

5.1 Algorithm MES

At the high level, the algorithm MES consists of two main stages, the preparation stage and routing stage.

Preparation Stage. At the beginning, the algorithm MES chooses a pair of two initial positions, denoted by A and B, satisfying $d(A, B) = \max_{\forall e_i, e_j \in E}\{d(e_i, e_j)\}$. If there are multiple such pairs, choose an arbitrary pair. Then, algorithm MES determines the direction vector $\overrightarrow{D} = (\overrightarrow{D_1}, \overrightarrow{D_2}, \overrightarrow{D_3}, \overrightarrow{D_4}) = (+\overrightarrow{y}, -\overrightarrow{y}, +\overrightarrow{x}, -\overrightarrow{x})$, where $+\overrightarrow{x} = \overrightarrow{AB}/|\overrightarrow{AB}|$, $-\overrightarrow{x} = \overrightarrow{BA}/|\overrightarrow{BA}|$, $+\overrightarrow{y}$ and $-\overrightarrow{y}$ are the unit vectors that are perpendicular to $+\overrightarrow{x}$ and $-\overrightarrow{x}$ respectively. Let G_A and G_B denote the group that evacuates from initial position A and B respectively. The MES divides each group into two subgroups arbitrarily, denoted by $G_A = (G_A^1, G_A^2)$ and $G_B = (G_B^1, G_B^2)$ respectively.

Routing Stage. Let $m = d(A, B)$ denote the maximum distance that evacuees can move towards a new direction, let n denote the total direction turning times, let $T_1T_2T_3T_4$ denote the rectangle with its length equal to 2 m and its width equal to m, with A and B as the midpoints of its 2 sides T_1T_2 and T_3T_4 respectively. The using superscripts and subscripts are $\{\alpha, i_1, i_2, j_1, j_2\}$. Let $i_1 \leftarrow 1, j_1 \leftarrow 1, i_2 \leftarrow 2, j_2 \leftarrow 2, \alpha \leftarrow 2, n \leftarrow 1$.

Firstly, subgroups $\{G_A^1, G_A^2, G_B^1, G_B^2\}$ move towards directions $\{\overrightarrow{D_{i_1}}, \overrightarrow{D_{i_2}}, \overrightarrow{D_{j_1}}, \overrightarrow{D_{j_2}}\}$ at once, with a distance equal to m respectively. The other evacuees move towards the nearest vertex of the rectangular $T_1T_2T_3T_4$, and join in the corresponding evacuation subgroup. During this evacuation, once someone reaches a boundary point of the affected region P, they can get complete boundary information by outside rescuers and send it to the other evacuees. In the meantime, evacuees still in P stop to move towards the nearest boundary. The algorithm MES terminates when all evacuees reach a boundary.

If no evacuee reaches a boundary of P during the previous process, algorithm MES updates $j_1 \leftarrow j_1 + (-1)^\alpha \cdot 2, j_2 \leftarrow j_2 + (-1)^\alpha, i_1 \leftarrow i_1 + (-1)^\alpha \cdot 3, i_2 \leftarrow i_2 + (-1)^\alpha \cdot 2,$ $n \leftarrow n + 1, \alpha \leftarrow \alpha + 1$. Subgroups $\{G_A^1, G_A^2, G_B^1, G_B^2\}$ turn to move towards directions $\{\overrightarrow{D_{i_1}}, \overrightarrow{D_{i_2}}, \overrightarrow{D_{j_1}}, \overrightarrow{D_{j_2}}\}$ at once, with a distance equal to m respectively. Repeat this process until someone reaches a boundary of P in some iteration, then the evacuees still in P stop to move towards the nearest boundary to their current position. The algorithm terminates as soon as all evacuees reach a boundary.

Algorithm 2. Multi-source Evacuation Strategy (MES)

Input: All evacuees and their initial positions, the affected region P.

Output: An evacuation route, the direction turning times n.

1. Determine the two initial positions A and B, the positions $T_1T_2T_3T_4$, and the subgroups $\{G_A^1, G_A^2, G_B^1, G_B^2\}$;

2. Determine direction vector $\overrightarrow{D} = (\overrightarrow{D_1}, \overrightarrow{D_2}, \overrightarrow{D_3}, \overrightarrow{D_4}) = (+\overrightarrow{y}, -\overrightarrow{y}, +\overrightarrow{x}, -\overrightarrow{x})$;

3. Let $i_1 \leftarrow 1, j_1 \leftarrow 1, i_2 \leftarrow 2, j_2 \leftarrow 2, \alpha \leftarrow 2, n \leftarrow 1, \beta \leftarrow 0, m \leftarrow d(A,B)$;

4. Let subgroups $\{G_A^1, G_A^2, G_B^1, G_B^2\}$ turn to move towards $\{\overrightarrow{D_{i_1}}, \overrightarrow{D_{i_2}}, \overrightarrow{D_{j_1}}, \overrightarrow{D_{j_2}}\}$ with a distance equal to m respectively. The other evacuees move towards the nearest vertex of $T_1T_2T_3T_4$, and join in the corresponding evacuation subgroup;

5. **If** someone reaches a boundary point of P **then**
      {The other evacuees stop to move towards the nearest boundary, algorithm terminates as soon as all evacuees reach a boundary of P;}
   else
      {Let $j_1 \leftarrow j_1 + (-1)^\alpha \cdot 2$, $j_2 \leftarrow j_2 + (-1)^\alpha$, $i_1 \leftarrow i_1 + (-1)^\alpha \cdot 3$,
      $i_2 \leftarrow i_2 + (-1)^\alpha \cdot 2$, $n \leftarrow n+1$, $\alpha \leftarrow \alpha+1$, go back to Step 4. }

5.2   Competitive Analysis on MES

Theorem 3. *For 2-source evacuation problem, the competitive ratio of the algorithm MES is no more than 4.*

Proof. In 2-source evacuation problem, all evacuees are located at 2 different initial evacuation positions. Clearly, the 2 positions can just be chosen as the special initial positions A and B by the algorithm MES. Suppose G_A^1 to be the subgroup that firstly reaches a boundary point (denoted by P_0) of the affected region P at an arbitrary time (denoted by t_0). Let l_0 denote the boundary line that P_0 lies on, let d_0 denote the distance that a subgroup covers from time 0 to time t_0, let d_K^i denote the shortest distance between subgroup G_K^i and a boundary of P at t_0, for $\forall K \in \{A, B\}, i \in \{1, 2\}$. Recall that $m = d(A, B)$ and F_2, F_3, F_4 denote the positions of G_A^1, G_B^2, G_B^1 at time t_0 respectively, n denote the direction turning times derived by the algorithm MES, let β denote the distance that a subgroup moves after turning the last (the nth) direction in the algorithm MES. Establish a rectangular coordinate system with vector $\overrightarrow{D_1}$, $\overrightarrow{D_3}$ and initial position P_0. Recall that $\overrightarrow{D_1}$ and $\overrightarrow{D_3}$ represent $+\overrightarrow{x}$ and $+\overrightarrow{y}$, respectively.

Let P_1 and P_2 denote the boundary points that have the shortest distance to initial positions B and A, respectively. Thus $l_{OPT}^A = d(A, P_2)$, $l_{OPT}^B = d(B, P_1)$. According to the positions of P_1 and P_2, we can study the competitive ratio of algorithm MES by analyzing the following 2 cases.

Case 1: P_1 and P_2 are both on the same boundary.
Case 2: P_1 and P_2 are not on the same boundary.

Case 1: P_1 and P_2 are both on the same boundary line l_0, see Fig. 2.

Suppose line l_0 satisfy the linear function $y = kx$, where $k \geq 0$ derived by Fact 1 and Fact 2. Clearly, l_0 is the boundary that is closest to every subgroup at t_0 in the worst case, meaning that F_2, F_3, F_4 lie on l_0. Let φ_2 denote the angle between $+\overrightarrow{x}$ and $\overrightarrow{P_0B}$, φ_1 denote the angle between $+\overrightarrow{x}$ and $\overrightarrow{P_0F_3}$, let $\theta = \arctan k \in [0, \frac{\pi}{2})$. Notice that $l_{MES} = d_0 + \max\limits_{K \in \{A,B\}i \in \{1,2\}} \{d_K^i\} = d_0 + d_B^2$, $l_{OPT} = \max\{l_{OPT}^A, l_{OPT}^B\}$, the competitive ratio can be formed as

$$c_{MES} = \frac{l_{MES}}{l_{OPT}} = \frac{(n-1)m + \beta + \sqrt{[(n-1)m + 2\beta]^2 + m^2 n^2} \cdot \sin(\varphi_1 + \theta)}{\sqrt{[\frac{(n-1)}{2}m + \beta]^2 + (m + \frac{n-1}{2}m)^2} \cdot \sin(\varphi_2 + \theta)}$$

$$= 2 + \frac{-m \sin\theta}{[\frac{(n-1)}{2}m + \beta] \cdot \cos\theta + \frac{n+1}{2}m \sin\theta} + \frac{(n-1)m + \beta}{[\frac{(n-1)}{2}m + \beta] \cdot \cos\theta + \frac{n+1}{2}m \sin\theta}$$

$$= 2 + c_1 + c_2$$

$$\leq 2 + c_1|_{\theta=0} + c_2|_{\theta=0}$$

$$\leq 4$$

Case 2: P_1 and P_2 are not on the same boundary line.

Let $d(K, l)$ denote the shortest distance from an arbitrary point $K \in \{A, B\}$ to an arbitrary line l. The boundary line that P_1 lies on satisfies linear equation l_P: $y = kx + b$. Clearly, there can't be any boundary point of the affected region P in the rectangular $P_0F_2F_3F_4$ derived by Fact 1, $BP_1 \perp l_P$ derived by Fact 3, and l_P can't intersect with rectangular $P_0F_2F_3F_4$ derived by Fact 2. The plane can be divided into 8 regions besides the rectangular $P_0F_2F_3F_4$, denoted by $\{R_1, R_2, \ldots, R_8\}$, see Fig. 3.

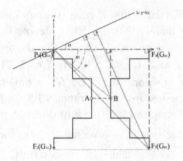

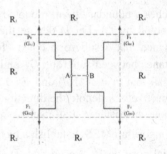

Fig. 2. The proof of Theorem 3 in Case 1

Fig. 3. Dividing the plane.

The competitive ratio can be formed as $c = (d_0 + \max\limits_{K\in\{A,B\},i\in\{1,2\}}\{d_K^i\})/$ $\max\{l_{OPT}^A, l_{OPT}^B\}$, and P_2 must lie on l_0 when analyzing the competitive ratio in the worst case. Clearly, l_{OPT}^B is no more than $d(B, l_0)$, and l_{OPT} is equal to l_{OPT}^A in case that $|BP_1| < d_{A,l_0}$. Thus, we just need to consider the possible positions of P_1 that satisfy $l_{OPT}^A < l_{opt}^B < d(B, l_0)$. When analyzing the competitive ratio in the worst case, we aim to consider the position of P_1 that l_{OPT}^B can approach to l_{OPT}^A as much as possible, and $\max\limits_{K\in\{A,B\},i\in\{1,2\}}\{d_K^i\}$ can be larger in the meantime. Noticing the geometrical features of Fig. 4, we can see that for any case in the situation that P_1 is located in R_1 (or R_2, R_3, R_4), there exist a worse case in the situation that P_1 is located in R_5 (or R_6, R_7, R_8). Besides, the situation that P_1 is located in R_6 (or R_7) is similar to the situation that P_1 is located in R_5 (or R_8). Thus, we only need to consider the situation that P_1 is located in R_6 (or R_5) and R_7 (or R_8).

Case 2.1: P_1 is located in R_7, see Fig. 4.

Derived by Fact 2, we know that $l_0 : y - kx + b$ satisfy $k \geq 0$, $b \geq 0$. It's clear that the points $F_0(m, 0)$, $B(\frac{n+1}{2}m, -\beta + \frac{1-n}{2}m)$ and $F_3(mn, -2\beta + (1 - n)m)$ are collinear. Line F_3F_0 intersect l_1 at point F_1, draw a line F_3P_1' perpendicular to l_0 at point P_1', then, ΔP_1F_1B and $\Delta P_1' F_1F_3$ are similar, let $\varepsilon = |F_1F_3| - |F_0F_3| = |F_0F_1|$. In the worst case, there doesn't exist any other boundary point of P that has a shorter distance to F_3 than $|F_3P_1'|$. Thus, the ratio can be formed as

$$c = \frac{d_0 + |F_3P_1'|}{l_{OPT}^B}$$

$$= \frac{d_0 + (\varepsilon + 2AP_0) \cdot \sin(\arctan k + \angle F_3F_0F_4)}{(\varepsilon + AP_0) \cdot \sin(\arctan k + \angle BF_0F_4)} = 1 + \frac{\sqrt{(\frac{n-1}{2}m)^2 + (\frac{n-1}{2}m + \beta)^2}}{\varepsilon + \sqrt{(\frac{n-1}{2}m)^2 + (\frac{n-1}{2}m + \beta)^2}}$$

$$+ \frac{(n - 1)m + \beta}{(\varepsilon + \sqrt{(\frac{n-1}{2}m)^2 + (\frac{n-1}{2}m + \beta)^2}) \cdot \sin(\arctan k + \arctan\frac{n-1}{2}m+\beta}{\frac{n-1}{2}m})}$$

in which, $0 \leq \beta < m$, $\frac{\partial c}{\partial k} < 0$, $\frac{\partial c}{\partial \varepsilon} < 0$. Hence, $c_{MES} < \lim\limits_{k,\varepsilon\to 0} c_{MES} \leq 4$.

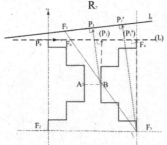

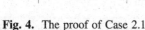

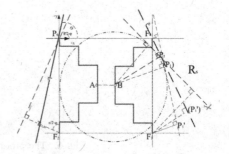

Fig. 4. The proof of Case 2.1   **Fig. 5.** The proof of Case 2.2

Case 2.2: P_1 is located in R_6, see the above Fig. 5.

Draw a line F_3P_1' that is perpendicular to l_0 at point P_1'. In the worst case, there doesn't exist any other boundary point of P that has a shorter distance to F_3 than $|F_3P_1'|$, and we only need to consider the case that $l_{OPT}^A \leq l_{OPT}^B < d(B, l_0)$, thus, the competitive ratio can be formed as $c = (d_0 + \max\{d_{F_2,l_0}, |F_3P_1'|\})/|BP_1|$ in the worst case, and the ratio c can get its maximum value when circle B (with its center at B, and its radius equal to $|BP_1|$) is tangent to the line F_4P_1. Let $l_0 : y = kx - kmn$, in which $k \leq 0$, let $\varphi = \arctan k - \pi/2$, thus $\varphi \in (0, \pi/2]$. We know $d_{A,l_0} \leq BP_1$, $F_3P_1' > d_{F_2,l_0}$. Then, the ratio can be further formed as

$$c = \frac{d_0 + |F_3P_1'|}{|BP_1|} = c_1 + c_2 < c_1|_{\varphi=0} + c_2\Big|_{\varphi=\arctan\frac{\frac{n-1}{2}m+\beta}{\frac{n-1}{2}m}} < 4$$

where $c_1 = \dfrac{(n-1)m+\beta}{(\frac{n-1}{2}m+\beta) \cdot \sin\varphi + \frac{n-1}{2}m \cdot \cos\varphi}$, $c_2 = \dfrac{(n-1)m+2\beta}{\frac{n-1}{2}m+\beta+\frac{n-1}{2}m \cdot \cot\varphi}$.

Case 2.3: P_1 is located in R_5 (or R_8). The analysis of the competitive ratio is similar to the situation that P_1 is located in R_6 (or R_7), and the competitive ratio in the worst case is also equal to 4.

The theorem follows.

Theorem 4. *For multi-source evacuation problem, the competitive ratio of the algorithm MES is no more than* $2 + \sqrt{5}$.

Proof. In multi-source evacuation problem, all evacuees are located at several different initial evacuation positions. the algorithm MES chooses a pair of two initial positions, denoted by A and B, that has the maximum distance between each other. If there are multiple such pairs, choose an arbitrary pair. Clearly, any other initial position must be in the intersection area of circle A (with its radius equal to m) and circle B (with its radius also equal to m), see the above Fig. 6.

Comparing the evacuation route derived by the algorithm MES in 2-source evacuation problem and in multi-source evacuation problem, we can see the difference is just in the first direction turning process, meaning that the direction turning time $n = 1$

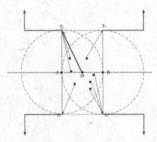

Fig. 6. The proof of Theorem 4

during the evacuation. During the first direction turning process ($n = 1$) by the algorithm MES, evacuees move a distance equal to m in *2-source evacuation problem*, while evacuees move a distance no more than $(\sqrt{5}/2) \cdot m$ in *multi-source problem*. Thus,

$$l_{MES}|_{multi-source} \leq l_{MES}|_{2-source} + (\frac{\sqrt{5}}{2} - 1)m$$

Note also that the l_{OPT} in multi-source evacuation problem is no less than the l_{OPT} in the corresponding *2-source evacuation problem*. Hence, the competitive ratio of the algorithm MES in *multi-source evacuation problem* satisfy

$$c_{MES} = \frac{l_{MES}|_{multi-source}}{l_{OPT}} \leq \frac{l_{MES}|_{2-source} + (\frac{\sqrt{5}}{2} - 1)m}{l_{OPT}} < 4 + \frac{(\frac{\sqrt{5}}{2} - 1)m}{\frac{n-1}{2}m + \beta} \leq 2 + \sqrt{5}.$$

Where β also denotes the distance that a subgroup covers after turning the last (the nth) direction in the algorithm MES.

This proves the theorem.

6   Performance Analysis

In this section, we analyze the performances of the algorithm SES and the algorithm MES respectively.

For *single-source evacuation problem*, we present the SES strategy, which improves the last step in the EET strategy proposed by Wei et.al. [6]. We compare the performance results among the SES strategy, the EET strategy and the lower bound, as shown in Fig. 7. Through the comparison, it can conclude that the SES strategy always performs a little better than the EET strategy, and both of the strategies can approach very closely to the lower bound as group number increases.

For *multi-source evacuation problem*, we prove the lower bound of this problem is 3, and we present the MES strategy which is derived by some observed facts about a convex polygon. The competitive ratio of the MES strategy is proved to be no more than $2 + \sqrt{5} \approx 4.237$, which is better than the best previous result of $4\sqrt{2} \approx 5.656$ proposed by Liu et.al. [7]. The performance of MES strategy, strategy proposed by Liu et al. [7], and the lower bound is shown in Table 1.

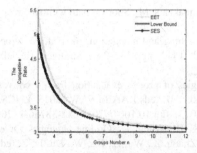

Fig. 7. The comparison on competitive ratio among the SES strategy, the EET strategy and lower bound

Table 1. The performance comparison

	Lower bound	The MES strategy	Liu's
The competitive ratio	3	$2 + \sqrt{5} \approx 4.237$	$4\sqrt{2} \approx 5.656$

7 Conclusions

In this paper, we study the problem that people evacuate from an affected convex region in the plane to a boundary of the region as soon as possible. For *multi-source evacuation problem*, we show that 3 is a lower bound on the competitive ratio. Based on the fact that with the help of helicopters or even satellite remote sensing technology, outside rescuers can easily get complete boundary information of the region, and they can share it with other evacuees once getting in touch with the evacuee who firstly reaches a boundary. Then we present the strategy MES derived by some observations about a convex polygon. The competitive ratio of MES is proved to be no more than $2 + \sqrt{5} \approx 4.237$, which is better than the best previous result of $4\sqrt{2} \approx 5.656$ proposed by Liu et.al. [7]. For *single-source evacuation problem*, we present the SES strategy by improving the last step in the EET strategy proposed by Wei et.al. [6]. The competitive ratio of the SES strategy is proved to be better than the EET strategy, and both can approach very closely to the lower bound as the groups number increases.

Further possible research could consider the case that outside rescuers are searching for the evacuees in the mean time during the evacuation. It would also be interesting to take evacuees' individual behaviors into account that people prefer to turn less directions during their evacuation.

Acknowledgments. This research is partially supported by the NSFC (Grant No. 71601152), and by the China Postdoctoral Science Foundation (Grant No. 2016M592811).

References

1. Deng, X.T., Kameda, T., Papadimitriou, C.: How to learn an unknown environment I: the rectilinear case. J. ACM **45**(2), 215–245 (1998)

2. Papadimitriou, C.H., Yannakakis, M.: Shortest path without a map. Theoret. Comput. Sci. **84** (1), 127–150 (1991)
3. Berman, P.: On-line searching and navigation. In: Fiat, A., Woeginger, G.J. (eds.) Online Algorithms. LNCS, vol. 1442, pp. 232–241. Springer, Heidelberg (1998). doi:10.1007/. BFb0029571
4. Xu, Y., Qin, L.: Strategies of groups evacuation from a convex region in the plane. In: Fellows, M., Tan, X., Zhu, B. (eds.) AAIM/FAW -2013. LNCS, vol. 7924, pp. 250–260. Springer, Heidelberg (2013). doi:10.1007/978-3-642-38756-2_26
5. Wei, Q., Tan, X., Jiang, B., Wang, L.: On-line strategies for evacuating from a convex region in the plane. In: Zhang, Z., Wu, L., Xu, W., Du, D.-Z. (eds.) COCOA 2014. LNCS, vol. 8881, pp. 74–85. Springer, Cham (2014). doi:10.1007/978-3-319-12691-3_7
6. Wei, Q., Wang, L., Jiang, B.: Tactics for evacuating from an affected area. Int. J. Mach. Learn. Comput. **3**(5), 435–439 (2013)
7. Liu, Y., Jiang, B., Zhang, H.: Online strategies for evacuating from a convex region by groups in the plane. In: 2015 Ninth International Conference on Frontier of Computer Science and Technology, pp. 178–183. IEEE (2015)
8. Qin, L., Xu, Y.: Fibonacci helps to evacuate from a convex region in a grid network. J. Comb. Optim. 1–16 (2016)
9. Lu, Q., George, B., Shekhar, S.: Capacity constrained routing algorithms for evacuation planning: a summary of results. In: Bauzer Medeiros, C., Egenhofer, Max J., Bertino, E. (eds.) SSTD 2005. LNCS, vol. 3633, pp. 291–307. Springer, Heidelberg (2005). doi:10. 1007/11535331_17
10. Zhang, H., Xu, Y.: The k-Canadian travelers problem with communication. In: Atallah, M., Li, X.-Y., Zhu, B. (eds.) Frontiers in Algorithmics and Algorithmic Aspects in Information and Management. LNCS, vol. 6681, pp. 17–28. Springer, Heidelberg (2011). doi:10.1007/ 978-3-642-21204-8_6

A Further Analysis of the Dynamic Dominant Resource Fairness Mechanism

Weidong Li[1,2], Xi Liu[1], Xiaolu Zhang[1], and Xuejie Zhang[1(✉)]

[1] Yunnan University, Kunming 650091, People's Republic of China
{weidong,zxl,xjzhang}@ynu.edu.cn
[2] Dianchi College of Yunnan University,
Kunming 650228, People's Republic of China

Abstract. Multi-resource fair allocation has been a hot topic in cloud computing. Recently, a *dynamic dominant resource fairness mechanism* (DDRF) is proposed for dynamic multi-resource fair allocation. In this paper, we develop a linear-time algorithm to find a DDRF solution at each step. Moreover, we give the competitive ratios of the DDRF mechanism under three widely used objectives.

Keywords: Multi-resource fair allocation · Dominant resource fairness · Dynamic dominant resource fairness · Competitive ratio

1 Introduction

Max-min fair allocation is a fundamental problem in combinatorial optimization. Several polynomial-time approximation algorithms are designed to find near-optimal solutions for max-min fair allocation with different constraints [1,2]. In 2011, Ghodsi et al. [13] proposed a novel max-min fair allocation problem, which is to allocation multiple resources to heterogeneous agents as fair as possible. They [13] advocated the *dominant resource fairness* (DRF) mechanism, which has a wide range of applications in cloud computing systems. Most importantly, agents can not be able to benefit by lying about their resource demands under DRF mechanism, i.e., DRF satisfies strategy-proof, which is also called incentive compatible or truthful.

Since then, DRF and its variants became a hot topic in cloud computing, computational economics, and computer networks, and has been extended to many dimensions. Joe-Wong et al. [24] designed a unifying multi-resource allocation framework that captures the trade-offs between fairness and efficiency, which generalizes the DRF measure. Gutman and Nisan [14] situated DRF in a common economics framework, obtaining a general economic perspective. Bhattacharya et al. [5] generalized DRF to a hierarchical scheduler that offers service isolations in a computing system with a hierarchical structure. Parkes et al. [18] extended DRF in several ways, including the presence of zero demands and the case of indivisible tasks. Wang et al. [22] generalized DRF into the cloud computing systems with heterogeneous servers. Psomans and Schwartz [20], and Friedman et al. [12] studied the multi-resource allocation of discrete tasks on multiple

M. Xiao and F. Rosamond (Eds.): FAW 2017, LNCS 10336, pp. 163–174, 2017.
DOI: 10.1007/978-3-319-59605-1_15

machines. Zarchy et al. [26] developed a framework for fair resource allocation that captures such implementation tradeoffs by allowing users to submit multiple resource demands. Chowdhury et al. [10] extended DRF from static and fixed demands to elastic demands and proposed an optimal allocation algorithm. More related results can be found in the recent survey [19].

There are several alternative mechanisms for multi-resource fair allocation. Notably, Dolev et al. [11] proposed an alternative notion of multi-resource fair allocation, called *bottleneck based fairness*, which guarantees that each user receives at least its entitlement on some bottleneck resource. Bonald and Roberts [6] argued that *proportional fairness* (PF) is preferable to DRF in certain environment. The same author [7] advocated *bottleneck max fairness* under a stochastic demand model. Jin and Hayashi [15] studied the efficiency of PF and DRF in the aspect of resource usage and total throughput when there are two different type resources are available. Zahedi and Lee [25] presented a Cobb-Douglas preferences based allocation mechanism, which satisfies fair properties similar to DRF. Wang and Varman [21] proposed *bottleneck aware allocation* mechanism, capturing trade-offs between fairness and system utilization in multi-tiered storage systems. Recently, Wang et al. [23] proposed *task share fairness* (TSF) mechanism for multi-resource fair allocation with placement constraints.

DRF uses complete information about the requirements of all agents in order to find the fair solution. However, in reality, agents arrive over time, and we do not know the requirements of forthcoming agents before allocating the resources to the arrived agents. Recently, Kash et al. [16] introduced a dynamic model of fair allocation and proposed a *dynamic dominant resource fairness* (DDRF) mechanism, which is generalized to a more general case [17]. They [16] mentioned that a DDRF solution can be found by using water-filling algorithm or solving the corresponding linear program. However, the running time of the water-filling algorithm is pseudo-polynomial in worst-case scenario. Although solving a linear program can be done within polynomial time, the running time is high. It is desired to design an efficient algorithm to find a DDRF solution.

In this paper, we further study the DDRF mechanism. We design an efficient algorithm to find a DDRF solution. In addition, motivated by [3, 4, 9], we discuss the efficiency of the DDRF mechanism from the worst-case perspective. We consider three important objectives, and give the tight ratios, by exploring the properties of DDRF. The rest of the paper is organized as follows. Section 2 describes the DDRF mechanism. Section 3 presents a polynomial-time algorithm, which can find a DDRF solution in $O(k)$ time at every step k. Section 4 gives the competitive ratios analysis of the DDRF mechanism. Finally, Sect. 5 concludes the paper and gives the future work.

2 The Description of DDRF

Throughout this paper, assume that resources are divisible. In a multi-resource environment, there are n agents and m resources. Each agent i requires D_{ir}-fraction of resource r for each task, assuming that $D_{ir} > 0$ for each resource r.

As defined in [13], the *dominant resource* of agent i is the resource r_i^* such that $D_{ir_i^*} = \max_r D_{ir}$. Following [16,18], the *normalized demand* of agent i is given by $\mathbf{d}_i = (d_{i1}, \ldots, d_{im})$, where $d_{ir} = D_{ir}/D_{ir_i^*}$ for each resource $r = 1, \ldots, m$. Clearly, $d_{ir} \leq 1$ and $d_{ir_i^*} = 1$ for each agent i.

In the dynamic resource allocation model considered in [16], agents arrive at different times and do not depart. Assume that agent 1 arrives first, and in general agent k arrives after agents $1, \ldots, k-1$, for $k \geq 2$. For convenience, we say that agent k arrives in *step k*. An agent reports its demand which does not change over time when it arrives. Thus, at step k, demand vectors $\mathbf{d}_1, \ldots, \mathbf{d}_k$ are known, and demand vectors $\mathbf{d}_{k+1}, \ldots, \mathbf{d}_n$ are unknown. At each step k, the DDRF mechanism [16] produces an allocation $\mathbf{A}^k$ over the agents present in the system, where $\mathbf{A}^k$ allocates A_{ir}^k-fraction of resource r to agent i, subject to the feasibility condition

$$\sum_{i=1}^{n} A_{ir}^k \leq \frac{k}{n}, \text{ for } r = 1, 2, \ldots, m. \tag{1}$$

Under the DDRF mechanism, assume that allocations are *irrevocable*, i.e., $A_{ir}^k \geq A_{ir}^{k-1}$, for every step $k \geq 2$, every agent $i \leq k-1$, and every resource r. At every step k, assume $\mathbf{A}^k$ is *non-wasteful*, which means that for every agent i there exists $y \in R^+$ such that for every resource r, $A_{ir}^k = y \cdot d_{ir}$. Let x_i^k be the *dominant share* of agent i, which means that agent i is allocated x_i^k-fraction of dominant resource. For a non-wasteful allocation $\mathbf{A}^k$, it is easy to verify that

$$A_{ir}^k = x_i^k \cdot d_{ir}, \text{ for } i = 1, 2, \ldots, k, \text{ and } r = 1, 2, \ldots, m. \tag{2}$$

At each step k, the DDRF mechanism [16] starts from the current allocation among the present agents $1, \ldots, k$ and keeps allocating resources to agents that have the minimum dominant share synchronously, until a k/n fraction of at least one resource is allocated. Formally, at every step k, the dominant share vector $(x_1^k, \ldots, x_k^k)$ of the DDRF allocation $\mathbf{A}^k$ can be obtained by solving the following linear program (LP):

$$\begin{cases} \text{Maximize } M^k \\ x_i^k \geq M^k, \forall i \leq k; \\ x_i^k \geq x_i^{k-1}, \forall i \leq k-1; \quad \text{(irrevocable)} \\ \sum_{i=1}^{k} d_{ir} x_i^k \leq \frac{k}{n}, \forall r. \quad \text{(capacity constraints)} \end{cases} \tag{3}$$

As shown in [16], the DDRF mechanism satisfies many desired properties. Especially, it satisfies *sharing incentives* (SI) and *dynamic Pareto optimality* (DPO). SI means that, for all steps k and all agents $i \leq k$, $x_i^k \geq 1/n$, i.e., when an agent arrives it receives an allocation that it likes at least as much as an equal split of the resources. DPO means that, for all steps k, there is a resource r such that $\sum_{i=1}^{k} d_{ir} x_i^k = k/n$, i.e., it should not be possible to increase the allocation of an agent without decreasing the allocation of at least another user, subject to not allocating more that k/n fraction of any resource.

3   A Linear-Time Optimal Algorithm

Although the water-filling algorithm can produce a DDRF solution [16], the running time is pseudo-polynomial [14]. Also, we can compute a DDRF solution by solving the LP (3). However, it is not a combinatorial optimal algorithm. At every step k, by modifying the linear-time algorithms for computing a DRF solution in [14,18] slightly, we can obtain a variant of water-filling algorithm with running time $O(k^2)$. In this section, we will design a linear-time algorithm to find a DDRF solution. In the proof below, M^k and x_i^k refer to the optimal solution of LP (3) at step k. The following two lemmas are very useful for designing the linear-time algorithm.

Lemma 1 [16]. *At any step $k \in \{1, \ldots, n\}$, it holds that $x_i^k = \max\{M^k, x_i^{k-1}\}$ for all agents $i \leq k$.*

Lemma 2 [16]. *At any step $k \in \{1, \ldots, n\}$, for all agents i, j such that $i < j$, it holds that $x_i^k \geq x_j^k$.*

Theorem 3. *At any step $k \in \{2, \ldots, n\}$, a DDRF solution can be found within $O(k)$ time.*

Proof. Consider an agent j. By Lemma 1, we have $x_j^k = \max\{M^k, x_j^{k-1}\}$. If $x_j^k = x_j^{k-1} > M^k$, by Lemma 2, for all agents $i \leq j$, we have $x_i^{k-1} \geq x_j^{k-1} > M^k$, which implies that $x_i^k = \max\{M^k, x_i^{k-1}\} = x_i^{k-1}$. If $x_j^k = M^k > x_j^{k-1}$, by Lemma 2, for all agents $i \geq j$, we have $x_i^{k-1} \leq x_j^{k-1} < M^k$, which implies that $x_i^k = \max\{M^k, x_i^{k-1}\} = M^k$. Therefore, at any step $k \geq 2$, there is an agent $\tau \leq k$ such that

$$\begin{cases} x_i^k = x_i^{k-1} > M^k, & \text{for } < \tau; \\ x_i^k = M^k \geq x_i^{k-1}, & \text{for } \tau \leq i \leq k. \end{cases} \tag{4}$$

Thus, if we know τ, M^k can be obtained by solving the following linear program

$$\begin{cases} \text{Maximize } M^k \\ \sum_{i:\tau \leq i \leq k} d_{ir} M^k + \sum_{i:i<\tau} d_{ir} x_i^{k-1} \leq \frac{k}{n}, & \text{for } r = 1, 2, \ldots, m. \end{cases} \tag{5}$$

As pointed in [18], this linear program can be rewritten as

$$M^k = \min_r \frac{k/n - \sum_{i:i<\tau} d_{ir} x_i^{k-1}}{\sum_{i:\tau \leq i \leq k} d_{ir}}. \tag{6}$$

We are now ready to describe our linear-time algorithm. Our main idea is to find τ by using a bisection method. At any step $k \geq 2$, consider the agent $l = \lceil (1+k)/2 \rceil$. Let

$$\begin{cases} x_i^k = x_i^{k-1}, & \text{for } i < l; \\ x_i^k = x_l^{k-1}, & \text{for } l \leq i \leq k, \end{cases} \tag{7}$$

For convenience, let

$$
\begin{cases}
\alpha_r = \sum_{i:i<l} d_{ir} x_i^{k-1}, \forall r; \\
\beta_r = \sum_{i:l \le i \le k} d_{ir}, \forall r.
\end{cases}
\tag{8}
$$

Clearly, if $\alpha_r + x_l^{k-1}\beta_r \le k/n$ for every resource r, i.e., $(x_1^k, \ldots, x_k^k)$ satisfies the capacity constraints in (1), we have $M^k \ge x_l^{k-1}$ and $l \ge \tau$. Otherwise, we have $M^k < x_l^{k-1}$ and $l < \tau$. We distinguish the following two cases:

Case 1. $l \ge \tau$. For every agent i satisfying $l \le i \le k$, we have $x_i^k = M^k$, as $i \ge l \ge \tau$. Let $\mathcal{AI} = \{i : l \le i \le k\}$ be set of known agents with identical dominant share in the optimal solution $(x_1^k, \ldots, x_k^k)$. Next, consider the agent $\lceil (1+l)/2 \rceil$ as before.

Case 2. $l < \tau$. For every agent i satisfying $i < l$, we have $x_i^k = x_i^{k-1}$, as $i < l < \tau$. Let $\mathcal{AS} = \{i : i < l\}$ be set of known agents with same dominant share as in step $k-1$ in the optimal solution $(x_1^k, \ldots, x_k^k)$. Next, consider the agent $\lceil (l+k)/2 \rceil$ as before.

At every step k, the number of unclassified agents in $\{i : i \notin \mathcal{AI}, i \notin \mathcal{AS}\}$ is reduced to half. Finally, all the agents are divided into two subsets $\mathcal{AI}$ and $\mathcal{AS}$, and we will find the τ and the optimal solution $(x_1^k, \ldots, x_k^k)$. Clearly, the running time of deciding whether $l \ge \tau$ at each iteration is linear in the number of unclassified agents. Thus, the total running time is $O(k+k/2+k/2^2+\cdots+1) = O(k)$, where m is seen as a constant. The complete algorithm is given as LINEAR-TIME DDRF ALGORITHM. ∎

4   Competitive Analysis of the DDRF Mechanism

In [16], the authors analyzed the performance of the DDRF mechanism on real data, for two objectives: the minimum dominant share (the maxmin objective) and the sum of dominant shares (the maxsum objective) of the agents present in the system. In this section, we analyze the performance of the DDRF mechanism in the worst-case scenario.

For a maximization problem, it is well known that the competitive ratio ρ of an online algorithm is the worst-case ratio between the cost of the solution found by the online algorithm and the cost of an optimal solution in an offline setting where all the demands of agents are known [8]. Clearly, $\rho \in [0, 1]$. Similarly, we define the *competitive ratio* of the DDRF mechanism as the worst-case ratio between the objective value of the DDRF solution $(x_1^k, \ldots, x_k^k)$ and the optimal solution $(\dot{x}_1^k, \ldots, \dot{x}_k^k)$ of instance I under certain objective function in the offline setting. Accordingly, the competitive ratio CR of the DDRF mechanism is defined as

$$
CR = \min_I \min_k \frac{\text{The objective value of } (x_1^k, \ldots, x_k^k)}{\text{The objective value of } (\dot{x}_1^k, \ldots, \dot{x}_k^k)}.
\tag{9}
$$

LINEAR-TIME DDRF ALGORITHM

1: **Data:** Demand $\mathbf{d}_i$, $1 \leq i \leq k$
2: **Result:** Allocation $\mathbf{A}^k$ at each step k
3: $x_1^1 \leftarrow 1/n$, $A_{1r}^1 \leftarrow x_1^1 \cdot d_{1r}$, $\forall r$;
4: $k \leftarrow 2$;
5: **while** $k \leq n$ **do**
6:     **if** $\sum_{i=1}^{k} d_{ir} x_1^{k-1} \leq k/n$, $\forall r$, **do**
7:       $\alpha_r \leftarrow 0$, $\beta_r \leftarrow \sum_{i=1}^{k} d_{ir}$, $\forall r$;
8:     **else, do**
9:       $LB \leftarrow 1$, $UB \leftarrow k$, $\tau \leftarrow \lceil (LB+UB)/2 \rceil$;
10:     $\alpha_r \leftarrow \sum_{i=1}^{\tau-1} d_{ir} x_i^{k-1}$, $\beta_r \leftarrow \sum_{i=\tau}^{k} d_{ir}$, $\forall r$;
11:      **while** $UB - LB > 1$, **do**
12:        **if** $\alpha_r + \beta_r x_\tau^{k-1} \leq k/n$, $\forall r$, **do**
13:        $LB \leftarrow LB$, $UB \leftarrow \tau$, $\tau \leftarrow \lceil (LB+UB)/2 \rceil$;
14:        $\alpha_r \leftarrow \alpha_r - \sum_{i=\tau}^{UB-1} d_{ir} x_i^{k-1}$, $\beta_r \leftarrow \beta_r + \sum_{i=\tau}^{UB-1} d_{ir}$;
15:        **else, do**
16:        $LB \leftarrow \tau$, $UB \leftarrow UB$, $\tau \leftarrow \lceil (LB+UB)/2 \rceil$;
17:        $\alpha_r \leftarrow \alpha_r + \sum_{i=LB}^{\tau-1} d_{ir} x_i^{k-1}$, $\beta_r \leftarrow \beta_r - \sum_{i=LB}^{\tau-1} d_{ir}$;
18:       **end if;**
19:      **end while;**
20:     **end if;**
21: $M^k \leftarrow \min_r (k/n - \alpha_r)/\beta_r$;
22: $x_i^k \leftarrow \max(x_i^{k-1}, M^k), \forall i \leq k$;
23: $A_{ir}^k \leftarrow x_i^k \cdot d_{ir}, \forall i \leq k$;
24: $k \leftarrow k + 1$;
25: **end while**

4.1 The Maxmin Objective

When the objective is minimum dominant share maximization (maxmin, for short), the optimal solution $(\tilde{x}_1^k, \ldots, \tilde{x}_k^k)$ at step k (≥ 2) in the offline setting can be obtained by solving the following linear program

$$\begin{cases} \text{Maximize } \min_i x_i^k \\ \sum_{i=1}^{k} d_{ir} x_i^k \leq \frac{k}{n}, \forall r. \end{cases} \tag{10}$$

Actually, $(\tilde{x}_1^k, \ldots, \tilde{x}_k^k)$ is a DRF solution [13,14,18], where the dominant shares of all agents are equal. Formally, for a given instance I, at every step k, $(\tilde{x}_1^k, \ldots, \tilde{x}_k^k)$ is obtained by

$$\tilde{x}_1^k = \cdots = \tilde{x}_k^k = \min_r \frac{k/n}{\sum_{i=1}^{k} d_{ir}}, \tag{11}$$

following from [14,18].

Therefore, the competitive ratio CR_1 of the DDRF mechanism for the maxmin objective can be defined as

$$CR_1 = \min_I \min_k \frac{\min_i x_i^k}{\min_i \tilde{x}_i^k} = \min_I \min_k \frac{\min_i x_i^k}{\tilde{x}_k^k} = \min_I \min_k \frac{x_k^k}{\tilde{x}_k^k}, \qquad (12)$$

where the last equality follows from the fact $\min_i x_i^k = x_k^k$, which can be obtained by Lemma 2.

Theorem 4. *When the objective is minimum dominant share maximization, the competitive ratio of the DDRF mechanism is $1/m$. Moreover, no mechanism satisfying DPO can do better than $1/(m-1)$.*

Proof. At every step $k \in \{2, \ldots, n\}$, since the DDRF mechanism satisfies the SI property, we have

$$x_k^k \geq \frac{1}{n}. \qquad (13)$$

By the pigeonhole principle, there exists a resource which is the dominant resource for at least $\lceil k/m \rceil$ agents. It implies that the DRF solution $(\tilde{x}_1^k, \ldots, \tilde{x}_k^k)$ satisfies

$$\tilde{x}_k^k \leq \frac{k/n}{\lceil k/m \rceil}. \qquad (14)$$

Thus, the competitive ratio of the DDRF mechanism satisfies

$$CR_1 = \frac{x_k^k}{\tilde{x}_k^k} \geq \frac{\lceil k/m \rceil}{k} \geq \frac{1}{m}. \qquad (15)$$

Consider a setting with m (>2) resources and $n = m^2 + 1$ agents. For $i = 1, 2, \ldots, m^2$, the demand of agent i is defined as

$$\mathbf{d}_i = \begin{cases} (1, \epsilon, \ldots, \epsilon), \text{if } i \equiv 1 (\text{mod } m) \\ (\epsilon, 1, \ldots, \epsilon), \text{if } i \equiv 2 (\text{mod } m) \\ \qquad \cdots \\ (\epsilon, \epsilon, \ldots, 1), \text{if } i \equiv 0 (\text{mod } m) \end{cases} \qquad (16)$$

where $\epsilon \to 0$ is a small enough number. The demand of agent $n = m^2 + 1$ is $\mathbf{d}_n = (1, 1, \ldots, 1)$. At step $k = m^2$, the DDRF solution is

$$x_i^k = \frac{m^2}{[m + \epsilon(m^2 - m)](m^2 + 1)} \to \frac{m}{m^2 + 1}, \text{for } i = 1, 2, \ldots, k, \qquad (17)$$

when $\epsilon \to 0$. Actually, after the first m^2 steps, at least $m^2/(m^2 + 1)$ share of at least one resource r^* must be exhausted for any dynamic mechanism satisfying the DPO property. It implies that at most $1/(m^2 + 1)$ share of resource r^* is left for the last agent n. Hence,

$$x_n^n \leq \frac{1}{m^2 + 1}, \qquad (18)$$

for any dynamic mechanism satisfying DPO, while the DRF solution $(\tilde{x}_1^n, \ldots, \tilde{x}_n^n)$ satisfies

$$\tilde{x}_i^n = \frac{1}{m+1+\epsilon(m^2-m)} \to \frac{1}{m+1}, \text{for } i = 1, 2, \ldots, n, \tag{19}$$

when $\epsilon \to 0$. It implies that, at step $k = n$, the competitive ratio of any dynamic mechanism satisfying DPO including the DDRF mechanism is at most

$$\frac{x_n^n}{\tilde{x}_n^n} \to \frac{m+1}{m^2+1} \leq \frac{1}{m-1}. \tag{20}$$

Thus, the theorem holds.   ∎

4.2 The Maxsum Objective

When the objective is the sum of dominant shares maximization (maxsum, for short), for a given instance I, the optimal solution $(\bar{x}_1^k, \ldots, \bar{x}_k^k)$ at step k (≥ 2) in the offline setting can be obtained by solving the following linear program

$$\begin{cases} \text{Maximize } \sum_{i=1}^{k} x_i^k \\ \sum_{i=1}^{k} d_{ir} x_i^k \leq \frac{k}{n}, \forall r. \end{cases} \tag{21}$$

Accordingly, the competitive ratio of the DDRF mechanism for the maxsum objective can be defined as

$$CR_2 = \min_{I} \min_{k} \frac{\sum_{i=1}^{k} x_i^k}{\sum_{i=1}^{k} \bar{x}_i^k}. \tag{22}$$

Theorem 5. *When the objective is the sum of dominant shares maximization, the competitive ratio of the DDRF mechanism is $1/m$, and the ratio is tight.*

Proof. Since the DDRF mechanism satisfies SI, we have $x_i^k \geq 1/n$ for every agent $i \leq k$ at step k, which implies that

$$\sum_{i=1}^{k} x_i^k \geq \frac{k}{n}. \tag{23}$$

Consider the optimal solution $(\bar{x}_1^k, \ldots, \bar{x}_k^k)$ obtained from (21). Clearly, at step k, for every resource r,

$$\sum_{i:r_i^*=r} \bar{x}_i^k = \sum_{i:r_i^*=r} d_{ir_i^*} \bar{x}_i^k \leq \sum_{i=1}^{k} d_{ir} \bar{x}_i^k \leq \frac{k}{n}, \tag{24}$$

following from the fact $d_{ir_i^*} = 1$ and the capacity constraint of (21). It implies that

$$\sum_{i=1}^{n} \bar{x}_i^k \leq \sum_{r=1}^{m} \sum_{i:r_i^*=r} \bar{x}_i^k \leq \frac{mk}{n}, \tag{25}$$

where the first inequality follows from the fact that each agent has at least one dominant resource. Thus, following (23) and (25), we have

$$\frac{\sum_{i=1}^{k} x_i^k}{\sum_{i=1}^{n} \bar{x}_i^k} \geq \frac{1}{m}, \tag{26}$$

i.e., the competitive ratio of the DDRF mechanism is at least $1/m$.

Next, we will prove that the competitive ratio is tight. Consider a setting with m (≥ 2) resources and n ($\gg m$) agents. For $i = 1, 2, \ldots, n - m$, the demand vector of agent i is $\mathbf{d}_i = (1, 1, \ldots, 1)$. For agents $i = n-m+1, n-m+2, \ldots, n$, the demand vectors are $(1, \epsilon, \ldots, \epsilon), (\epsilon, 1, \ldots, \epsilon), \cdots, (\epsilon, \epsilon, \ldots, 1)$, respectively, where ϵ is a small enough number. It is easy to verify that the DDRF mechanism produces a solution with

$$x_i^n = \frac{1}{n - m + 1 + \epsilon(m - 1)} \rightarrow \frac{1}{n - m + 1}, \text{for } i = 1, 2, \ldots, n, \tag{27}$$

at step n. The optimal solution will allocate all resources to the last m agents, obtaining a solution with

$$\bar{x}_i^n = \frac{1}{1 + \epsilon(m - 1)} \rightarrow 1, \text{ for } i = n - m + 1, n - m + 2, \ldots, n, \tag{28}$$

and $\bar{x}_i^n = 0$ for other agents. Thus, the competitive ratio is

$$\frac{\sum_{i=1}^{k} x_i^n}{\sum_{i=1}^{n} \bar{x}_i^n} \rightarrow \frac{n}{m(n - m + 1)} = \frac{1}{m} \frac{1}{1 - m/n + 1/n}. \tag{29}$$

when $\epsilon \rightarrow 0$. If n is large enough, the ratio approaches $1/m$. Thus, the theorem holds. ∎

4.3   Resource Utilization Maximization

In cloud computing systems, the resource managers care more about the resource utilization. For example, Ghodsi et al. [13] shows CPU and memory utilization for the small workload when using DRF compared to Hadoop's fair scheduler (slot). Given an allocation $(x_1, \ldots, x_n)$, let $c_r = \sum_{i=1}^{n} d_{ir} x_i$ be the utilization rate (or consumption) of resource r. Define *utilization* of $(x_1, \ldots, x_n)$ as $\min_r c_r$, which is the minimum utilization rate of m resources.

When the objective is resource utilization maximization, the optimal solution $(\hat{x}_1^k, \ldots, \hat{x}_k^k)$ at step k (≥ 2) in the offline setting can be obtained by solving the following linear program

$$\begin{cases} \text{Maximize } c \\ c \leq \sum_{i=1}^{k} d_{ir} x_i^k \leq k/n, \forall r. \end{cases} \tag{30}$$

Accordingly, the competitive ratio of the DDRF mechanism for utilization maximization can be defined as

$$CR_3 = \min_I \min_k \frac{\min_r \sum_{i=1}^k d_{ir} x_i^k}{\min_r \sum_{i=1}^k d_{ir} \hat{x}_i^k}. \tag{31}$$

Theorem 6. *When the objective is utilization maximization, the competitive ratio of the DDRF mechanism is 1/n, and the ratio is tight.*

Proof. Let $(\hat{x}_1, \ldots, \hat{x}_n)$ be a utilization maximized solution with utilization $\hat{c}$. Since DDRF satisfies the SI property, for each agent i, we have

$$d_{ir_i^*} x_i^k \geq \frac{1}{n}. \tag{32}$$

Consider the resource r_i^*, by the constraint in (30), we have

$$d_{ir_i^*} \hat{x}_i^k \leq \sum_{i=1}^k d_{ir} \hat{x}_i^k \leq \frac{k}{n} \leq 1, \tag{33}$$

Therefore, for $i = 1, \ldots, k$ at step k, we have

$$x_i^k \geq \frac{\hat{x}_i^k}{n}. \tag{34}$$

Let r' be the resource such that $\sum_{i=1}^k d_{ir} \hat{x}_i^k$ is minimized, which implies that the utilization of $(\hat{x}_1^k, \ldots, \hat{x}_k^k)$ is $\hat{c} = \sum_{i=1}^k d_{ir'} \hat{x}_i^k$. Consider the DDRF solution $(x_1^k, \ldots, x_k^k)$. Obviously, for each resource $r = 1, \ldots, m$, we have

$$\sum_{i=1}^k d_{ir} x_i^k \geq \frac{1}{n} \sum_{i=1}^k d_{ir} \hat{x}_i^k \geq \frac{1}{n} \sum_{i=1}^k d_{ir'} \hat{x}_i^k \geq \frac{\hat{c}}{n}, \tag{35}$$

where the first inequality follows from (34), and the last two inequalities follow from the definitions of r' and $\hat{c}$. Thus, the competitive ratio of the DDRF mechanism for the utilization maximization objective is at least $1/n$.

Consider a setting with two resources and n agents. The demand of agent 1 is $(1,1)$. For $i = 2, \ldots, n$, the demand of agent i is $(1, 1/n^K)$, where K is a large positive number. The optimal allocation with utilization maximization will allocate the entire resource to agent 1, for a utilization 1. In contrast, under the DDRF mechanism, at step n, each agent will receive a $1/n$-fraction of the resource 1 for a utilization $1/n + (n-1)/n^K$. When K grows larger, the competitive ratio of the DDRF mechanism approaches

$$\lim_{K \to \infty} \frac{1/n + (n-1)/n^K}{1} = \lim_{K \to \infty} \frac{n^{K-1} + (n-1)}{n^K} = \frac{1}{n}. \tag{36}$$

Thus, the competitive ratio of the DDRF mechanism is at most $1/n$. Therefore, the theorem holds. ∎

5   Discussions and Future Work

We have described a non-trivial polynomial-time algorithm to find a DDRF allocation, whose running time is linear in the number of present agents at every step, improving the result in [16]. We also analyzed the competitive ratios of the DDRF mechanism, which shows that the DDRF mechanism is a nearly optimal mechanism satisfying DPO for the maxmin objective.

Note that another fair allocation mechanism, called *cautious LP*, is proposed in [16]. Cautious LP achieves near optimal maxmin value at the last step. However, since cautious LP violates the DPO property and allocates too many resources at the last several steps, it is unfair to compare cautious LP with DDRF for the maxmin objective. It is interesting to analyze the competitive ratio of the cautious LP mechanism under different objectives. Since solving the linear program takes too much time, it is challenging to develop a combinatorial algorithm to find a cautious LP solution as in Sect. 3.

On the other hand, our results show that the DDRF mechanism is not good enough for the maxsum (or utilization maximization) objective. It is interesting to design other dynamic fair allocation mechanisms, which satisfy necessary fair properties and have higher welfare (i.e. the sum of dominant shares) or resource utilization in the worst-case scenario.

Acknowledgment. The work is supported in part by the National Natural Science Foundation of China [Nos. 61662088, 11301466], the Natural Science Foundation of Yunnan Province of China [No. 2014FB114], and IRTSTYN.

References

1. Annamalai, C., Kalaitzis, C., Svensson, O.: Combinatorial algorithm for restricted max-min fair allocation. In: Proceedings of the Twenty-Sixth Annual ACM-SIAM Symposium on Discrete Algorithms, pp. 1357–1372 (2015)
2. Asadpour, A., Saberi, A.: An approximation algorithm for max-min fair allocation of indivisible goods. SIAM J. Comput. **39**(7), 2970–2989 (2012)
3. Aumann, Y., Dombb, Y.: The efficiency of fair division with connected pieces. ACM Trans. Econ. Comput. **3**(4) (2015). Article No. 23
4. Bertsimas, D., Farias, V.F., Trichakis, N.: The price of fairness. Oper. Res. **59**(1), 17–31 (2011)
5. Bhattacharya, A.A., Culler, D., Friedman, E., Ghodsi, A., Shenker, S., Stoica, I.: Hierarchical scheduling for diverse datacenter workloads. In: Proceedings of the 4th Annual Symposium on Cloud Computing, SOCC 2013 (2013). Article No. 4
6. Bonald, T., Roberts, J.: Enhanced cluster computing performance through proportional fairness. Perform. Eval. **79**, 134–145 (2014)
7. Bonald, T., Roberts, J.: Multi-resource fairness: objectives, algorithms and performance. ACM SIGMETRICS Perform. Eval. Rev. **43**(1), 31–42 (2015)
8. Borodin, A., El-Yaniv, R.: Online Computation and Competitive Analysis. Cambridge University, Cambridge (1998)
9. Caragiannis, I., Kaklamanis, C., Kanellopoulos, P., Kyropoulou, M.: The efficiency of fair division. Theory Comput. Syst. **50**(4), 589–610 (2012)

10. Chowdhury, M., Liu, Z., Ghodsi, A., Stoica, I.: HUG: multi-resource fairness for correlated and elastic demands. In: Proceedings of the 13th USENIX Symposium on Networked Systems Design and Implementation (NSDI 2016), pp. 407–424 (2016)
11. Dolev, D., Feitelson, D.G., Halpern, J.Y., Kupferman, R., Linial, N.: No justified complaints: on fair sharing of multiple resources. In: Proceedings of the 3rd Innovations in Theoretical Computer Science Conference, pp. 68–75 (2012)
12. Friedman, E., Ghodsi, A., Psomas, C.-A.: Strategyproof allocation of discrete jobs on multiple machines. In: Proceedings of the Fifteenth ACM Conference on Economics and Computation, pp. 529–546 (2014)
13. Ghodsi, A., Zaharia, M., Hindman, B., Konwinski, A., Shenker, S., Stoica, I.: Dominant resource fairness: fair allocation of multiple resource types. In: Proceedings of the 8th USENIX Conference on Networked Systems Design and Implementation, pp. 24–37 (2011)
14. Gutman, A., Nisan, N.: Fair allocation without trade. In: Proceedings of the 11th International Conference on Autonomous Agents and Multiagent Systems, pp. 719–728 (2012)
15. Jin, Y., Hayashi, M.: Efficiency comparison between proportional fairness and dominant resource fairness with two different type resources. In: 2016 Annual Conference on Information Science and Systems (CISS), pp. 643–648 (2016)
16. Kash, I., Procaccia, A.D., Shah, N.: No agent left behind: dynamic fair division of multiple resources. J. Artif. Intell. Res. 51, 579–603 (2014)
17. Li, W., Liu, X., Zhang, X., Zhang, X.: Dynamic fair allocation of multiple resources with bounded number of tasks in cloud computing systems. Multiagent Grid Syst. Int. J. 11, 245–257 (2015)
18. Parkes, D.C., Procaccia, A.D., Shah, N.: Beyond dominant resource fairness: extensions, limitations, and indivisibilities. ACM Trans. Econ. Comput. 3(1) (2015). Article No. 3
19. Procaccia, A.D.: Cake cutting: not just child's play. Commun. ACM 56(7), 78–87 (2013)
20. Psomas, C.-A., Schwartz, J.: Beyond beyond dominant resource fairness: indivisible resource allocation in clusters. Technical report Berkeley (2013)
21. Wang, H., Varman, P.J.: Balancing fairness and efficiency in tiered storage systems with bottleneck-aware allocation. In: Proceedings of the 12th USENIX Conference on File and Storage Technologies, pp. 229–242 (2014)
22. Wang, W., Liang, B., Li, B.: Multi-resource fair allocation in heterogeneous cloud computing systems. IEEE Trans. Parallel Distrib. Syst. 26(10), 2822–2835 (2015)
23. Wang, W., Li, B., Liang, B., Li, J.: Towards multi-resource fair allocation with placement constraints. In: Proceedings of the 2016 ACM SIGMETRICS International Conference on Measurement and Modeling of Computer Science, pp. 415–416 (2016)
24. Wong, C.J., Sen, S., Lan, T., Chiang, M.: Multi-resource allocation: fairness efficiency tradeoffs in a unifying framework. IEEE/ACM Trans. Netw. 21(6), 1785–1798 (2013)
25. Zahedi, S.M., Lee, B.C.: REF: resource elasticity fairness with sharing incentives for multiprocessors. ACM SIGARCH Comput. Architect. News 42(1), 145–160 (2014)
26. Zarchy, D., Hay, D., Schapira, M.: Capturing resource tradeoffs in fair multi-resource allocation. In: 2015 IEEE Conference on Computer Communications (INFOCOM), pp. 1062–1070 (2015)

A 42k Kernel for the Complementary Maximal Strip Recovery Problem

Wenjun Li[1], Haiyan Liu[1], Jianxin Wang[2(✉)], Lingyun Xiang[1], and Yongjie Yang[2]

[1] Hunan Provincial Key Laboratory of Intelligent Processing of Big Data on Transportation, Changsha University of Science and Technology, Changsha, China
[2] School of Information Science and Engineering, Central South University, Changsha, China
jxwang@csu.edu.cn

Abstract. In the COMPLEMENTARY MAXIMAL STRIP RECOVERY problem (CMSR), we are given two strings S_1 and S_2 of distinct letters, where each letter appears either in the positive form or the negative form. The question is whether there are k letters whose deletion results in two matched strings. String S_1 matches string S_2 if there are partitions of S_1 and S_2, such that, for each component S_1^i of the partition of S_1, there is a unique component S_2^j in the partition of S_2 which is either equal to S_1^i or can be obtained from S_1^i by firstly reversing the order of the letters and then negating the letters. The CMSR problem is known to be NP-hard and fixed-parameter tractable with respect to k. In particular, a linear kernel of size $74k + 4$ was developed based on 8 reduction rules. Very recently, by imposing 3 new reduction rules to the previous kernelization, the linear kernel was improved to $58k$. We aim to simplify the kernelization, yet obtain an improved kernel. In particular, we study 7 reduction rules which lead to a linear kernel of size $42k + 24$.

1 Introduction

A major task in comparative genomic is to compare genetic maps of different organisms. In many cases, the genetic maps to be compared have noise and ambiguities. Hence, a preprocessing to eliminate these noise and ambiguities is carried out before the comparison. In this paper, we study the COMPLEMENTARY MAXIMAL STRIP RECOVERY problem (CMSR), which models the preprocessing scenario of deleting minimum number of gene markers of two genomes so that the remaining genomes can be decomposed into syntonic blocks. The CMSR problem and numerous related problems were proposed and studied in [4,17]

This work is supported by the National Natural Science Foundation of China (Grants No. 61672536, 61502054, 61420106009), the Natural Science Foundation of Hunan Province, China (Grant No. 2017JJ3333), the Scientific Research Fund of Hunan Provincial Education Department (Grant No. 17C0047), and the China Postdoctoral Science Foundation (Grant No. 2017M612584).

© Springer International Publishing AG 2017
M. Xiao and F. Rosamond (Eds.): FAW 2017, LNCS 10336, pp. 175–186, 2017.
DOI: 10.1007/978-3-319-59605-1_16

from the biological point of view, and since then have received a considerable amount of attention from biological and computer science communities [2,3,15,16]. Recently, the CMSR problem has been studied from the parameterized complexity point of view [5,7], which is also the main focus of this paper.

Recall that a parameterized problem is *fixed-parameter tractable* (FPT) if it can be solved in $O(f(k)|I|^{O(1)})$ time, where k is the parameter and $|I|$ is the size of the main part. A *kernelization* of a parameterized problem Q is an algorithm that transforms each instance (I, k) of Q in time $(|I| + k)^{O(1)}$ into an instance (I', k') of Q such that (1) $(I, k) \in Q$ if and only if $(I', k') \in Q$; (2) $k' \leq f(k)$ for some computable function f; and (3) $|I'| \leq g(k)$ for some computable function g. Here, the new instance (I', k') is called the *kernel* and $g(k)$ is the *size* of the kernel. Moreover, if g is a polynomial (linear) function of k, we say that (I', k') is a *polynomial kernel (linear kernel)* of the problem Q. We refer to [11] for further discussion on parameterized complexity.

Once a kernel of an FPT problem is developed, the next question which is of particular importance is to improve the quality of the kernel (see, e.g., [9,12, 14]). In this paper, we derive a kernel of size $42k + 24$ for the CMSR problem based on 7 reduction rules. The previous kernelization is built upon 11 reduction rules and yields a kernel of size $58k$ [5]. So, our kernelization outperforms the previous kernelization in [5] not only in terms of the size of the kernel but also in terms of simplicity. Before further discussion, let's give the formal definition of the problem. To not distract the attention, we describe the problem in a mathematic language, and refer to [4,17] and references therein for detailed biological background of the CMSR problem.

Problem Statement. Unless stated otherwise, all numerical data mentioned hereinafter are integers. Let $\mathcal{L}$ be a set of *letters*, each of which occurs either in the *positive form* or the *negative form*. In particular, for a letter $a \in \mathcal{L}$, $+a$ denotes its positive form and $-a$ denotes its negative form. When it is clear from the context we remove the plus symbol from "$+a$". Moreover, *negating a* gives us $-a$ and negating $-a$ gives us a.

A *string* is a sequence of distinct letters, each of which is either in its positive form or its negative form. We denote a sequence of elements by embracing its elements in a brace according to their relative positions in the sequence and separating them by the comma. For instance, a string consisting of "a_1", "$-a_3$", "a_2" in the order of their appearances is denoted by $(a_1, -a_3, a_2)$. In a sequence $(a_1, a_2, \ldots, a_n), a_1$ is the *first element* and a_n the *last element*. The *length* of a string is the number of letters in the string. For a string S, let $\mathcal{L}(S)$ be the set of letters appearing in S. For instance, for $\mathcal{L} = \{a_1, a_2, a_3, a_4\}$ and $S = (a_1, -a_3, a_2)$, we have that $\mathcal{L}(S) = \{a_1, a_2, a_3\}$. In addition, let $-\overleftarrow{S}$ be the string obtained from S by first reversing the order of the letters, and then negating each letter. For instance, for $S = (a_1, a_2, -a_3, a_4, -a_5)$, we have that $-\overleftarrow{S} = (a_5, -a_4, a_3, -a_2, -a_1)$. Moreover, for a subset $L \subseteq \mathcal{L}$ of letters, $S \backslash L$ is the string obtained from S by deleting all letters in L. For instance, if $S = (a_1, -a_2, -a_3, a_4, a_5)$ and $L = \{a_2, a_4\}, S \backslash L = (a_1, -a_3, a_5)$. A *substring* of a string is a sequence of consecutive letters in the string.

A $(t, 2_\geq)$-*partition* of a string S is a t-tuple $(S^1, S^2, \ldots, S^t)$ such that (1) each S^x where $1 \leq x \leq t$ is a substring of S; (2) each S^x where $1 \leq x \leq t$ consists of at least 2 letters; (3) $\mathcal{L}(S^x) \cap \mathcal{L}(S^y) = \emptyset$ for all $1 \leq x \neq y \leq t$; and (4) $\bigcup_{x \in [t]} \mathcal{L}(S^x) = \mathcal{L}(S)$, where $[t] = \{1, 2, \ldots, t\}$. Two (sub)strings S_1 and S_2 are *syntonic* if $S_1 = S_2$, or $S_2 = -\overleftarrow{S_1}$. A string S_1 *matches* another string S_2 if there is a $(t, 2_\geq)$-partition $(S_i^1, S_i^2, \ldots, S_i^t)$ of S_i for both $i = 1, 2$, where t is a positive integer, and an one-to-one mapping f from $\{S_1^1, S_1^2, \ldots, S_1^t\}$ to $\{S_2^1, S_2^2, \ldots, S_2^t\}$ such that for each $1 \leq x \leq t$, S_1^x and $f(S_1^x)$ are syntonic. Now we are ready to give the formal definition of the CMSR problem.

COMPLEMENTARY MAXIMAL STRIP RECOVERY (CMSR)

Input: Two strings S_1 and S_2 of length n such that $\mathcal{L}(S_1) = \mathcal{L}(S_2)$, an integer $0 \leq k \leq n - 2$.

Parameter: k.

Question: Is there an $L \subseteq \mathcal{L}(S_1)$ such that $|L| \leq k$ and $S_1 \backslash L$ matches $S_2 \backslash L$?

Due to space limitation, many proofs are deferred to the full version.

2 Related Work and Our Contribution

If we adopt $n - k$ as the parameter in the definition of the CMSR problem, we have the MAXIMAL STRIP RECOVERY Problem (MSR). Zheng et al. [17] set up the line of research on the MSR and CMSR problems. As a matter of fact, they studied the optimization version of the MSR problem, where the objective is to find an $L \subseteq \mathcal{L}(S_1)$ of maximum size such that $S_1 \backslash L'$ matches $S_2 \backslash L'$, where $L' = \mathcal{L}(S_1) \backslash L$. The optimization version of the CMSR problem is to find an L of minimum size such that $S_1 \backslash L$ matches $S_2 \backslash L$. Apparently, the optimization versions of the MSR problem and the CMSR problem are polynomial-time equivalent. In fact, both problems are NP-hard [13]. Concerning approximation algorithms, Jiang et al. [6] developed a factor-3 polynomial-time approximation algorithm for the optimization version of the CMSR problem, which was later improved to a factor-2.33 polynomial-time approximation algorithm by Lin et al. [10]. For the optimization version of the MSR problem, a factor-4 polynomial-time approximation algorithm was proposed by Chen et al. [3]. On the negative side, both problems were shown to be APX-hard [2, 8].

From the parameterized complexity point of view, the CMSR problem was shown to be FPT. In particular, Jiang et al. [6] devised an FPT-algorithm with running time $O(3^k n^2)$. Shortly later, the FPT-algorithm was improved to $O(2.36^k n^2)$ by Bulteau et al. [1]. Given these positive results, an intriguing question is whether the CMSR problem admits a polynomial kernel, or even a linear kernel. This line of research has been initiated by Jiang and Zhu [7], who crafted a linear kernel of size $74k + 4$ for the CMSR problem[1]. To obtain this linear kernel, Jiang and Zhu [7] devised 8 reduction rules. Following the work of Jiang and Zhu [7], Hu et al. [5] recently reported a kernelization which leads to

---

[1] In the paper [7], Jiang and Zhu claimed $74k + 4 = 78k$ as the kernel size.

an improved linear kernel of size $58k$. The kernelization of Hu et al. [5] relies on the previous 8 reduction rules studied in [7] together with 3 additional reduction rules.

In this paper, we further improve the kernelization of the CMSR problem. Instead of complicating the kernelization by imposing new reduction rules to deal with further specific structures into the previous kernelization, we simplify the kernelization by utilizing 7 reduction rules. Our reduction rules do not only cover all the previous reduction rules but also are able to deal with more new structures that are beyond the ability of the previous reduction rules studied in [5,7]. More importantly, our reduction rules result in a kernel of size $42k + 24$. Hence, for $k > 1$ the bound of our kernel improves the previous result $58k$. We believe that our simplification of the kernelization provides a better understanding of the CMSR problem, and shed some light on further studies on the CMSR problem.

Finally, we would like to point out that in a sharp contrast to the positive results for the CMSR problem discussed above, the investigation on the parameterized complexity of the MSR problem moves slowly—it is not even known whether the problem is FPT (A generalization of the MSR problem, where instead of two strings there are d strings in the input, has been studied by Bulteau et al. [1] from the parameterized complexity. In particular, they proved that the generalization is W[1]-hard for every $d \geq 4$).

3   Reduction Rules

In this section, we study the kernelization of the CMSR problem. Let $I = (S_1, S_2, k)$ be a given instance of the CMSR problem. A substring b of S_i where $i \in \{1, 2\}$ is a *block* if there is a substring b' of S_{3-i} such that b and b' are syntonic and, moreover, b is maximal under this condition. We denote $syn(b) = b'$ and analogously $syn(b') = b$, and call $\{b, syn(b)\}$ a *block pair*, or say b and b' are *paired blocks*. So, blocks must appear in pairs, and two blocks in a block pair are from different strings. An *isolator* is a block consisting of one letter. For convenience, we call a block of length i an *i-block*, and a block of length at least i an *$i_{\geq}$-block*. If $\{b, syn(b)\}$ is a block pair and b is an i-block, we call $\{b, syn(b)\}$ an *i-block pair*. A *super block* of S_i where $i = 1, 2$ is a maximal sequence of consecutive $2_{\geq}$-blocks in S_i. Consecutive blocks in a string are blocks without any other block between them in the string.

A *graph* G is a tuple (V, E) where V is the set of vertices and E the set of edges. We also use $V(G)$ and $E(G)$ to denote the vertex set and edge set respectively. An *edge* between two vertices v, u is denoted by (v, u). For a vertex v of G, let $N_G(v)$ be the set of neighbors of v in G, i.e., $N_G(v) = \{u \in V(G) \mid (v, u) \in E(G)\}$. For $B \subseteq V(G)$, let $N_G(B) = \bigcup_{v \in B} N(v) \backslash B$. By *merging* a set A of vertices, we mean to first create a new vertex v_A and create an edge between v_A and every vertex $u \in N_G(A)$, and then delete all vertices in A from the graph. Notice that merging a subset of vertices may result in multiple edges between two vertices. A *path* is a sequence $(v_0, v_1, \ldots, v_{t-1})$ of vertices such that there is an edge between v_i and v_{i+1} for every $i = 0, \ldots, t - 2$. If there is an edge between v_0 and v_{t-1} in a path $(v_0, v_1, \dots, v_{t-1})$, then we have a *cycle*. An *induced subgraph* of G is a graph with vertex set V' for some $V' \subseteq V(G)$ and edge set $\{(v, u) \mid v, u \in V', (v, u) \in E(G)\}$. A graph is *connected* if there is a path between every pair of vertices. A *connected component* is a maximal induced subgraph of G which is connected. A *tree* is a connected graph without cycles. A *forest* is a graph with each connected component being a tree.

The terms "block" and "super block" have been studied in [5,7] with the same meanings. In particular, the kernelizations in [5,7] create an auxiliary graph whose vertices correspond to super blocks. Moreover, there is an edge between two vertices if the corresponding super blocks contain a block pair. Then, based on the auxiliary graph, the kernelizations reduce the size of the super blocks by several reduction rules.

Our kernelization follows the recipe of that studied in [5,7], but with more fine-grained ingredients. First, same as the kernelizations in [5,7], we first find in polynomial-time all $2_{\geq}$-blocks and super blocks. Then, we create an auxiliary graph G as follows. For each $2_{\geq}$-block pair $(b, syn(b))$ we create two vertices $v(b)$ and $v(syn(b))$ correspondingly, and create an edge between $v(b)$ and $v(syn(b))$. In addition, we connect the vertices corresponding to blocks in each super block one by one in a path, according to the relative positions of the blocks in the super block, i.e., if b and b' are two consecutive blocks in the super block, then there is an edge between $v(b)$ and $v(b')$. Hence, our auxiliary graph is different from the one used in [5,7]. See Fig. 1 for an illustration. Finally, based on the auxiliary graph G, we utilize 7 reduction rules to shrink the instance.

In our reduction rules, we will mark some blocks. Informally speaking, we mark a block as we know that there is an optimal solution excluding the letters of the block. Note that it is possible for the letters of a marked block to be included in some optimal solution, but the point is that there exists at least one optimal solution that does not include any letter of the marked block. Hence, marking blocks are just auxiliary operations and not part of the problem definition.

A reduction rule is *sound* if each application of the reduction rule does not affect the answer to the instance. An instance is *reduced* by a set of reduction rules if no reduction rule in the set applies to the instance. We assume that before the i-th reduction rule is introduced where $i \geq 2$, the instance is reduced by the j-th reduction rule for every $1 \leq j < i$. The following lemma is due to [6].

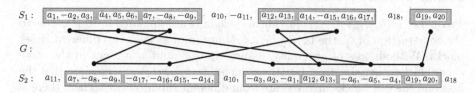

Fig. 1. An illustration of blocks, super blocks and the auxiliary graph G. Each $2_{\geq}$-block is bounded in a white-background rectangle, and each super block is bounded in a gray-background rectangle.

Lemma 1. *There exists an optimal solution L_{opt} such that*

1. *for every $2_\geq$-block b either $L_{opt} \cap \mathcal{L}(b) = \emptyset$ or $\mathcal{L}(b) \subseteq L_{opt}$; and*
2. *for every $4_\geq$-block b, it holds that $L_{opt} \cap \mathcal{L}(b) = \emptyset$.*

By and large, the above lemma states that every instance has a special optimal solution such that each $2_\geq$-block is either completely disjoint with the solution or completely included in the solution and, moreover, every $4_\geq$-block is completely disjoint from the solution.

Now we describe the reduction rules. The first 3 reduction rules are to identify a number of $2_\geq$-blocks that are completely disjoint with a certain optimal solution. If such a block is identified, we *mark* it. Moreover, *if we mark a block b, we simultaneously mark $syn(b)$ as well.* Lemma 1 suggests that we can mark all $4_\geq$-blocks.

Rule 1. *Mark all $4_\geq$-blocks.*

Rule 1 has been studied in [7] (Rule 2.1 in [7]). The second reduction rule marks $2_\geq$-blocks corresponding to vertices forming a cycle in the graph G.

Rule 2. *If there is a cycle in G with a vertex whose corresponding block is unmarked, then mark all unmarked blocks corresponding to the vertices in the cycle.*

To prove the soundness of Rule 2, we need the following lemma. For a subset B of blocks, let $\mathcal{L}(B) = \bigcup_{b \in B} \mathcal{L}(b)$.

Lemma 2. *Let $L \subseteq \mathcal{L}(S_1)$ be a subset of letters such that $S_1 \backslash L$ matches $S_2 \backslash L$. Let B be a subset of $2_\geq$-blocks in S_1 and S_2 such that $\mathcal{L}(B) \subseteq L$. Let I be the set of isolators if we bring back all blocks in B and their paired blocks into $S_1 \backslash L$ and $S_2 \backslash L$ correspondingly. Then, there exists an $L' \subseteq L$ such that $S_1 \backslash L'$ matches $S_2 \backslash L'$ and $|L'| = |L| - |\mathcal{L}(B)| + |\mathcal{L}(I)|$.*

Now, we are ready to prove the soundness of Rule 2. For a sequence of consecutive blocks $P = (b_1, \ldots, b_t)$, let $first(P)$ and $last(P)$ denote the first and the last blocks in P, respectively, i.e., $first(P) = b_1$ and $last(P) = b_t$.

Lemma 3. *Rule 2 is sound.*

Proof. Let $(v(b_1), v(b_2), \ldots, v(b_t))$ be a cycle in G as stipulated in Rule 2. Due to the construction of G, the blocks $b_1, \ldots, b_t$ are from at least two different super blocks. Without loss of generality, assume that b_1 and b_t are not in the same super block (otherwise, we rotate the sequence $(v(b_1), v(b_2), \ldots, v(b_t))$ so that the first and the last blocks involved are from different super blocks). Then, there is a partition $(P_0 = (b_1, \ldots, b_x), P_1 = (b_{x+1}, \ldots, b_y), \ldots, P_{\ell-1} = (b_z, b_{z+1}, \ldots, b_t))$ of $(b_1, \ldots, b_t)$ such that for every $i = 0, 1, \ldots, \ell - 1$ (1) all blocks in P_i are from the same super block; (2) $last(P_{i \bmod \ell}) = syn(first(P_{(i+1) \bmod \ell}))$; and (3) P_i consists of at least two blocks. Let B be the set of all blocks appearing in the first or last positions in some P_i, i.e., $B = \bigcup_{0 \leq i \leq \ell-1} \{last(P_i), first(P_i)\}$.

Clearly, $|B|$ is even. More precisely, according to the above discussion, if $b \in B$, then $syn(b) \in B$. We first show that we can mark each block b in B. Let L_{opt} be an optimal solution that does not contain any letters of already marked blocks. Let B' be the set of blocks $b \in B$ such that $\mathcal{L}(b) \subseteq L_{opt}$. Clearly, if $b \in B'$, then $syn(b) \in B'$. As a result, $|B'|$ is even. Let d be the number of P_is where $0 \le i \le \ell - 1$ such that $\{first(P_i), last(P_i)\} \subseteq B'$, and e the number of P_is such that $|\{first(P_i), last(P_i)\} \cap B'| = 1$. If we bring back all blocks in B' into $S_1 \backslash L_{opt}$ and $S_2 \backslash L_{opt}$ correspondingly, then there will result in at most $2d + e$ new 1-block pairs, 2 for each P_i such that $\{first(P_i), last(P_i)\} \subseteq B'$, and 1 for each P_i such that $|\{first(P_i), last(P_i)\} \cap B'| = 1$. Since $|\mathcal{L}(B')| \ge 2 \cdot \frac{|B'|}{2} = |B'| = 2d + e$, due to Lemma 2, we can get another optimal solution $L_{opt'}$ obtained from L_{opt} by deleting all letters of blocks in B' and adding all letters in the 1-block pairs, arising after bringing back all blocks in B' into $S_1 \backslash L_{opt}$ and $S_2 \backslash L_{opt}$. Clearly, $L_{opt'} \cap \mathcal{L}(B') = \emptyset$ and $L_{opt'}$ does not contain any letters of already marked blocks.

Now consider other blocks in each P_i. Let b be an intermediate block in P_i. Due to symmetry, assume that b is from S_1. If $\mathcal{L}(b) \in L'_{opt}$, then bringing the block b back into $S_1 \backslash L'_{opt}$ does not result in any isolator. Bringing back $syn(b)$ in $S_2 \backslash L'_{opt}$, however, may result in at most two 1-block pairs. Then due to Lemma 2, $L_{opt''} = L_{opt'} \backslash \mathcal{L}(B) \cup I$ is another optimal solution that does not contain any letters of already marked blocks, where I is the set of letters in the isolators arising after bringing back b and $syn(b)$ into the strings. By iteratively applying the above argument, we can achieve at an optimal solution that does not include any letter in $\mathcal{L}(B)$ and all already marked blocks. As a consequence, we can mark all blocks in B. The lemma follows.   □

Now we study another reduction rule which marks all blocks involved on a path between two vertices.

Rule 3. *If there is a path $(v(b_1), v(b_2), \ldots, v(b_n))$ in G such that*
1. *$n \ge 3$;*
2. *for each $i \in \{1, n\}$, b_i is either a marked block or a 3-block;*
3. *every b_i, $2 \le i \le n - 1$, is unmarked,*
then mark all blocks $b_1, \ldots, b_n$.

Lemma 4. *Rule 3 is sound.*

Previous reduction rules only mark blocks but never shrinks the size of the instance. Now we study reduction rules to shrink the size of the marked blocks. For a $5_{\ge}$-block pair $\{b, syn(b)\}$ where $b = (a_1, a_2, \ldots, a_t)$, by *contracting* $\{b, syn(b)\}$ we mean to reset b as (a_1, a_2, a_3, a_t), and reset $syn(b)$ as (a_1, a_2, a_3, a_t) if $syn(b) = (a_1, a_2, \ldots, a_t)$ in advance; and reset $syn(b) = (-a_t, -a_3, -a_2, -a_1)$ otherwise. What's important in the contraction operation is to remain the rightmost and leftmost letters of b and $syn(b)$, and keep the length of b and $syn(b)$ to be 4. A marked block b is a *boundary marked block* if b is either the leftmost or the rightmost block of a maximal sequence of consecutive marked blocks.

Rule 4. *Let b be a marked block. If b is a $5_{\ge}$-block, contract $\{b, syn(b)\}$; and if both b and $syn(b)$ are non-boundary marked blocks, delete $\mathcal{L}(b)$.*

Lemma 5. *Rule 4 is sound.*

For two sequences $A = (a_1, a_2, \ldots, a_x)$ and $B = (b_1, b_2, \ldots, b_y)$, let (A, B) be the concatenation of A and B, i.e., $(A, B) = (a_1, \ldots, a_x, b_1, \ldots, b_y)$. We further develop several reduction rules to reduce marked blocks. When we create a new block pair $\{\bar{b}, syn(\bar{b})\}$ we mean that the letters of $\bar{b}$ (and $syn(\bar{b})$) are disjoint with the letters of all existing blocks in the current instance. To utilize these reduction rules, we first create two new 4-block pairs $(b_1^*, syn(b_1^*))$ and $(b_2^*, syn(b_2^*))$ and concatenate them with S_1 and S_2 in a certain way. In particular, we reset $S_1 := (S_1, (b_1^*, syn(b_2^*)))$ and $S_2 := (S_2, (b_2^*, syn(b_1^*)))$. Clearly, this does not change the answer to the instance. Due to Rule 1, all these newly created blocks are marked.

Rule 5. *If there is a boundary marked block b in S_i lying before b_i^* for $i \in \{1, 2\}$ such that $syn(b)$ is a non-boundary marked block lying before b_{3-i}^*, then put $syn(b)$ after the last block in S_{3-i}.*

See Fig. 2 for an illustration of Rule 5.

Lemma 6. *Rule 5 is sound.*

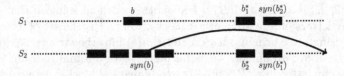

Fig. 2. An illustration of Reduction Rule 5. Black blocks are marked blocks.

Note that after an exhaustive application of Rule 5 (and other rules described above), every maximal sequence of marked block before b_i^*, $i = 1, 2$, consists of at most 2 blocks. Moreover, all non-boundary marked blocks lie after b_1^* or b_2^*.

Rule 6. *If there are two consecutive boundary marked blocks b, b' in S_i lying before b_i^* for $i \in \{1, 2\}$ such that $syn(b)$ and $syn(b')$ lie after b_{3-i}^* then,*

- *create a new 4-block pair $\{\bar{b}, syn(\bar{b})\}$, put $\bar{b}$ immediately after b', and put $syn(\bar{b})$ after the last block in S_{3-i};*
- *mark $\bar{b}$ and $syn(\bar{b})$; and*
- *delete the letters of b and b' in the instance.*

See Fig. 3 for an illustration of Rule 6.

Lemma 7. *Rule 6 is sound.*

Proof. Let b, b' and $\bar{b}$ be as stipulated in Rule 6, and I' the instance obtained from I after applying Rule 6. Observe that in I, b and b' do not form new blocks with other blocks after deleting any letters not in the marked blocks. Moreover, $\bar{b}$ does not form new blocks with other blocks after deleting any letters not in the marked blocks in I'. Hence, any optimal solution of I which is complete disjoint with marked blocks is a solution of I' which is complete disjoint with marked blocks, and vice versa. □

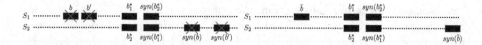

Fig. 3. An illustration of Rule 6. Black blocks are marked blocks. The left-hand instance is the original instance, and the right-hand one is the instance after the reduction rule.

For a block b, let $\overrightarrow{\mathcal{L}(b)}$ be the order of letters in $\mathcal{L}(b)$ according to their relative orders in b. For instance, if $b = (-x, y - z, w)$ then $\overrightarrow{\mathcal{L}(b)} = (x, y, z, w)$. For each marked block b, we define a 3-tuple $\mathbf{x}_b \in \{0, 1\} \times \mathcal{L} \cup \{\bot\} \times \mathcal{L} \cup \{\bot\}$. The first component indicates whether b is a boundary marked block. In particular, $\mathbf{x}_b[1] = 1$ means b is a boundary marked block; and $\mathbf{x}_b[1] = 0$ means b is a nonboundary marked block. The second and third components are determined as follows. A boundary marked block is a *left (resp. right) boundary marked block* if it is the leftmost (resp. rightmost) block of a maximal sequence of marked blocks. If b is not a left boundary marked block, then $\mathbf{x}_b[2] = \bot$; otherwise $\mathbf{x}_b[2]$ is the first letter in $\overrightarrow{\mathcal{L}(b)}$. Moreover, if b is not a right boundary marked block, then $\mathbf{x}_b[3] = \bot$; otherwise $\mathbf{x}_b[3]$ is the last letter in $\overrightarrow{\mathcal{L}(b)}$. For instance, if the block $b = (a_4, a_5, a_6)$ of S_1 in Fig. 1 is marked, then $\mathbf{x}_b = (1, a_4, a_6)$ if both $(a_1, -a_2, a_3)$ and $(a_7, -a_8, -a_9)$ are unmarked, and $\mathbf{x}_b = (1, a_4, \bot)$ if $(a_1, -a_2, a_3)$ is unmarked and $(a_7, -a_8, -a_9)$ is marked. According to the definition, for each marked block b if $\mathbf{x}_b[1] = 0$, then $\mathbf{x}_b[2] = \mathbf{x}_b[3] = \bot$. Now we introduce the last reduction rule. We call b_1^* and b_2^* *dividing blocks*.

Rule 7. *If there are two consecutive marked blocks b^1 and b^2 in S_i lying before b_i^* where $i \in \{1, 2\}$ such that*

1. *$syn(b^j)$ is a boundary marked block, i.e., $\mathbf{x}_{syn(b^j)}[1] = 1$, where $j \in \{1, 2\}$;*
2. *there is another boundary marked block b' lying consecutively with $syn(b^j)$ in S_{3-i};*
3. *at least one of $\{syn(b'), syn(b^{3-j})\}$ lies after the dividing blocks, then,*

- *If $\{\mathbf{x}_{b^j}[2], \mathbf{x}_{b^j}[3]\} \cap \{\mathbf{x}_{syn(b^j)}[2], \mathbf{x}_{syn(b^j)}[3]\} \setminus \{\bot\} \neq \emptyset$, then, delete the letters of exactly one of $\{syn(b'), syn(b^{3-j})\}$ which lies after the dividing blocks (if both lie after the dividing blocks, choose arbitrarily one).*
- *If $\{\mathbf{x}_{b^j}[2], \mathbf{x}_{b^j}[3]\} \cap \{\mathbf{x}_{syn(b^j)}[2], \mathbf{x}_{syn(b^j)}[3]\} \setminus \{\bot\} = \emptyset$, then, we create two new 4-block pairs $(\bar{b}, syn(\bar{b}))$ and $(\ddot{b}, syn(\ddot{b}))$, mark these new blocks, put $\bar{b}$ between b^j and b^{3-j} in S_i, put $syn(\bar{b})$ after the last block in S_{3-i}, put $\ddot{b}$ between b' and $syn(b^j)$ in S_{3-i}, and put $syn(\ddot{b})$ after the last block in S_i. Then, if $syn(b^{3-j})$ lies after the dividing blocks, delete the letters of b^j and b^{3-j}; otherwise, delete the letters of b^j and b'.*

Lemma 8. *Rule 7 is sound.*

4   Analysis of the Kernel

Let $I = (S_1, S_2, k)$ be an instance such that none of the reduction rules studied in the previous section applies. Let G be the auxiliary graph corresponding to

I as discussed in the previous section. We analyze the size of I. To this end, we create a graph G', which is obtained from G by first deleting all edges between two vertices $v(b), v(syn(b))$ such that b is a marked block, and then for each super block merging all vertices corresponding to blocks in the super block. Hence, edges in G' one-to-one correspond to block pairs $\{b, syn(b)\}$ where b is an unmarked $2_\geq$-block. For a vertex sv in G', let $supblock(sv)$ be the super block corresponding to sv, and $blocks(sv)$ the set of all blocks in $supblock(sv)$. We first show that G' is a forest. In fact, if this was not the case, then there will be unmarked blocks in a cycle in G, which contradicts with the fact that the instance is reduced by Rule 2.

Lemma 9. *The graph G' is a forest.*

Let $H_1, H_2, \ldots, H_m$ be the connected components of G', and q_i the number of edges in each H_i. Due to Lemma 9, each H_i is a tree. The following lemma was studied in [7] (see proofs of Lemmas 7, 8 and Theorem 1 in [7]). As our reduction rules cover all of the ones studied in [7], the lemma holds here.

Lemma 10. $m \leq 4k - \sum_{i=1}^{m} q_i + 2.$

Now we are ready to analyze the size of I, i.e., $2|\mathcal{L}(S_1)|$. Assume that I is a YES-instance and $L \subseteq \mathcal{L}(S_1)$ is a solution of I. We define four sets A, B, C, D as follows: A consists of all blocks b such that $\mathcal{L}(b) \subseteq L$; B consists of all isolators except the ones in A; C consists of all marked blocks; and D consists of all unmarked $2_\geq$-blocks. Clearly, $A \cup B \cup C \cup D$ is the set of all blocks in S_1 and S_2 except $b_1^*, syn(b_1^*), b_2^*, syn(b_2^*)$.

Obviously, $|\mathcal{L}(S_1)| \leq |\mathcal{L}(A)| + |\mathcal{L}(B)| + |\mathcal{L}(C)| + |\mathcal{L}(D)| + 8$ (notice that it may be that $\mathcal{L}(A) \cap \mathcal{L}(D) \neq \emptyset$). The correctness of the following lemma is independent of the reduction rules, and has been studied in [7].

Lemma 11. $|\mathcal{L}(A)| + |\mathcal{L}(B)| \leq 5k.$

Now we investigate the size of C and D. For $V' \subseteq V(G')$, let $supblock(V') = \{supblock(sv) \mid sv \in V'\}$, and $blocks(V') = \bigcup_{sv \in V'} blocks(sv)$.

Lemma 12. *If for some H_i, where $1 \leq i \leq m$, there is a vertex $sv \in V(H_i)$ such that $C \cap blocks(sv) \neq \emptyset$, then all blocks in $blocks(V(H_i)) \setminus (C \cap blocks(sv))$ are unmarked 2-blocks.*

Lemma 13. *If for some H_i, where $1 \leq i \leq m$, all blocks in $blocks(V(H_i))$ are unmarked and there is a 3-block b in $blocks(V(H_i))$, then all blocks in $blocks(V(H_i)) \setminus \{b, syn(b)\}$ are 2-blocks.*

Now we are ready to investigate the size of C and D.

Lemma 14. $|\mathcal{L}(C)| + |\mathcal{L}(D)| \leq 16k + 8.$

Proof. For each H_i, $i \in \{1, 2, \ldots, m\}$, let C_i and D_i be the sets of marked and unmarked $2_\geq$-blocks in $blocks(V(H_i))$ lying before b_1^* or b_2^*, respectively. Notice that only marked blocks lie after b_i^* where $i \in \{1, 2\}$. Moreover, if a marked block b lies after b_i^* where $i \in \{1, 2\}$, then $syn(b)$ must lie before b_{3-i}^*. Hence, $|\mathcal{L}(C)| + |\mathcal{L}(D)| = \sum_{i=1}^{m} |\mathcal{L}(C_i)| + |\mathcal{L}(D_i)|$. For each H_i where $i \in \{1, 2, \ldots, m\}$, let X_i be the set of all super blocks in subblocks($V(H_i)$) containing some marked block, and let $x_i = |X_i|$. Due to Rule 4, each marked block is of size at most 4. Due to Rule 3, each super block in X_i has exactly one maximal sequence of consecutive marked blocks. Moreover, due to Rules 5–7, each C_i consists of no more than two blocks. Furthermore, if some C_i consists of two blocks b, b', then there is a $\bar{b} \in \{b, b'\}$ such that both $\bar{b}$ and $syn(\bar{b})$ are boundary marked blocks. It then follows that

$$|\mathcal{L}(C)| = \sum_{i=1}^{m} |\mathcal{L}(C_i)| \leq \sum_{i=1}^{m} 4x_i \tag{1}$$

Due to Lemma 12, it holds that $x_i \in \{0, 1\}$ for every $1 \leq i \leq m$. Moreover, if $x_i = 1$ for some $1 \leq i \leq m$, each edge of H_i corresponds to an unmarked 2-block pair. On the other hand, if $x_i = 0$, due to Lemma 13 at most one edge of H_i corresponds to an unmarked 3-block pair, and every other edge corresponds to an unmarked 2-block pair. Due to Lemma 9, each H_i is a tree. Then, as each unmarked $2_{\geq}$-block pair corresponds to an edge in G', we have that

$$|\mathcal{L}(D)| = \sum_{i=1}^{m} 2q_i + (1 - x_i) \tag{2}$$

In summary, we have

$$
\begin{aligned}
|\mathcal{L}(C)| + |\mathcal{L}(D)| &= \sum_{i=1}^{m} (|\mathcal{L}(C_i)| + |\mathcal{L}(D_i)|) \\
&\leq \sum_{i=1}^{m} (4x_i + 2q_i + 1 - x_i) && \text{due to (1) and (2)} \\
&= 2 \sum_{i=1}^{m} q_i + m + 3 \sum_{i=1}^{m} x_i \leq 2 \sum_{i=1}^{m} q_i + 4m && x_i \in \{0, 1\} \\
&\leq 2 \sum_{i=1}^{m} q_i + 4(4k + 2 - \sum_{i=1}^{m} q_i) && \text{Lemma 10} \\
&= 16k + 8 - 2 \sum_{i=1}^{m} q_i \leq 16k + 8
\end{aligned}
$$

$\square$

Theorem 1. *The CMSR problem admits a kernel of size $42k + 24$.*

Proof. Let $I = (S_1, S_2, k)$ be a given instance. The kernelization first creates the auxiliary graph G as illustrated in Sect. 3. Then, based on the graph G, the reduction rules are used until none of them is applicable. Due to the soundness of the reduction rules (see Lemmas 3–8), the original instance is a YES-instance if and only if the reduced instance is a YES-instance. Moreover, due to Lemmas 11 and 14, if the reduced instance is a YES-instance, then each of S_1 and S_2 without $\{b_1^*, b_2^*, syn(b_1^*), syn(b_2^*)\}$ consists of at most $|\mathcal{L}(A)| + |\mathcal{L}(B)| + |\mathcal{L}(C)| + |\mathcal{L}(D)| \leq 5k + 16k + 8 = 21k + 8$ letters. Due to this, if the reduced instance has at most $2(21k + 8) + 8 = 42k + 24$ letters, we return the reduced instance; otherwise, we return a trivial NO-instance. The last number 8 here is $|\mathcal{L}(\{b_1^*, b_2^*\})|$.   □

References

1. Bulteau, L., Fertin, G., Jiang, M., Rusu, I.: Tractability and approximability of maximal strip recovery. Theor. Comput. Sci. **440–441**, 14–28 (2012)
2. Bulteau, L., Fertin, G., Rusu, I.: Maximal strip recovery problem with gaps: hardness and approximation algorithms. J. Discret. Algorithms **19**, 1–22 (2013)
3. Chen, Z., Fu, B., Jiang, M., Zhu, B.: On recovering syntenic blocks from comparative maps. J. Comb. Optim. **18**(3), 307–318 (2009)
4. Choi, V., Zheng, C., Zhu, Q., Sankoff, D.: Algorithms for the extraction of synteny blocks from comparative maps. In: Giancarlo, R., Hannenhalli, S. (eds.) WABI 2007. LNCS, vol. 4645, pp. 277–288. Springer, Heidelberg (2007). doi:10.1007/978-3-540-74126-8_26
5. Hu, S., Li, W., Wang, J.: An improved kernel for the complementary maximal strip recovery problem. In: Xu, D., Du, D., Du, D. (eds.) COCOON 2015. LNCS, vol. 9198, pp. 601–608. Springer, Cham (2015). doi:10.1007/978-3-319-21398-9_47
6. Jiang, H., Li, Z., Lin, G., Wang, L., Zhu, B.: Exact and approximation algorithms for the complementary maximal strip recovery problem. J. Comb. Optim. **23**(4), 493–506 (2012)
7. Jiang, H., Zhu, B.: A linear kernel for the complementary maximal strip recovery problem. J. Comput. Syst. Sci. **80**(7), 1350–1358 (2014)
8. Jiang, M.: Inapproximability of maximal strip recovery. Theor. Comput. Sci. **412**(29), 3759–3774 (2011)
9. Kowalik, L., Pilipczuk, M., Suchan, K.: Towards optimal kernel for connected vertex cover in planar graphs. Discret. Appl. Math. **161**(7–8), 1154–1161 (2013)
10. Lin, G., Goebel, R., Li, Z., Wang, L.: An improved approximation algorithm for the complementary maximal strip recovery problem. J. Comput. Syst. Sci. **78**(3), 720–730 (2012)
11. Niedermeier, R.: Invitation to Fixed-Parameter Algorithms. Oxford University Press Inc., Oxford (2006)
12. Wang, J., Yang, Y., Guo, J., Chen, J.: Planar graph vertex partition for linear problem kernels. J. Comput. Syst. Sci. **79**(5), 609–621 (2013)
13. Wang, L., Zhu, B.: On the tractability of maximal strip recovery. J. Comput. Biol. **17**(7), 907–914 (2010)
14. Yang, Y.: Towards optimal kernel for edge-disjoint triangle packing. Inf. Process. Lett. **114**(7), 344–348 (2014)
15. Zheng, C.: Pathgroups, a dynamic data structure for genome reconstruction problems. Bioinformatics **26**(13), 1587–1594 (2010)
16. Zheng, C., Zhu, Q., Adam, Z., Sankoff, D.: Guided genome halving: hardness, heuristics and the history of the hemiascomycetes. In: ISMB, pp. 96–104 (2008)
17. Zheng, C., Zhu, Q., Sankoff, D.: Removing noise and ambiguities from comparative maps in rearrangement analysis. IEEE/ACM Trans. Comput. Biol. Bioinform. **4**(4), 515–522 (2007)

On-line Scheduling with a Monotonous Subsequence Constraint

Kelin Luo[1(✉)], Yinfeng Xu[1,2], Huili Zhang[1], and Wei Luo[3]

[1] School of Management, Xi'an Jiaotong University, Xi'an 710049, China
{luokelin,zhang.huilims}@stu.xjtu.edu.cn, yfxu@xjtu.edu.cn
[2] The State Key Lab for Manufacturing Systems Engineering,
Xi'an 710049, China
[3] Department of Geography, Santa Barbara, University of California,
Berkeley, USA
wei.luo@ucsb.edu

Abstract. In this paper, we study a new on-line scheduling problem that each server has to process a monotonous request subsequence. The customer requests are released over-list, and the operator has to decide whether or not to accept the current request and arrange it to a server immediately. The goal of this paper is to find a strategy which accepts the maximal requests. When the number of servers k is less than that of the request types m, we give several lower bounds for this problem. Also, we present the optimal strategy for $k = 1$ and $k = 2$ respectively.

Keywords: Scheduling · On-line algorithm · Competitive analysis · Monotonous subsequence

1 Introduction

In recent years, "Sharing Economy" is making traveling easier and cheaper [1]. To thumb a lift to destination may hinge on luck, Uber COMMUTE meets your needs for sure. People can take some passengers with the same destination or on the way to the destinations. How to schedule requests to the potential drivers efficiently is a tough problem, especially in peak hours. In this paper, we use "monotonous subsequence [2]" to represent the "path with a fixed destination" feature. By labeling all the possible requests as 1 to m, and the destination as 1 or m based on various factors, like request and driver locations, as shown in Fig. 1. Each driver (server) has two optional paths, which can be identified as a monotonous subsequence (an increasing or decreasing subsequence) in the request set, as shown in Fig. 2.

The problem that we solve is this: Suppose there are k servers and the request r_j are released over-list and non-preemptively. Given information of all possible request labels ($r_j \in \{1, 2, \ldots m\}$), we seek to find a strategy, along with a monotonous subsequence constraint, that maximizes the total accepted requests. Denote this problem as *On-line scheduling problem with a monotonous subsequence constraint* $(k - MS)$.

© Springer International Publishing AG 2017
M. Xiao and F. Rosamond (Eds.): FAW 2017, LNCS 10336, pp. 187–195, 2017.
DOI: 10.1007/978-3-319-59605-1_17

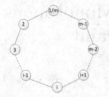
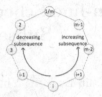

Fig. 1. A request set   **Fig. 2.** An example of subsequence

This problem is close to the longest monotonous subsequence problem [6]. A decision maker observes a subsequence of independent variables $\{X_1, X_2, \ldots, X_n\}$ with distribution F (n is unknown in advance), and the task is to select a subsequence $\{X_{r1}, X_{r2}, \ldots, X_{rn}\}$ such that $X_{r1} \leqslant X_{r2} \leqslant X_{rn}$ or $X_{r1} \geqslant X_{r2} \geqslant X_{rn}$. In early year, researchers [3,4] studied the off-line longest monotonous circular subsequence problem based on an exist integer set. Albert et al. [5] came up with the on-line version problem and analyzed the expected situation. In this paper, we investigate this problem from competitive analysis aspect and apply it to practical matters. Our approach allows to select more than one monotonous subsequence. We thereby answer this problem by referring to the on-line scheduling problem [7–10].

The rest of this paper is organized as follows. Section 2 introduces the $k-MS$ problem and some preliminaries. In Sect. 3, we develop several lower bounds for the $k-MS$ problems. A description of the algorithm with competitive analysis follows in Sect. 4. Final results and remarks are given in Sect. 5.

2   Model

We consider a k servers scheduling problem, with each server serving an increasing or a decreasing request subsequence. For example, $\{1, 3, 4, 7\}$ is an increasing subsequence, $\{9, 5, 2\}$ is a deceasing subsequence. Denote server s_i, $i \in K$ and $K = \{1, 2, \ldots k\}$. The request r_j are released over-list, denote $r_j \in M$ and $M = \{1, 2, \ldots m\}$. Define variable x_{ij} as follows:

$$x_{ij} = \begin{cases} 1, & \text{if for request } r_j, \text{ we arrange it to server } s_i, \\ 0, & \text{otherwise.} \end{cases}$$

The $k-MS$ problem is formulated as follows:

$$Max \quad \sum_{i=1}^{k} \sum_{j=1}^{m} x_{ij} \tag{2.1}$$

$$s.t \quad r_{j_1} > r_{j_2}, \quad \forall j_1, j_2, j_1 > j_2, \overrightarrow{s_i} = +, x_{ij_1} = x_{ij_2} = 1 \quad i \in K. \tag{2.2}$$

$$r_{j_1} < r_{j_2}, \quad \forall j_1, j_2, j_1 > j_2, \overrightarrow{s_i} = -, x_{ij_1} = x_{ij_2} = 1 \quad i \in K. \tag{2.3}$$

$$\sum_{i=1}^{k} x_{ij} \in \{0, 1\}, \quad j \in M. \tag{2.4}$$

$$x_{ij} \in \{0, 1\}. \tag{2.5}$$

The objective, (2.1), of the $k - MS$ problem is to maximize the number of requests accepted in the schedule. Constraint (2.2) and Constraint (2.3) ensure that each server serves either an increasing request subsequence or a decreasing request subsequence. $\overrightarrow{s_i} = +$ represents the server serves an increasing subsequence, and $\overrightarrow{s_i} = -$ represents the server serves a decreasing subsequence. Integer restrictions are imposed in constraint (2.4) and (2.5), meaning that either a request is accepted in the schedule or it is rejected.

The performance is measured by competitive ratio (see [11]). Translated into our problem terminology, for any input request sequence R, let $A(R)$ be the objective value produced by an on-line algorithm A, and $OPT(R)$ be that obtained by one of the optimal scheduler OPT who has the request information in advance. The competitive ratio of A is define as $c_A = sup_R \frac{OPT(R)}{A(R)}$. Let ON be the set of all the on-line strategies. The lower bound β is defined as $\beta = inf_{A \in ON} c_A$. We say A is an optimal on-line strategy if $c_A = \beta$.

3   Lower Bound for k-MS Problem

In this section we discuss the lower bound of $k - MS$ problem when $m > k > 2$, $k = 1$ and $k = 2$, respectively. For the ease of expression, we denote ALG as any one of the on-line algorithm and OPT as one of the optimal scheduler.

Theorem 1. *For $k - MS$ problem, there exists no on-line algorithm that is $\frac{3}{2}$-competitive for all $m > k > 2$.*

Proof. Consider OPT releases the following 1st sequence of at most $2k$ requests: $\underbrace{\{1, m, 1, m, \ldots 1, m\}}_{2k}$. According to ALG response, we can classify this problem into two cases.

Case 1. ALG accepts all these requests on k servers, OPT will release and accept 2nd sequence $\underbrace{\{1, 2, \ldots m; \ldots; 1, 2, \ldots m\}}_{k \cdot m}$, yielding a ratio on this sequence of $\frac{OPT(R)}{A(R)} = \frac{k \cdot m}{2k} > \frac{m}{2}$.

Case 2. ALG rejects some requests of the 1st sequence, saying accepts k_{sum} ($\frac{k_{sum}}{2} \le k$) requests in all, which k_1 ($k_1 \le k$) servers accepts $\{1, m\}$, k_2 ($k_2 \le k$) servers accepts $\{1\}$ and k_3 ($k_3 \le k$) servers accepts $\{m\}$.

Case 2.1 $k_{sum} \le k$. OPT will not release requests any more and accepts the 1st sequence, yielding a ratio on this sequence of $\frac{OPT(R)}{A(R)} = \frac{2k}{k_{sum}} \ge 2$.

Case 2.2 $k_{sum} > k$. OPT will release the 3rd sequence. If $k_2 \ge k_3$, the 3rd sequence is $\underbrace{\{m, m, \ldots, m; m-1, m-1, \ldots, m-1; \ldots; 1, 1, \ldots, 1\}}_{k \cdot m}$. If $k_2 < k_3$, the 3rd sequence is $\underbrace{\{1, 1, \ldots, 1; 2, 2, \ldots, 2; \ldots; m, m, \ldots, m\}}_{k \cdot m}$. OPT accepts the requests $\{1, 2, \ldots m\}$ on each of the servers, yielding a ratio on this sequence of $\frac{OPT(R)}{A(R)} = \frac{k \cdot m}{2k_1 + 2max\{k_2, k_3\} + m \cdot (k - k_1 - max\{k_2, k_3\})}$. The ratio of $\frac{OPT(R)}{A(R)}$ is no greater than $\frac{3}{2}$ (Further analysis of the ratio in Appendix A).

Thus, $c \geq \frac{3}{2}$ and the theorem follows.   □

3.1   Unique Server

In this part, we exhibit the lower bound of $k - MS$ problem when $k = 1$.

Theorem 2. *For $k - MS$ problem, there exists no on-line algorithm with competitive ratio less than $\frac{m}{2}$ for $k = 1$.*

Proof. Consider OPT releases the 1st request $\lceil \frac{m}{2} \rceil$. According to ALG response, we can classify this problem into two cases.

Case 1. ALG rejects this request, OPT will not release any more, yielding a ratio on this sequence of $\frac{OPT(R)}{A(R)} = \infty$.

Case 2. ALG accepts this request, OPT then releases the following 1st sequence of at most $\lceil \frac{m}{2} \rceil$ requests: $\{1, 2, \ldots, \lceil \frac{m}{2} \rceil\}$.

Case 2.1. ALG rejects all the 1st sequence requests, OPT will not release any more. OPT accepts all the 1st sequence requests, yielding a ratio on this sequence of $\frac{OPT(R)}{A(R)} = \lceil \frac{m}{2} \rceil$.

Case 2.2. Once ALG accepts a request in the 1st sequence, OPT will release the rest requests of the 1st sequence sequentially, and then release the 2nd sequence of at most $\lfloor \frac{m}{2} \rfloor$ requests: $\{\lceil \frac{m}{2} \rceil + 1, \lceil \frac{m}{2} \rceil + 2, \ldots, m\}$. OPT accepts all the m requests, yielding a ratio on this sequence of $\frac{OPT(R)}{A(R)} = \frac{m}{2}$.

Thus, The lower bound of $1 - OIS$ problem is $\frac{m}{2}$.   □

3.2   Two Servers

In this part, we exhibit the lower bound of $k - MS$ problem when $k = 2$.

Theorem 3. *For $k - MS$ problem, there exists no on-line algorithm with competitive ratio less than $\frac{m}{3}$ for $k = 2$.*

Proof. Consider OPT releases a request $\lceil \frac{m}{2} \rceil$. According to ALG, we can classify this problem into two cases.

Case 1. ALG rejects $\lceil \frac{m}{2} \rceil$, OPT will not release any more, yielding a ratio on this sequence of $\frac{OPT(R)}{A(R)} = \infty$.

Case 2. ALG accepts $\lceil \frac{m}{2} \rceil$. Without loss of generality, suppose this request is accepted by server s_1, and let r_{11} represent this request. OPT then releases the following 1st sequence of at most $2 \cdot \lceil \frac{m}{3} \rceil$ requests: $\{1, 1, 2, 2, \ldots, \lceil \frac{m}{3} \rceil, \lceil \frac{m}{3} \rceil\}$.

Case 2.1. ALG rejects all the 1st sequence requests, OPT will not release any more. OPT accepts all the 1st sequence requests, yielding a ratio on this sequence of $\frac{OPT(R)}{A(R)} = 2 \cdot \lceil \frac{m}{3} \rceil$.

Case 2.2. ALG accepts a request of the 1st sequence by server s_1 , let r_{12} denote this request. OPT will release the rest requests of the 1st sequence sequentially, and then release the 2nd sequence of at most $2 \cdot \lfloor \frac{2m}{3} \rfloor$ requests: $\{\lceil \frac{m}{3} \rceil + 1, \lceil \frac{m}{3} \rceil + 1, \lceil \frac{m}{3} \rceil + 2, \lceil \frac{m}{3} \rceil + 2, \ldots, m, m\}$.

Case 2.2.1. ALG rejects all the rest of the 1st sequence and the 2nd sequence. OPT accepts all the $2m$ requests (1st sequence and 2nd sequence), yielding a ratio on this sequence of $\frac{OPT(R)}{A(R)} = \frac{2m}{2} = m$.

Case 2.2.2. Once ALG accepts one more request, this request must be accepted by server s_2 because this request is greater than r_{12} and $\vec{s_1} = -$. Without loss of generality, denote the third request as r_{21}.

(i) If $r_{21} \geq \lceil \frac{m}{2} \rceil$, OPT will not release requests any more. OPT accepts the requests $\{1, 2, \ldots, r_{21}\}$ on each of the server, yielding a ratio on this sequence of $\frac{OPT(R)}{A(R)} = \frac{2r_{21}}{3} \geq \frac{m}{3}$.

(ii) If $r_{21} < \lceil \frac{m}{2} \rceil$ ($r_{21} \geq r_{12}$), OPT will not release the above sequence, and release the 3rd sequence: $\{m, m, m-1, m-1, \ldots, r_{21}, r_{21}\}$.

ALG rejects all request of the 3rd sequence, OPT accepts all request of the 3rd sequence, yielding a ratio on this sequence of $\frac{OPT(R)}{A(R)} = \frac{2(m-r_{21}+1)}{3} > \frac{m}{3}$.

ALG accepts a request of the 3rd sequence, which denoted as r_{22}, OPT will release the rest requests of the 3rd sequence sequentially, and then release the 4th sequence of at most $\lfloor \frac{m}{2} \rfloor$ requests: $\{r_{21} - 1, r_{21} - 1, r_{21} - 2, r_{21} - 2, \ldots, r_{12}, r_{12}\}$. ALG could not accept the request of 4th sequence any more because the last request of the server s_1 is r_{12} and $\vec{s_1} = -$, and the last request of the server s_2 is r_{22} and $\vec{s_1} = +$. OPT accepts all the $2 \cdot (m - r_{12} + 1)$ requests, yielding a ratio on this sequence of $\frac{OPT(R)}{A(R)} = \frac{2 \cdot (m - r_{12} + 1)}{4} = \frac{m - r_{12} + 1}{2} > \frac{2/3m}{2} = \frac{m}{3}$.

Case 2.3. ALG accepts a request of the 1st sequence by server s_2, let r_{21} denote this request. OPT will not release the above 1st sequence any more, and then release the 2nd sequence of at most $2 \cdot \lceil \frac{m}{3} \rceil$ requests: $\{m, m, m-1, m-1, \ldots, m - \lfloor \frac{m}{3} \rfloor, m - \lfloor \frac{m}{3} \rfloor\}$.

Case 2.3.1. ALG rejects all the 2nd sequence, OPT will not release request any more and accept all the 2^{nrd} sequence, yielding a ratio on this sequence of $\frac{OPT(R)}{A(R)} = \frac{2 \cdot \lceil \frac{m}{3} \rceil}{2} = \lceil \frac{m}{3} \rceil$.

Case 2.3.2. ALG accepts the third request, there are 2 cases.

(i) The third request is accepted by server s_1, which denoted as r_{12}. OPT will release the rest request of the 2nd sequence sequentially, and then release the 3rd sequence requests: $\{m - \lfloor \frac{m}{3} \rfloor - 1, m - \lfloor \frac{m}{3} \rfloor - 1, m - \lfloor \frac{m}{3} \rfloor - 2, m - \lfloor \frac{m}{3} \rfloor - 2, \ldots, r_{21}, r_{21}\}$.

ALG rejects all the above sequence (the rest request of the 2nd sequence and the 3rd sequence), OPT will not release request any more and accept $\{m, m-1, \ldots, r_{12}\}$ on each of the two servers, yielding a ratio on this sequence of $\frac{OPT(R)}{A(R)} = \frac{2(m-r_{21})}{3} > \frac{2 \cdot 2/3 \cdot m}{3} = \frac{4m}{9}$.

ALG accepts one of the requests, it must be accepted by server s_2 and $\vec{s_2} = +$, because the last request of the server s_1 is r_{12} and $\vec{s_1} = +$. OPT will release the rest of the above sequence and the on-line algorithm could not accept any more request because the last request of the server s_1 is r_{12} and $\overrightarrow{s_1} = +$, and the last request of the server s_2 is r_{22} and $\overrightarrow{s_1} = +$. OPT accepts all the $2(m - r_{21})$ requests, yielding a ratio on this sequence of $\frac{OPT(R)}{A(R)} = \frac{2(m-r_{21})}{4} = \frac{m}{3}$.

(ii) The third request is accepted by server s_2, which denoted as r_{22}. OPT will not release the 2nd sequence any more, and then release the 3rd sequence of at most $2 \cdot \frac{m}{2}$ requests: $\{1, 1, 2, 2, \ldots, \frac{m}{2}, \frac{m}{2}\}$.

ALG rejects all the above sequence since then, OPT will not release request any more and accept $\{1, 2, \ldots, \frac{m}{2}\}$ on each of the two servers, yielding a ratio on this sequence of $\frac{OPT(R)}{A(R)} = \frac{2 \cdot \frac{m}{2}}{3} > \frac{m}{3}$.

ALG accepts one of the request, the fourth request must be accepted by server s_1 and $\overrightarrow{s_2} = -$, because the last request of the server s_2 is r_{22} and $\overrightarrow{s_1} = +$. OPT will release the rest of the 3rd sequence and then release the 4th sequence $\{\lceil\frac{m}{2}\rceil + 1, \lceil\frac{m}{2}\rceil + 1, \lceil\frac{m}{2}\rceil + 2, \lceil\frac{m}{2}\rceil + 2, \ldots, r_{22}, r_{22}\}$. The on-line algorithm could not accept the request of 4th sequence any more because the last request of the server s_1 is r_{12} and $\overrightarrow{s_1} = -$, and the last request of the server s_2 is r_{22} and $\overrightarrow{s_1} = +$. OPT accepts $2r_{22}$ requests $\{1, 2, \ldots, r_{22}\}$ on each of the two servers, yielding a ratio on this sequence of $\frac{OPT(R)}{A(R)} \geq \frac{2 \cdot 2/3 \cdot m}{4} = \frac{m}{3}$.

Thus, The lower bound of $2 - OIS$ problem is $\frac{m}{3}$.   □

4   On-line Scheduling by K-Interval Algorithm

In the k-interval algorithm $(k - IA)$, we create two types of intervals denoted I_i and FI_i $(i \in K)$. Let S_i represent all request accepted by s_i. r_{ij} is the jth request of S_i. I_i represents the acceptable request interval of s_i when $|S_i| < 2$ $(a = \frac{m}{2(k+1)})$. FI_i represents the feasible request interval of s_i when $|S_i| \geq 2$.

$$I_i = \begin{cases} [r_{i,1}, r_{i,1} + (m - 2a)], & r_{i,1} \leq a \\ [a, m - a], & a < r_{i,1} < m - a \\ [r_{i,1} - (m - 2a), r_{i,1}], & r_{i,1} \geq m - a \end{cases} \qquad FI_i = \begin{cases} [r_{i,|S_i|}, m], & \overrightarrow{s_i} = +, \\ [1, r_{i,|S_i|}], & \overrightarrow{s_i} = -. \end{cases}$$

Algorithm 1. k-Interval Algorithm

Input: k (k=1 or 2) identical servers, m types of possible requests, $r_j \in M$;

Initialize: Set $i = 1$, $j = 1$, schedule r_j to s_i, go to Step 3.

Step 1: If $i < k$, let $i = i + 1$ and schedule r_j to s_i, goto Step 3.
　　　　Otherwise, go to Step 2.

Step 2: If $\exists i^* \in K_i, |S_{i^*}| \geq 2$, and $r_j \in FI_{i^*}$, schedule r_j to s_{i^*}, go to Step 3.
　　　　If $\exists i^* \in K_i, |S_{i^*}| < 2$, and $r_j \in I_{i^*}$, schedule r_j to s_{i^*}, go to Step 3.
　　　　Otherwise, go to Step 3.

Step 3: If all the requests have already been arranged, the game terminates;
　　　　otherwise, j=j+1, go to Step 1.

Note 1: $K_i = \{s_1, s_2, \ldots s_i\}$.

Observation. The k-Interval Algorithm 1 $(k = 1)$ $1 - IA$ has a performance guarantee of $\frac{m}{2}$ for one server with m types of requests.

Note that when $k = 1$, any non-trivial algorithm will take in at least 2 integers from the given sequence, and therefore is optimal.

Theorem 4. *The k-Interval Algorithm 1 (k = 2) $2 - IA$ has a performance guarantee of $\frac{m}{3}$ for 2 servers and m $(m > 2)$ types of requests.*

Proof. For all schedules, $2 - IA$ has a performance guarantee of $\frac{m}{3}$. There are six cases based on the schedule produced by $2 - IA$.

Case 1: $|S_1| = 0$, $|S_2| = 1$. The schedule produced by $2 - IA$ includes one request ($\sum_{i \in K} |S_i| = 1$). According to the Step 1 of $2 - IA$, because $2 - IA$ accepts two requests without constraint, so $|M| = 1$, and OPT at most accepts one request, too. Therefore, yielding a ratio on this sequence of $\frac{OPT(R)}{A(R)} = \frac{1}{1} = 1 \leq \frac{1}{3}m$.

Case 2: $|S_1| = 1$, $|S_2| = 1$. Each server accepts one request, respectively. According to the Step 2 of $2 - IA$, s_i rejects all the requests outside of I_i, so OPT at most accepts $2a$ requests on each of servers. Therefore, $OPT(R) \leq \frac{2}{3}m$ and $A(R) \geq 2$, yielding a ratio on this sequence of $\frac{OPT(R)}{A(R)} \leq \frac{2 \cdot 2 \cdot a}{2} = \frac{1}{3}m$ $(a = \frac{m}{2(k+1)} = \frac{1}{6}m)$.

Case 3: $|S_1| = 1$, $|S_2| >= 2$. One server accepts one request, and the other one accepts more than 2 requests. According to the Step 2 of $2 - IA$, OPT at most accepts $m - 2a$ requests on server s_2 and at most accepts $2a$ requests on server s_1. Therefore, $OPT(R) \leq m$ and $A(R) \geq 3$, yielding a ratio on this sequence of $\frac{OPT(R)}{A(R)} \leq \frac{m}{3} = \frac{1}{3}m$.

Case 4: $|S_1| = 2$, $|S_2| = 2$. Each server accepts two requests, respectively. According to the Step 2 of $2 - IA$, OPT at most accepts $m - 2a$ requests on each of servers. Therefore, $OPT(R) \leq 2(m - 2a)$ and $A(R) = 4$, yielding a ratio on this sequence of $\frac{OPT(R)}{A(R)} \leq \frac{2(m-2a)}{4} = \frac{1}{3}m$.

Case 5: $|S_1| = 2$, $|S_2| >= 3$. One server accepts 2 requests, and the other one accepts more than 2 requests. According to the Step 2 of $2 - IA$, OPT at most accepts $m - a$ requests on each of servers. Therefore, $OPT(R) \leq 2(m - a)$ and $A(R) \geq 5$, yielding a ratio on this sequence of $\frac{OPT(R)}{A(R)} \leq \frac{2(m-a)}{5} = \frac{1}{3}m$.

Case 6: $3 <= |S_1| <= |S_2|$. According to the Step 2 of $2 - IA$, each server accepts requests within its request interval, so OPT at most accepts $2m$ requests on each of servers. Therefore, $OPT(R) \leq 2m$ and $A(R) \geq 6$, yielding a ratio on this sequence of $\frac{OPT(R)}{A(R)} \leq \frac{2m}{6} = \frac{1}{3}m$.

Using Case 1-6 we derive $\frac{OPT(R)}{A(R)} \leq \frac{1}{3}m$. □

Both of $1 - IA$ and $2 - IA$ are optimal because the lower bound of the problem is equal to the algorithm competitive ratio.

5   Conclusions

Our solutions of $1 - OMS$ and $2 - OMS$ are optimal in expectation. This result is useful to scheduling idle drivers if their destinations are fixed, but their traveling are flexible. The optimal algorithm of $k-OMS$ $(k < m)$ remains open. Another, interesting future research is to analyze the relationship of k and m for most of the requests. Moreover, it would be interesting to study the problem when k fluctuates over time.

Acknowledgement. This work was partially supported by the NSFC (Grant No. 71601152), and by the China Postdoctoral Science Foundation (Grant No. 2016M592811).

A   Appendix

The analysis of ratio $\frac{OPT(R)}{A(R)}$ in Theorem 1. Case 2.2.

As we already know, $A(R) = 2k_1 + 2max\{k_2, k_3\} + m \cdot (k - k_1 - max\{k_2, k_3\})$ and $OPT(R) = k \cdot m$ for all $m > k > 2$, where $k_2 + k_3 + 2k_1 = k_{sum}$ ($k < k_{sum} \leq 2k$). To make the ratio of $\frac{OPT(R)}{A(R)}$ easier to solve, we analyse

$$\frac{A(R)}{OPT(R)} = \frac{2k_1 + 2max\{k_2, k_3\} + m \cdot (k - k_1 - max\{k_2, k_3\})}{k \cdot m} \quad (5.1)$$

$$= \frac{k_1 \cdot (2 - m) + max\{k_2, k_3\} \cdot (2 - m) + m \cdot k}{k \cdot m}$$

$$= 1 + C$$

$$C = \frac{k_1 \cdot (2 - m) + max\{k_2, k_3\} \cdot (2 - m)}{k \cdot m}. \quad (5.2)$$

We analyse the change of C with respect to k_1, $max\{k_2, k_3\}$ and $k_1 + max\{k_2, k_3\}$, respectively.

$$\frac{\partial C}{\partial k_1} = \frac{\partial C}{\partial max\{k_2, k_3\}} = \frac{\partial C}{\partial k_1 + max\{k_2, k_3\}} = \frac{(2 - m) \cdot k \cdot m}{(k \cdot m)^2} \quad (5.3)$$

$$< 0 \text{ (Because } m > k > 2).$$

Because $k_2 + k_3 + 2k_1 = k_{sum}$ ($k < k_{sum} \leq 2k$) and 5.3, we have:

$$C \leq \frac{\frac{k+1}{2} \cdot (2 - m)}{m \cdot k} = -\frac{1}{3} \text{ (Because } m > k > 2),$$

and

$$\frac{OPT(R)}{A(R)} \geq \frac{1}{1 - \frac{1}{3}} \quad (5.4)$$

$$\geq \frac{3}{2} \text{ (Because } m > k > 2),$$

References

1. Zervas, G., Proserpio, D., Byers, J.: The rise of the sharing economy: estimating the impact of Airbnb on the hotel industry. 18 November 2016. Boston U. School of Management Research Paper No. 2013–2016
2. Fredman, M.L.: On computing the length of longest increasing subsequences. Discret. Math. **11**(1), 29–35 (1975)
3. Deorowicz, S.: An algorithm for solving the longest increasing circular subsequence problem. Inf. Process. Lett. **109**(12), 630–634 (2009)
4. Romik, D.: The Surprising Mathematics of Longest Increasing Subsequences. Cambridge University Press, Cambridge (2014)
5. Albert, M.H., Golynski, A., Hamel, A.M., et al.: Longest increasing subsequences in sliding windows. Theoret. Comput. Sci. **321**(2–3), 405–414 (2004)
6. Arlotto, A., Nguyen, V.V., Steele, J.M.: Optimal online selection of a monotone subsequence: a central limit theorem. Stochast. Process. Appl. **125**(9), 3596–3622 (2014)
7. Nagarajan, V., Sviridenko, M.: Tight bounds for permutation flow shop scheduling. Math. Oper. Res. **34**(2), 417–427 (2009)
8. Sitters, R.: Competitive analysis of preemptive single-machine scheduling. Oper. Res. Lett. **38**(6), 585–588 (2010)
9. Nther, E., Maurer, O., Megow, N., et al.: A new approach to online scheduling: approximating the optimal competitive ratio. In: Twenty-Fourth ACM-SIAM Symposium on Discrete Algorithms, pp. 118–128. Society for Industrial and Applied Mathematics (2012)
10. Karhi, S., Shabtay, D.: On the optimality of the TLS algorithm for solving the online-list scheduling problem with two job types on a set of multipurpose machines. J. Comb. Optim. **26**(1), 198–222 (2013)
11. Borodin, A., El-Yaniv, R.: Online Computation and Competitive Analysis. Cambridge University Press, Cambridge (1998)

A 1.4-Approximation Algorithm for Two-Sided Scaffold Filling

Jingjing Ma, Haitao Jiang, Daming Zhu[✉], and Shu Zhang

School of Computer Science and Technology, Shandong University,
Jinan, People's Republic of China
majingjing.sdu@gmail.com, {htjiang,dmzhu}@sdu.edu.cn,
zhangshu365@163.com

Abstract. Scaffold Filling aims at getting what can be used as whole genomes from scaffolds by computation. Two-Sided Scaffold Filling is given by two scaffolds, asks respectively, to fill one scaffold with those genes in the other but the scaffold itself, so that the two new produced scaffolds have as many as possible common adjacencies. This problem has long been learnt to be NP-Hard and can be approximated to a constant performance ratio. In this paper, we devise an approximation algorithm which can achieve a performance ratio $1.4 + \varepsilon$. This improves upon the so far best approximation algorithm proposed by Liu et al.

1 Introduction

With the genomes being published in scaffold or contig forms increasingly [1], people often run into scaffolds or contigs in their genomic analysis related practice. The use of draft genomes often makes analyses and interpretations tentative and prone to error, and leads to particular problems. Thus the scaffold filling problem is motivated by extracting whole genomes from scaffolds via computation [5].

Muñoz *et al.* pioneered to investigate the One-Sided Scaffold Filling of signed permutations, and devised a polynomial time algorithm to minimize the DCJ distance [5]. Subsequently, Jiang and Zhu proposed a polynomial time algorithm for Two-Sided Scaffold Filling of signed permutations under the objective of minimizing the DCJ distance.

Many genome similarity measures can be used as an optimization objective in filling scaffolds with duplicated genes, where the *breakpoint distance* as well as the (*common*) *adjacency number* seems the most basic. Although the breakpoint distance or the adjacency number is computationally easy for genomes with duplicated genes, it has been shown that One-Sided Scaffold Filling is NP-Hard under breakpoint distance [2], which means NP-Hard under adjacency number. Two-Sided Scaffold Filling under breakpoint distance or adjacency number is NP-Hard consequently.

Scaffold filling for genomes with duplicated genes admits approximation algorithms with constant performance ratios, if adjacency number is used as the

© Springer International Publishing AG 2017
M. Xiao and F. Rosamond (Eds.): FAW 2017, LNCS 10336, pp. 196–208, 2017.
DOI: 10.1007/978-3-319-59605-1_18

maximization objective. For One-Sided Scaffold Filling, Jiang *et al.* designed a greedy 1.33-approximation algorithm [3,4]. Liu *et al.* then improved the performance ratio to $\frac{5}{4}$ by the so called local improvement followed by greedy [6]. Recently, Ma *et al.* presented a $\frac{6}{5}$-approximation local search algorithm for One-Sided Scaffold Filling [10]. For Two-Sided Scaffold Filling, Liu *et al.* proposed an approximation algorithm which can achieve a performance ratio 1.5 [9], which is the best as we know by now.

In this paper, we present an approximation algorithm for Two-sided Scaffold Filling with performance ratio $1.4+\varepsilon$. This improves upon the 1.5-approximation algorithm proposed by Liu *et al.*. Section 2 will present on those notations of the two-sided scaffold filling problem formally. In Sect. 3, we bound on how many adjacencies scaffold filling can increase firstly. Then by finding an approximated maximum independent set in a 5-claw free graph and a 7-claw free graph, we propose the algorithm which can carry out the two-sided scaffold filling in polynomial time. Section 5 will show that the performance ratio of the algorithm is $1.4+\varepsilon$.

2   Preliminaries

Let Σ be an alphabet in which each element represents a gene family. A *scaffold* on Σ is a gene *sequence*, where a gene is an occurrence of a gene family in Σ. In the absence of exception, we will represent a gene with the same symbol as with which we represent the gene's family. Let $S = s_1 s_2 \cdots s_n$ be a scaffold on Σ, then all genes in S form a multi-set which will be denoted as $c(S)$. A substring in S with m genes is referred as an *m-string*. A 2-string in S is particularized as a *pair*. Let $P[S]$ be the multi-set of pairs in S. Then $P[S] = \{s_1 s_2, s_2 s_3, \ldots, s_{n-1} s_n\}$. A pair is *in* S, if and only if it is in $P[S]$.

Let $A = a_1 a_2 \cdots a_n$ and $B = b_1 b_2 \cdots b_m$ be two scaffolds on Σ, $P[A]$ and $P[B]$ the sets of pairs in A and B respectively, $a_i a_{i+1} \in P[A]$, $b_j b_{j+1} \in P[B]$. If $a_i a_{i+1} = b_j b_{j+1}$ (or $b_{j+1} b_j$), then $a_i a_{i+1}$ and $b_j b_{j+1}$ form a *match* $(a_i a_{i+1}, b_j b_{j+1})$ between $P[A]$ and $P[B]$. Let $\mathcal{R}$ be the set of all matches between $P[A]$ and $P[B]$. Then a subset $R \subseteq \mathcal{R}$ is referred to as a *maximum matching* between $P[A]$ and $P[B]$ (or A and B) if, (1) any pair in A or B does not occur in two or more matches in R; (2) the cardinality of R is maximized over all subsets of $\mathcal{R}$ which subject to (1).

According to the maximum matching between $P[A]$ and $P[B]$ (resp. A and B), an *adjacency* as well as a *breakpoint* in A or B can be identified by,

Definition 1. *Let R be a maximum matching between $P[A]$ and $P[B]$. A pair in A (resp. B) is an* adjacency *relative to B (resp. A) with respect to R, if it forms a match in R with a pair in B (resp. A), otherwise, a* breakpoint.

No matter with respect to which maximum matching those pairs in A (resp. B) have been identified as adjacencies or breakpoints, A must have as many adjacencies (resp. breakpoints) as B has. So in the situation without any need of stressing, we will mention an adjacency or a breakpoint in A or B

regardless of the maximum matching. We denote by $a(A, B)$ the set of adjacencies in A relative to B. Note that there must be as many adjacencies in $a(A, B)$ as the matches in a maximum matching between A and B.

An *insertion* of a gene into a scaffold refers to the operation to insert the gene between two genes of a pair, onto the left side of the first or the right side of the last gene in the scaffold. Let $S = s_1 \ldots s_i \, s_{i+1} \ldots s_n$ be a scaffold on Σ and x a gene with its family in Σ. Then inserting x between s_i and s_{i+1} transforms S into $S' = s_1 \ldots s_i \, x \, s_{i+1} \ldots s_n, 1 \le i \le n - 1$, inserting x onto the left (resp. right) side of s_1 (resp. s_n) transforms S into $S' = x \, s_1 \ldots s_n$ (resp. $S' = s_1 \ldots s_n \, x$). For a gene x with family in Σ, let $S + x \equiv S + \{x\}$ denote the set of scaffolds resulted from inserting x into S. For a gene set X, let $S + X$ denote the set of scaffolds resulted from inserting all those genes in X into S, which can be formulated recursively as,

$$
S + X = \begin{cases} S, & X = \emptyset; \\ \displaystyle\bigcup_{S' \in S+x} (S' + (X - \{x\})), & x \in X. \end{cases} \tag{1}
$$

We also refer to the operation to insert a gene x into a scaffold S as an insertion to *fill S with x*.

The scaffold filling problem is originated from Muñoz *et al.* [5], who suggested to get a seemingly no-gene-missing genome by filling a scaffold with those genes it has missed, where the adjacency number has been most commonly used to measure how alike two scaffolds are, as well as how optimal the resulting scaffolds are. Two-Sided Scaffold Filling has first been proposed in [9], which is given by two scaffolds, asks to fill each one with the genes in the other but it, such that the number of adjacencies between the two resulting scaffolds is maximized.

Instance: Two scaffolds A and B on Σ.

Objective: Find $A' \in A + (c(B) - c(A)), B' \in B + (c(A) - c(B))$ such that $|a(A', B')| - |a(A, B)|$ is maximized.

A gene is *missing* if it is in $c(B) - c(A)$ or $c(A) - c(B)$. Following [9], the first and the last gene of A (resp. B) both be #, which is different from any other one in A (resp. B). Thus we only accept those insertions each of which happens to insert missing genes between two genes of a pair in A or B.

3   An Adjacency Number Bound in the Optimal Solution

Let A, B be two scaffolds on Σ which are given in a two-sided scaffold filling instance, $X = c(A) - c(B), Y = c(B) - c(A)$.

Let $X' \subseteq X, Y' \subseteq Y, A' \in A + Y', B' \in B + X'$. A substring of A' (resp. B') is *missing*, if all its genes are missing. A missing substring (or string) of A' (resp. B') is *maximal*, if either of the gene on the left side of its first gene and the gene on the right side of its last gene is not missing. A maximal missing substring of A' (resp. B') is abbreviated as an MMS, and a k-MMS, if it has k genes.

Let I be an arbitrary MMS of A' (resp. B'), A'' (resp. B'') the scaffold resulted from removing those genes in I from A' (resp. B'). Usually, it suffices to treat I as one made by one insertion to fill A'' (resp. B''). Thus, the insertion to fill A'' (resp. B'') with I *increases* k *adjacencies* between A'' (B'') and B' (A'), if $|a(A', B')| - |a(A'', B')| = k$ (resp. $\lceil a(A', B')\rceil - |a(A', B'')| = k$). We relax to say the insertion to fill A'' with I increases k adjacencies between A and B, provided it increases k adjacencies between A'' and B'.

A substring *involves* a pair, if it shares a gene with the pair. A k-MMS of A (resp. B) can involve at most $k + 1$ adjacencies. Thus it can increase at most $k + 1$ adjacencies between A and B to fill A (resp. B) with a k-string. In [9], a polynomial time algorithm has been proposed with guarantee as follows,

Lemma 1. *For any two scaffolds A and B on Σ, it can find string insertions in polynomial time to fill A and B with genes in $c(B) - c(A)$ and $c(A) - c(B)$ respectively, such that each insertion can fill the scaffold with a missing gene string to increase at least as many adjacencies as those genes in the string.*

Let M' be the maximum matching with respect to which the adjacencies in A' and B' have been identified. A pair in A' (resp. B') is *boundary* for an MMS, if it shares exactly one gene with the MMS. A match in M' is *boundary*, if both of its pairs in A' and B' are boundary.

To evaluate on at most how many adjacencies A' and B' can have, we take two MMSs into consideration together, if they involve two pairs forming a boundary match in M'. Thus we set a bipartite graph $\mathfrak{G}(M') = (V_1, V_2, E)$ with $V_1 = \{u \mid u$ corresponds to an MMS in $A'\}$, $V_2 = \{v \mid v$ corresponds to an MMS in $B'\}$, $E = \{(u, v) \mid$ those two MMSs corresponding to u and v involve two pairs forming a boundary match in $M'\}$.

Since an MMS can share pairs with at most two boundary matches, a vertex in $\mathfrak{G}(M')$ is incident with at most two edges. Thus a connected component in $\mathfrak{G}(M')$, must be a path, which can take the form of an isolated vertex, a path with two 1-degree vertices, or a cycle. In what follows, a *path* (resp. *vertex*, *cycle*) *component* in $\mathfrak{G}(M')$ refers to a connected component which is a path in $\mathfrak{G}(M')$. For an arbitrary vertex v in $\mathfrak{G}(M')$, we denote by $mms(v)$ that MMS corresponding to which v has been set. We assign a weight to a vertex v by $|mms(v)|$, which is the gene number of $mms(v)$. Let P be a path component in $\mathfrak{G}(M')$. Then we denote by $mms(P)$ the set of MMSs corresponding to which the vertices on P have been set. Moreover, the weight of P will be mentioned as $|mms(P)| = \sum_{v \in V(P)} |mms(v)|$. A path component of weight i is abbreviated as an i-path for $i > 0$.

Let $A^* \in A + Y, B^* \in B + X$ with $|a(A^*, B^*)| = \max \{|a(A', B')| : A' \in A + Y, B' \in B + X\}$, M^* the maximum matching between $P[A^*]$ and $P[B^*]$ with respect to which those pairs in A^* and B^* have been identified as adjacencies or breakpoints. Then those vertices on a path component in $\mathfrak{G}(M^*)$ must correspond to a set of MMSs which can involve as many adjacencies as,

Lemma 2. *Let P be an i-path in $\mathfrak{G}(M^*)$, removing from A^*, B^* those strings in $mms(P)$ transform A^* and B^* into A^{**} and B^{**}. Then $|mms(P)| \leq |a(A^*, B^*)| - |a(A^{**}, B^{**})| \leq |mms(P)| + 1$.*

Proof. The inequality $|mms(P)| \leq |a(A^*, B^*)| - |a(A^{**}, B^{**})|$ holds because otherwise, by Lemma 1, one can find insertions to fill A^{**} and B^{**} with those genes in the strings in $mms(P)$ to increase at least $|mms(P)|$ adjacencies between A and B. Since P is a path component in $\mathfrak{G}(M^*)$, those MMSs in $mms(P)$ can involve at most two boundary pairs which do not happen to any boundary matches in M^*. Thus if each pair involved by a string in $mms(P)$ is an adjacency with respect to M^*, there are at most $|mms(P)| + 1$ less adjacencies between A^{**} and B^{**} than between A^* and B^*. This leads to $|a(A^*, B^*)| - |a(A^{**}, B^{**})| \leq |mms(P)| + 1$.

A path component P in $\mathfrak{G}(M')$ is *good*, if removing from A' and B' all those MMSs in $mms(P)$ transforms A' and B' into A'' and B'' respectively, such that $|a(A', B')| - |a(A'', B'')| = |mms(P)| + 1$.

Let $PC^*[k]$ be the set of good k-paths in $\mathfrak{G}(M^*)$ for $k \geq 1$. Then the number of matches in M^* can be accounted by,

Lemma 3

$$|M^*| = |a(A, B)| + |X| + |Y| + \sum_{k=1}^{|X|+|Y|} |PC^*[k]| \qquad (2)$$

Proof. Since $X \cap Y = \emptyset$, a match in M^* is either boundary, or has at least one pair no MMS in A^* or B^* can involve. Thus for any two path components, say P_1, P_2 in $\mathfrak{G}(M^*)$, no pair involved by an MMS in $mms(P_1)$ can form a match in M^* with a pair involved by an MMS in $mms(P_2)$. Let P be a path component in M^*. By Lemma 2, those pairs involved by the MMSs in $mms(P)$ must join in $|mms(P)| + 1$ matches in M^*, if P is good; or $|mms(P)|$ matches otherwise. Thus those matches in M^* each with at least a pair involved by an MMS can be summed up as $|X| + |Y| + \sum_{k=1}^{|X|+|Y|} |PC^*[k]|$. The lemma equation follows. □

Lemma 3 implies $|a(A^*, B^*)| = |a(A, B)| + |X| + |Y| + \sum_{k=1}^{|X|+|Y|} |PC^*[k]|$. To bound the number of those more adjacencies in $a(A^*, B^*)$ than in $a(A, B)$, it suffices to take into account of those good paths with weight no more than 2. That is,

Lemma 4

$$|a(A^*, B^*)| - |a(A, B)| \leq \frac{4}{3}(|X| + |Y|) + \frac{2}{3}|PC^*[1]| + \frac{1}{3}|PC^*[2]| \qquad (3)$$

Proof. The lemma follows from Lemma 3 and $\sum_{k=3}^{|X|+|Y|} |PC^*[k]| \leq \frac{1}{3}(|X| + |Y| - |PC^*[1]| - 2|PC^*[2]|)$. □

4   Approximation Algorithm for Two-Sided Scaffold Filling

Let $X' \subseteq X, Y' \subseteq Y, A' \in A + Y', B' \in B + X', M'$ the maximum matching with respect to which the adjacencies and breakpoints in $P(A')$ and $P(B')$ have been identified. An MMS set, say S, is *good*, if there exists a good path component, say P in $\mathfrak{G}(M')$, such that $S = mms(P)$. A missing string is an *MMS candidate* for A' and B' if all its genes are in $(X - X')$ or $(Y - Y')$. An *MMS candidate set* will be mentioned with no two strings sharing a gene. Let S be an MMS candidate set. Then the *length* of S, denoted as $L(S)$, refers to the number of genes all those strings in S have. An *insertion of* S into A' and B' refers to a group of insertions each of which can insert a string in S between two genes of a pair in A' or B', such that provided they transform A' and B' into A'' and B'', each string in S will turn into an MMS in A'' or B''. Every time we mention to insert S into A' and B' or fill A' and B' with S, it refers to an insertion of S into A' and B'.

An insertion of S into A' and B' is *good*, if it transforms A' and B' into A'' and B'' such that S turns into a good MMS set in A'' and B''.

For a good insertion of S which transforms A' and B' into A'' and B''. Let $position(S)$ denote the multiset of pairs in $P(A')$ or $P(B')$, between those two genes of which the elements are inserted into turn into MMSs as members in S.

Lemma 5. *For $p_r \in position(S), p_r \in P(B')$ (resp. $P(A')$), if p_r is an adjacency in $P(B')$ with respect to M', there must exist at least one breakpoint p_r or its reversal in $P(B')$ (resp. $P(A')$) with respect to M'.*

For a good insertion of S, let the multiset of pairs in $P(A'')$ and $P(B'')$ involved by MMSs in S denote as $I(S)$, the pairs in $P(A')$ and $P(B')$ matched by those in $I(S)$ which do not form boundary matches in M'' denote as NBMP(S).

Lemma 6. *For $c_z c_{z+1} \in$ NBMP(S), $c_z c_{z+1} \in p(B')$ (resp. $P(A')$), if $c_z c_{z+1}$ is an adjacency in $P(B')$ (resp. $P(A')$) with respect to M', there must be at least one breakpoint $c_z c_{z+1}$ or $c_{z+1} c_z$ in $P(B')$ (resp. $P(A')$) with respect to M'.*

Lemma 7. *If there exists a good insertion of S into A' and B', then there must exist a good insertion of S into A' and B' such that: (1) the MMSs in S each is inserted into a breakpoint in $P(A')$ or $P(B')$; (2) the pairs in $P(A')$ or $P(B')$ matched by those involved by MMSs in S are all breakpoints.*

From Lemma 7, the good insertion of MMS set S into A' and B' discussed below is a good insertion of S, which by default inserts the MMSs in S each into a breakpoint in $P(A')$ or $P(B')$, and the pairs in $P(A')$ or $P(B')$ matched by those involved by MMSs in S are all breakpoints.

An MMS candidate set S is *good* for A' and B', if there exists a good insertion of S into A' and B'. Let S be a good MMS candidate set for A' and B', inserting those strings in S into A' and B' transform A' and B' into A'', B''. Each element in S must *correspond to* a breakpoint in $P(A')$ or $P(B')$, between those two genes of which the element can be inserted to turn into an MMS as a member in a good MMS set.

Let S_1 and S_2 be two good MMS candidate sets for A' and B'. Then S_1 *conflicts with* S_2 if one of the following situations occurs: (1) an element in S_1 shares a gene with an element in S_2; (2) an element in S_1 and an element in S_2 happen to correspond to one and the same breakpoint in $P(A')$ or $P(B')$, or the breakpoint in $P(A')$ or $P(B')$ which an element in S_2 corresponds to is matched by a pair involved by one element in S_1; (3) the breakpoint in $P(A')$ or $P(B')$ matched by a pair involved by one element in S_2 is also matched by a pair involved by one MMS in S_1; or the breakpoint in $P(A')$ or $P(B')$ matched by the one involved by one element in S_2 is also a breakpoint which one MMS in S_1 corresponds to.

To obtain more good MMS sets, we set a *conflict graph* $G = (V, E)$ in accordance with the conflict relationships of good MMS candidate sets, where a vertex in V stands for a good MMS candidate set, and the edge set $E = \{(u,v) \mid$ the good MMS candidate set u stands for conflicts with the one which v stands for,$u,v \in V$ }.

A graph is called k-claw free if it contains no independent set of k vertices, all adjacent to a common vertex [7]. Based on the conflict relations, we know,

Lemma 8. *A conflict graph G, where each vertex stands for a good MMS candidate set on a path of length at most k, is $(2k + 3)$-claw free.*

Proof. A good MMS candidate set on an i-path, say S_i, consksts of i genes from X and Y, and corresponds to j_1 breakpoknts in $P(A)$, and j_2 breakpoints in $P(B)$, where $1 \le j_1 + j_2 \le i$. By the definition of conflict, another good MMS candidate set could conflict with S_i, if (1) one of its MMSs shares one gene in S_i; (2) for the breakpoints in $P(A)$ or $P(B)$ which the MMSs in S_i correspond to, one of its MMSs corresponds to one of them, or one of them is matched by a pair involved by one of its MMSs; (3) for the breakpoints in $P(A)$ and $P(B)$ matched by pairs involved by MMSs in S_i, one of them is matched by a pair involved by one of its MMSs; or one of them is a breakpoint which one of its MMSs corresponds to.

Note that there are i genes consisting of S_i from X and Y, j_1+j_2 breakpoints the MMSs in S_i correspond to in $P(A)$ and $P(B)$, and $(i+2-(j_1+j_2))$ breakpoints in $P(A)$ and $P(B)$ matched by pairs involved by MMSs in S_i. We treat them as $2i + 2$ different factors of S_i. S_i cannot be obtained without anyone of them. So, good MMS candidate sets conflicting with S_i could be divided into $2i + 2$ types for different factors. It means that, at most $2i + 2$ types of good MMS candidate sets conflict with S_i, and among them, those conflicting with S_i for the same factor must conflict with each other, and those conflicting with S_i for distinct factors could not have conflicts.

Thus, in conflict graph G, there are at most $2i + 2$ types of vertices standing for the good MMS candidate sets, which are adjacent to a common vertex which S_i stands for. Among them, vertices standing for the same type of good MMS candidate sets conflicting with S_i must be adjacent to each other pairwiseand vertices standing for the same type of good MMS candidate sets conflicting with S_i could not connect to each other. Thus, there are no independent set of at most $2i + 3$, among those vertices which are adjacent to a common vertex standing for a good MMS candidate set on an i-path.

Note that in G, each vertex stands for a good MMS candidate set on a path of length at most k, so every vertex in G can not have a $(2k + 3)$-claw as an induced subgraph, G is a $(2k + 3)$-claw free graph.   □

By Lemma 4, we focus on searching for certain number of good MMS sets on a 1-path or a 2-path in our algorithm, by constructing conflict graphs based on their conflict relationships. Since any good MMS candidate set can probably conflict with another one, we set a 5-claw free graph only with good MMS candidate sets on a 1-path, and a 7-claw free graph with good MMS candidate sets on a path of length at most 2. Then, we compute the maximum independent set in the 5,7-claw free graph respectively. We choose better results by comparing the number of good MMS sets obtained respectively, and fill Scaffold A and B with them. Finally, we fill the scaffolds with the remaining genes, guaranteeing that it can increase at least as many adjacencies as genes inserted by the algorithm in [9].

A good MMS set on a 1-path must be a 1-MMS from X or Y. Thus, by enumerating 1-MMSs in X or Y, we identify all the good MMS candidate sets on a 1-path in polynomial time. Let the set of the good MMS candidate sets on a 1-path in X be X_1, and in Y be Y_1, then $V_g = X_1 \bigcup Y_1$. By enumeration, we also identify all the good MMS candidate sets on a 2-path. A good MMS set on a 2-path can be a 2-MMS from X or Y, or a 1-MMS from X and a 1-MMS from Y. Let the set of good MMS candidate sets on a 2-path only from X be X_2, the set of good MMS candidate sets on a 2-path only from Y be Y_2, For the set of good MMS candidate sets on a 2-path, where one 1-MMS from X and the other 1-MMS from Y, we denote it by Z. Let $V'_g = X_1 \bigcup X_2 \bigcup Y_1 \bigcup Y_2 \bigcup Z$.

4.1   Searching for Good MMS Sets in a 5-claw Free Graph

In order to obtain as many good MMS sets on a 1-path as possible, we set a conflict graph $G' = (V', E')$, where a vertex in V' stands for a good MMS candidate set in V_g, and the edge set is $E' = \{(u, v)|$ if the good MMS candidate set which u stands for conflicts with the one that v stands for$\}$. Then, from Lemma 8, G' is a 5-claw free graph. Using the approximation algorithm in [8], we can compute good MMS sets on a 1-path in polynomial time by seeking the approximative maximum independent set in G'.

Lemma 9. *There exists a polynomial approximation algorithm with ratio $\frac{1}{2+\varepsilon} (\varepsilon > 0)$ for the maximum independent set problem in a 5-claw free graph [8].*

By computing the maximum independent set of the conflict graph using the approximation algorithm in [8], we can obtain certain number of good MMS sets on a 1-path. The algorithm is described as follows:

Algorithm 1. MIS-1(A, B, X, Y)

1: identify all the good MMS candidate sets on a 1-path for A and B.
2: construct the conflict graph $G' = (V', E')$.
3: compute the approximative maximum independent set in G'.(Lemma 9).
4: **return** the approximative maximum independent set D_1.

Let D_1 be the result returned by MIS-1(A, B, X, Y). For scaffold A and B, we obtain $|D_1|$ good MMS sets on a 1-path, then we have

Lemma 10. *In some optimal solution, if there are* $|PC^*[1]|$ *good 1-paths in* A^* *and* B^*, *then*

$$|D_1| \geq \frac{1}{2+\varepsilon_1}|PC^*[1]| \tag{4}$$

Proof. Let D' be the maximum independent set in G'. By definition, the $|PC^*[1]|$ good 1-paths stand for $|PC^*[1]|$ good MMS sets in the optimal solution. They also construct an independent set in G' because of their optimality, so we have $|D'| \geq |PC^*[1]|$. Moreover, from Lemma 9, we can obtain $|D_1| = \frac{1}{2+\varepsilon_1}|D'|$. The lemma follows. $\square$

4.2   Searching for Good MMS Sets in a 7-claw Free Graph

For the sake of more good MMS sets on a 1-path and on a 2-path, we set a conflict graph $G'' = (V'', E'')$, where a vertex in V'' stands for a good MMS candidate set in V_g', and the edge set is $E'' = \{(w, z)|$ if the good MMS candidate set which w stands for conflicts with the one that z stands for.$\}$. Then, from Lemma 8, G' is a 7-claw free graph. We can compute good MMS sets on a 1-path, or on a 2-path in polynomial time by seeking the approximative maximum independent set in G'', using the algorithm in [8].

Lemma 11. *There exists a polynomial approximation algorithm with ratio* $\frac{1}{3+\varepsilon}(\varepsilon > 0)$ *for the maximum independent set problem in a 7-claw free graph* [8].

By computing the maximum independent set of the conflict graph G'' by an approximation algorithm, we can obtain an approximative maximum independent setwhich stands for certain number of good MMS sets on a 1-path or on a 2-path. The algorithm is described as follows:

Algorithm 2. MIS-2(A, B, X, Y)

1: identify all the good MMS candidate sets on a 1, 2-path for A and B.
2: construct the conflict graph $G'' = (V'', E'')$.
3: compute the approximative maximum independent set in G''.(Lemma 11).
4: **return** the approximative maximum independent set D_2.

Let D_2 be the result returned by MIS-2(A, B, X, Y). For scaffold A and B, we obtain $|D_2|$ good MMS sets on a 1-path or on a 2-path, then we have

Lemma 12. *In some optimal solution, if there are* $|PC^*[1]|$ *good 1-paths in* A^* *and* B^*, *and* $|PC^*[2]|$ *good 2-paths in* A^* *and* B^*, *then*

$$|D_2| \geq \frac{1}{3+\varepsilon_2}(|PC^*[1]| + |PC^*[2]|) \tag{5}$$

Proof. Let D'' be the maximum independent set in G''. In the optimal solution, we can know by definition, that the $|PC^*[1]|$ good 1-paths stand for $|PC^*[1]|$ good MMS sets, the $|PC^*[2]|$ good 2-paths stand for $|PC^*[2]|$ good MMS sets in G''. They also construct an independent set in G'' because of their optimality. So $|D''| \geq (|PC^*[1]| + |PC^*[2]|)$. Moreover, from Lemma 11, we can obtain $|D_2| = \frac{1}{3+\varepsilon_2}|D''|$. The lemma follows. □

4.3   The Algorithm for Two-Sided Scaffold Filling

Now we describe our approximation algorithm for the Two-sided Scaffold Filling as a whole.

Algorithm 3. TwoSideScaffoldFill(A, B, X, Y)

Input: Two scaffolds A and B, $X = c(A) - c(B)$,$Y = c(B) - c(A)$.
Output: $A' \in A + Y$, $B' \in B + X$.
1: $\alpha = \beta = \gamma = \phi$.
2: $\alpha \leftarrow$ MIS-1(A, B, X, Y).
3: $\beta \leftarrow$ MIS-2(A, B, X, Y).
4: If $(|\alpha| \geq |\beta|)$ $\gamma = \alpha$.
5: Else $\gamma = \beta$.
6: fill A and B with the good MMS sets in γ, obtain A^+ and B^+.
7: fill A^+ and B^+ with the remaining genes, obtain A' and B' by the algorithm in [9].
8: **return** A',B'.

---

5   Proof of the Approximation Ratio

To analyze the performance of our algorithm, we need to compare the number of good MMS sets obtained by our algorithm with that in some optimal solution. So we assume that A' and B' be the result scaffolds returned by TwoSideScaffoldFill(A, B, X, Y), and M' is maximum matching with respect to which the adjacencies are identified in A' and B'. In graph $\mathfrak{G}(M')$, let $PC[k]$ be the set of good k-paths. Then in A' and B', there are $|PC[k]|$ good MMS sets on a k-path.

By the definition and Lemma 1, we have

$$|a(A', B')| - |a(A, B)| = |X| + |Y| + \sum_{i=1}^{|X|+|Y|} |PC[k]| \tag{6}$$

Let A^+ and B^+ be the scaffolds returned by running TwoSideScaffoldFill(A, B, X, Y) ended in Line 6. Let $PC^+[1]$ be the set of good MMS sets on a 1-path, $PC^+[2]$ be the set of good MMS sets on a 2-path in A^+ and B^+.

Lemma 13

$$\sum_{i=1}^{|X|+|Y|} |PC[k]| \geq |PC^+[1]| + |PC^+[2]| \tag{7}$$

Proof. Note that there are $\sum_{i=1}^{|X|+|Y|} |PC[k]|$ good MMS sets in A' and B', and there are $|PC^+[1]| + |PC^+[2]|$ good MMS sets in A^+ and B^+. Let $X^+ = c(B') - c(B^+)$, and $Y^+ = c(A') - c(A^+)$. If X^+ and Y^+ are empty, obviously there are as many good MMS sets in A' and B' as those in A^+ and B^+. Otherwise, there are some insertions in Line 7 by running TwoSideScaffoldFill(A, B, X, Y), and assume λ is an arbitrary insertion. From the algorithm in [9], if λ insert an MMS set between two genes of a breakpoint, it will be a good MMS set or not without destroying the original ones in the scaffold, so λ will increase good MMS sets by number zero or one; if λ insert an MMS set between two genes of an adjacency, the number of good MMS sets will not increase. Since those insertions like λ transform A^+ into A', and transform B^+ into B', there are at least as many good strings in A' and B' as in A^+ and B^+.   □

Let s be a number such that $\frac{4}{3} < \frac{1}{s} < \frac{3}{2}$. Thus from Lemma 4, we have

$$|a(A^*, B^*)| - |a(A, B)|$$
$$\leq \frac{4}{3}(|X| + |Y|) + \frac{2}{3}|PC^*[1]| + \frac{1}{3}|PC^*[2]|$$
$$= \frac{1}{s}((|X| + |Y|) + (\frac{4s}{3} - 1)(|X| + |Y|) + \frac{2s}{3}|PC^*[1]| + \frac{1s}{3}|PC^*[2]|)$$

Since the number of genes in good MMS sets is not larger than that of X and Y, we have $|X| + |Y| \geq |PC^*[1]| + 2|PC^*[2]|$. What's more, $\frac{4s}{3} - 1 < 0$, thus the bound of optimal solution can be transformed as,

$$|a(A^*, B^*)| - |a(A, B)|$$
$$\leq \frac{1}{s}((|X| + |Y|) + (\frac{4s}{3} - 1)(|PC^*[1]| + 2|PC^*[2]|) + \frac{2s}{3}|PC^*[1]| + \frac{1s}{3}|PC^*[2]|)$$
$$= \frac{1}{s}((|X| + |Y|) + (2s - 1)|PC^*[1]| + (3s - 2)|PC^*[2]| \tag{8}$$

Now from the above formula 6 and 8, we can obtain the approximation ratio $\frac{1}{s}$, if and only if the formula below always holds,

$$\sum_{i=1}^{|X|+|Y|} |PC[k]| \geq (2s - 1)|PC^*[1]| + (3s - 2)|PC^*[2]| \tag{9}$$

Moreover, by Lemma 13, if the formula

$$|PC^+[1]| + |PC^+[2]| \geq (2s - 1)|PC^*[1]| + (3s - 2)|PC^*[2]| \tag{10}$$

holds, we will have the approximation ratio $\frac{1}{s}$.

Let α be the result obtained by running the TwoSideScaffoldFill(A, B, X, Y) ended in Line 2, β be the result obtained by running Line 3 in the TwoSideScaffoldFill (A, B, X, Y), γ the returned by running TwoSideScaffoldFill(A, B, X, Y) ended in Line 5, then $|\gamma| = \max\{|\alpha|, |\beta|\}$.

Note by Lemma 10, we obtain

$$|\alpha| \geq \frac{1}{2 + \varepsilon_1}|PC^*[1]| \tag{11}$$

By Lemma 12, we obtain

$$|\beta| \geq \frac{1}{3 + \varepsilon_2}(|PC^*[1]| + |PC^*[2]|) \tag{12}$$

By Algorithm 3, we obtain

$$|PC^+[1]| + |PC^+[2]| = |\gamma| = max\{|\alpha|, |\beta|\} \tag{13}$$

Let $|\phi| = (2s - 1)|PC^*[1]| + (3s - 2)|PC^*[2]|$. We should assign s to satisfy $|\gamma| \geq |\phi|$. In what follows, we aim at acquiring the value of s, by comparing the bound of $|\alpha|$ in formula (11) and $|\beta|$ in formula (12).

If $\frac{1}{2+\varepsilon_1}|PC^*[1]| \geq \frac{1}{3+\varepsilon_2}(|PC^*[1]| + |PC^*[2]|)$ holds, then $|PC^*[2]| \leq \frac{1+\varepsilon_2-\varepsilon_1}{2+\varepsilon_1}$ $|PC^*[1]|$. Moreover, $3s - 2 > 0$. So we have $|\phi| = (2s - 1)|PC^*[1]| + (3s - 2)$ $|PC^*[2]| \leq (2s - 1)|PC^*[1]| + (3s - 2)\frac{1+\varepsilon_2-\varepsilon_1}{2+\varepsilon_1}|PC^*[1]| = ((2s - 1) + (3s - 2)$ $\frac{1+\varepsilon_2-\varepsilon_1}{2+\varepsilon_1})|PC^*[1]|$.

From formula(11) and (13), $|\gamma| \geq |\alpha| \geq (\frac{1}{2+\varepsilon_1})|PC^*[1]|$. $|\gamma| \geq |\phi|$ always holds, iff $(2s - 1) + (3s - 2)\frac{1+\varepsilon_2-\varepsilon_1}{2+\varepsilon_1} \leq \frac{1}{2+\varepsilon_1}$ holds, then $\frac{1}{s} \leq \frac{7+3\varepsilon_2-\varepsilon_1}{5+2\varepsilon_2-\varepsilon_1}$.

If $\frac{1}{2+\varepsilon_1}|PC^*[1]| < \frac{1}{3+\varepsilon_2}(|PC^*[1]| + |PC^*[2]|)$ holds, then $|PC^*[1]| < \frac{2+\varepsilon_1}{3+\varepsilon_2}$ $(|PC^*[1]| + |PC^*[2]|)$. Moreover, $1 - s > 0$. So we have $|\phi| = (2s - 1)|PC^*[1]| +$ $(3s - 2)|PC^*[2]| = (3s - 2)(|PC^*[1]| + |PC^*[2]|) + (1 - s)|PC^*[1]| < (3s - 2)$ $(|PC^*[1]| + |PC^*[2]|) + (1 - s)\frac{2+\varepsilon_1}{3+\varepsilon_2}(|PC^*[1]| + |PC^*[2]|) = ((3s - 2) + (1 - s)$ $\frac{2+\varepsilon_1}{3+\varepsilon_2})(|PC^*[1]| + |PC^*[2]|)$.

From formula (12) and (13), $|\gamma| \geq |\beta| \geq \frac{1}{3+\varepsilon_2}(|PC^*[1]| + |PC^*[2]|)$. $|\gamma| \geq |\phi|$ always holds, iff $(3s - 2) + (1 - s)\frac{2+\varepsilon_1}{3+\varepsilon_2} \leq \frac{1}{3+\varepsilon_2}$ holds, then $\frac{1}{s} \leq \frac{7+3\varepsilon_2-\varepsilon_1}{5+2\varepsilon_2-\varepsilon_1}$.

Thus, when $\frac{1}{s} = \frac{7+3\varepsilon_2-\varepsilon_1}{5+2\varepsilon_2-\varepsilon_1}$, the performance ratio of our algorithm is $\frac{1}{s} = 1.4 + \varepsilon'$.

6 Conclusion

In this paper, we conduct a further research on the Two-sided Scaffold Filling problem, and improve the approximation algorithm to $1.4 + \varepsilon$ based on a new lower bound. We devise a new approximation algorithm with $1.4 + \varepsilon$-ratio, which applies an approximative maximum independent set algorithm in a 5-claw free and a 5-claw free graph. Exploring new algorithms with smaller approximation ratio is still a valuable future work.

Acknowledgments. This research is partially supported NSF of China under grant 61472222.

References

1. Huson, D.H., Reinert, K., Myers, E.W.: The greedy path-merging algorithm for contig scaffolding. J. ACM **49**(5), 603–615 (2002)
2. Jiang, H., Zheng, C., Sankoff, D., Zhu, B.: Scaffold filling under the breakpoint distance. In: Tannier, E. (ed.) RECOMB-CG 2010. LNCS, vol. 6398, pp. 83–92. Springer, Heidelberg (2010). doi:10.1007/978-3-642-16181-0_8
3. Jiang, H., Zhong, F., Zhu, B.: Filling scaffolds with gene repetitions: maximizing the number of adjacencies. In: Giancarlo, R., Manzini, G. (eds.) CPM 2011. LNCS, vol. 6661, pp. 55–64. Springer, Heidelberg (2011). doi:10.1007/978-3-642-21458-5_7
4. Jiang, H., Zheng, C., Sankoff, D., Zhu, B.: Scaffold filling under the breakpoint and related distances. IEEE/ACM Trans. Bioinform. Comput. Biol. **9**(4), 1220–1229 (2012)
5. Muñoz, A., Zheng, C., Zhu, Q., Albert, V., Rounsley, S., Sankoff, D.: Scaffold filling, contig fusion and gene order comparison. BMC Bioinform. **11**, 304 (2010)
6. Liu, N., Jiang, H., Zhu, D., Zhu, B.: An improved approximation algorithm for scaffold filling to maximize the common adjacencies. IEEE/ACM Trans. Comput. Biol. Bioinform. **10**(4), 905–913 (2013)
7. Yu, G., Goldschmidt, O.: Local optimality and its application on independent sets for k-claw free graphs. J. Comb. Optim. **1**(2), 151–164 (1997)
8. Halldórsson, M.M.: Approximating discrete collections via local improvements. In: Proceedings of the 6th Annual ACM-SIAM Symposium on Discrete Algorithms (SODA 1995), pp. 160–169 (1995)
9. Liu, N., Zhu, D., Jiang, H., Zhu, B.: A 1.5-approximation algorithm for two-sided scaffold filling. Algorithmica **74**(1), 91–116 (2016)
10. Ma, J., Jiang, H.: Notes on the 6/5-approximation algorithm for one-sided scaffold filling. In: Zhu, D., Bereg, S. (eds.) FAW 2016. LNCS, vol. 9711, pp. 145–157. Springer, Cham (2016). doi:10.1007/978-3-319-39817-4_15

FPT Algorithms for FVS Parameterized by Split and Cluster Vertex Deletion Sets and Other Parameters

Diptapriyo Majumdar$^{(\boxtimes)}$ and Venkatesh Raman

The Institute of Mathematical Sciences, HBNI, Chennai, India
{diptapriyom,vraman}@imsc.res.in

Abstract. A feedback vertex set in an undirected graph is a subset of vertices whose deletion results in an acyclic graph. The problem (which we call FVS) of finding a minimum (or k sized) feedback vertex set is NP-hard in general graphs, while it is polynomial time solvable in some classes of graphs including split graphs and cluster graphs. The current best fixed-parameter tractable (FPT) algorithm for determining whether a given undirected graph has a feedback vertex set of size at most k has a runtime of $\mathcal{O}^*(3.618^k)$ ($\mathcal{O}^*$ notation hides polynomial factors). We consider the parameterized complexity of feedback vertex set parameterized by (vertex deletion) distance to some polynomially solvable classes of graphs including cluster and split graphs. We call a graph G a (c,i)-graph if its vertex set can be partitioned into c cliques and i independent sets. When $c = 0$ and $i = 2$, such a graph is simply a bipartite graph where FVS is NP-hard. It can be deduced easily that FVS is NP-hard even for constant c when $i \geq 2$. When $c \leq 1$ and $i \leq 1$, then the graph is a split graph where FVS is solvable in polynomial time. Given a graph, let k be the size of the modulator whose deletion results in a (c,i)-graph. We address the parameterized complexity of FVS parameterized by k when $i \leq 1$. Specifically we show that

1. FVS admits an FPT algorithm that runs in $\mathcal{O}^*(3.148^k)$ time, when $c \leq 1$ and $i \leq 1$ (i.e. when the modulator is a deletion set to a split graph). When $c \geq 2$, we generalize the algorithm to one with run-time $\mathcal{O}(3.148^{k+c} \cdot n^{\mathcal{O}(c)})$. We also show that FVS is $W[1]$-hard when parameterized by c (i.e. the c in the exponent of n is unavoidable) if $i \leq 1$ extending a known hardness reduction for the case when $i = 0$.
2. For the special case when $i = 0$ and $c \geq 1$, and when there are no edges across vertices in different parts (i.e. the modulator is a deletion set to a cluster graph), we give an $\mathcal{O}^*(5^k)$ algorithm.

1 Introduction and Motivation

A feedback vertex set in an undirected graph is a subset of vertices whose deletion results in an acyclic graph. It finds applications in multiple domains where one needs to identify a small number of nodes that cover loops or cycles. It is a well-studied parameter in algorithmic paradigms. The problem of finding a minimum

© Springer International Publishing AG 2017
M. Xiao and F. Rosamond (Eds.): FAW 2017, LNCS 10336, pp. 209–220, 2017.
DOI: 10.1007/978-3-319-59605-1_19

feedback vertex set is one of the classical NP-hard problems in general graphs. It is closely related to the other well-studied parameter VERTEX COVER as it has a simple reduction from VERTEX COVER, and it also has a ratio 2 approximation algorithm [1].

However the parameterized complexity of feedback vertex set (FVS) appears harder than that of vertex cover. While vertex cover has an efficient $O^*(1.28^k)$ fixed-parameter algorithm [6] (see Sect. 2 for definitions) when parameterized by the solution size, the current best fixed-parameter tractable (FPT) algorithm for FVS has a runtime of $\mathcal{O}(3.618^k \cdot n^{\mathcal{O}(1)})$ [12] after a series of initial algorithms [4, 5, 12, 17]. While vertex cover has a $O(k^2)$ sized kernel with at most $2k - \lg k$ vertices [15], an $O(k^2)$ sized kernel for FVS [18] is considerably involved.

There has also been a number of structural parameterizations of vertex cover [11], where the parameter is the size of some structure in the input, rather than the solution size. In fact, it is easy to see that if $\mathcal{F}$ is a hereditary class of graphs where vertex cover is polynomial time solvable, then vertex cover is fixed-parameter tractable in a graph that is k vertices away from a graph in $\mathcal{F}$ when parameterized by k. However the parameterized complexity of FVS is varied when parameterized by deletion distance to a class of graphs where it is polynomially solvable [10]. See [16] for kernel results on some structural parameterizations of FVS. We continue this line of work on structural parameterizations of FVS by giving efficient fixed-parameter tractable algorithms when parameterized by deletion distance to split and cluster graphs.

Our Results: For a given $c, i \in \mathbb{N}$, an undirected graph G is called (c, i)-graph if $V(G)$ can be partitioned into c cliques and i independent sets. We consider the parameterized complexity of the following problem.

FVS-DELETION TO (c, i)-GRAPH
Input: An undirected graph $G = (V, E), S \subseteq V(G)$ such that $G \setminus S$ is a graph for which $V(G) \setminus S = A_1 \dot{\cup} \ldots \dot{\cup} A_c \dot{\cup} B_1 \dot{\cup} \ldots \dot{\cup} B_i$ where $A_1, \ldots, A_c$ are cliques, $B_1, \ldots, B_i$ are independent sets and an integer ℓ.
Question: Is there a feedback vertex set of G with at most ℓ vertices?

We study the parameterized complexity of the problem when parameterized by $|S|$ and sometimes by c and $|S|$.

When $c \leq 1$ and $i \leq 1$, a (c, i)-graph is a split graph where FEEDBACK VERTEX SET is polynomial time solvable. In this case S is the *split vertex deletion set* and we give the following special name to the problem.

FVS-SPLIT-VERTEX-DELETION   **Parameter:** k
Input: An undirected multi-graph G, $S \subseteq V(G)$ of size at most k such that $G \backslash S$ is a split graph and an integer ℓ
Question: Does G have a feedback vertex set of size at most ℓ?

Our algorithm for this problem that appears in Sect. 4.1 takes $\mathcal{O}^*(3.148^k)$ time. Note that the base of the exponent is smaller than 3.618 which is the base for the best known FVS when parameterized by solution size.

When $i = 2$ and $c = 0$, a (c,i)-graph is a bipartite graph where FVS is **NP**-Complete and this result can easily be extended for $i \geq 2$ and $c = 0$. Thus the FVS-DELETION TO (c,i)-GRAPH is para-**NP**-hard when $i \geq 2$ even when parameterized by c. Jansen et al. [10] proved that FEEDBACK VERTEX SET is $W[1]$-hard but contained in XP when parameterized by the number of cliques in a vertex clique cover. In Sect. 4.2), we extend these results for $i \leq 1$; I.e. we show that FVS is $W[1]$-hard in a (c,i)-graph when parameterized by c, but has an algorithm of running time $\mathcal{O}(n^{f(c)})$.

In Sect. 5, we consider deletion distance to a special case of the (c,i)-graph where $i = 0$. For this special case, each part of the c partitions is a connected component. So, there is no edge between two vertices of different parts. Such a (c,i)-graph is simply a cluster graph.

FVS-CVD   **Parameter:** k
Input: An undirected multi-graph G, $S \subseteq V(G)$ of size at most k such that every component of $G\backslash S$ is a clique and an integer ℓ
Question: Does G have a feedback vertex set of size at most ℓ?

We provide an algorithm with running time $\mathcal{O}^*(5^k)$ for this problem.

2 Preliminaries

We use $\mathbb{N}$ to denote the set of natural numbers. For $r \in \mathbb{N}$, we use $[r]$ to denote the set $\{1, \ldots, r\}$. We use standard graph theoretic notations from Diestel [9]. For a vertex $u \in V(G)$ we use $N_G(u) = \{v \in V(G)|(u,v) \in E(G)\} \setminus \{u\}$ to denote the open neighborhood of u and $N_G[u] = \{v \in V(G)|(u,v) \in E(G)\} \cup \{u\}$ to denote the close neighborhood of u. By $deg_G(u)$, we denote the number of edges incident on u in the graph G. Let F be a subgraph of G. Then by $deg_F(u)$, we denote the number of edges incident on u in F. Let $S \subseteq V(G)$. By $G[S]$, we denote the graph induced on the vertex set S. By $G\backslash S$, we denote the graph that we obtain after deleting the vertices of S and the edges incident on S. Equivalently, $G\backslash S$ can be represented as $G[V(G) \setminus S]$. We use these two notations ($G\backslash S$ and $G[V(G) \setminus S]$) interchangeably to denote the same thing. By *multiplicity of an edge* (u,v) we mean the number of edges present between u and v. For a subset S and for an integer d, we denote $\binom{S}{\leq d}$ as the collection of subsets of S with at most d elements. And by $\binom{S}{\geq d}$, we denote the collection of subsets of S with at least d elements.

Definition 1 (Fixed Parameter Tractability). *Let $L \subseteq \Sigma^* \times \mathbb{N}$ is a parameterized language. L is said to be fixed parameter tractable (or FPT) if there exists an algorithm $\mathcal{B}$, a constant c and a computable function f such that $\forall x, \forall k, \mathcal{B}$ on input (x,k) runs in at most $f(k) \cdot |x|^c$ time and outputs $(x,k) \in L$ iff $\mathcal{B}([x,k]) = 1$. We call the algorithm $\mathcal{B}$ as fixed-parameter algorithm.*

Definition 2 (Slice-Wise Polynomial (XP)). *Let $L \subseteq \Sigma^* \times \mathbb{N}$ is a parame-terized language. L is said to be Slice-Wise Polynomial (or in XP) if there exists an algorithm $\mathcal{B}$, a constant c and computable functions f, g such that $\forall x, \forall k$, $\mathcal{B}$ on input (x, k) runs in at most $f(k).|x|^{g(k)+c}$ time and outputs $(x, k) \in L$ iff $\mathcal{B}([x, k]) = 1$. We call the algorithm $\mathcal{B}$ as XP Algorithm.*

There is a notion of hardness in parameterized complexity captured by the weft hierarchy, W-hard complexity classes and parameterized reduction. W-hierarchy is defined as $FPT \subseteq W[1] \subseteq W[2] \subseteq \cdots \subseteq XP$ using boolean circuits and parameterized reductions. It is believed that the subset relations are strict in this sequence and a parameterized problem that is hard for some complexity class above FPT is unlikely to be FPT.

Definition 3 (Parameterized Reduction). *Let P_1 and P_2 be two parameter-ized languages. We say that there is a parameterized reduction from P_1 to P_2 if there exists an algorithm $\mathcal{B}$, a constant c and computable functions $f, g : \mathbb{N} \to \mathbb{N}$ such that $\forall x, \forall k$ such that given an instance (x, k) of P_1, $\mathcal{B}$ runs in at most $f(k) \cdot |x|^c$ time and outputs an instance (x', k') of P_2 such that $(x, k) \in P_1$ if and only if $(x', k') \in P_2$ and $k' \leq g(k)$.*

Theorem 1 *(See Chap. 13, [7]). Let P_1, P_2 be two parameterized languages and let there is a parameterized reduction from P_1 to P_2. Then, if P_2 is fixed-parameter tractable, then P_1 is fixed-parameter tractable. Equivalently, for a given $i \geq 1$, if P_1 is $W[i]$-hard, then P_2 is $W[i]$-hard (unlikely to be FPT). We call the algorithm $\mathcal{B}$ a parameterized reduction.*

$L \subseteq \Sigma^* \times \mathbb{N}$ is called *para-NP-hard* if L is NP-hard even when the parameter is a constant. We refer the reader to the recent monograph [7] for more on parameterized complexity.

3   FVS-DELETION TO (c, i)-GRAPH when $i \geq 2$

As FEEDBACK VERTEX SET on bipartite graphs is **NP**-Complete, we have the following observation.

Observation 1. *When $c = 0, i \geq 2$, FEEDBACK VERTEX SET is NP-Complete on (c, i) graphs. So, when the parameter is either c or k and $i \geq 2$, FVS-DELETION TO (c, i)-GRAPH is Para-NP-Hard.*

4   FVS-DELETION TO (c, i)-GRAPH when $i \leq 1$

4.1   When $c \leq 1$

A graph G is called a *Split Graph* if $V(G) = C \dot{\cup} I$ where $G[C]$ is a clique and I is an independent set. A set of vertices $S \subseteq V(G)$ is called *split vertex deletion set* if $G \backslash S$ is a *split graph*. In this section, we provide a fixed-parameter algorithm when it is parameterized by the size (number of vertices) of a *split vertex deletion set*. Note that the size of the *split vertex deletion set* is incomparable to the solution size. We assume that the *split vertex deletion set* is also given with the input. Otherwise we use an algorithm by Cygan and Pilipczuk [8] that runs in $\mathcal{O}(1.2738^k \cdot k^{O(\log_2 k)} \cdot n^{\mathcal{O}(1)})$ time to determine the existence of a split vertex deletion set of size at most k. Now, we are ready to describe our algorithm. We use (G, S, ℓ) to denote the input instance. We denote $F = G \backslash S$. F is a split graph whose vertices can be partitioned into a clique and an independent set. So, we denote $F = (C, I)$ where $V(F) = C \dot\cup I$. Note that any FVS must contain at least $|C| - 2$ vertices from F. In particular, all vertices of S with $|C|$ vertices from F form a feedback vertex set of G. So, if $\ell \geq |S| + |C|$, then (G, S, ℓ) is YES-INSTANCE as vertices of I is an independent set. So, we have the following reduction rule whose correctness is also easy to see.

Reduction Rule 1. *If $\ell \geq |S| + |C|$, then (G, S, ℓ) is* YES-INSTANCE.

So, we can assume that $\ell \leq |S| + |C| - 1$. Now we apply the following standard rules to make the graph minimum degree three. It is easy to see that these rules can be implemented in polynomial time. For correctness, refer to Chap. 3 in [7].

Reduction Rule 2. *If there exists $u \in V(G)$ where there is a self loop in u, then delete u and reduce ℓ by 1.*

Reduction Rule 3. *If $\exists u \in V(G)$ such that $deg_G(u) \leq 1$, then delete u.*

Reduction Rule 4. *If there exists $u \in V(G)$ such that $deg_G(u) = 2$, $N_G(u) = \{v, w\}$, then delete u and add an extra-edge between v and w.*

The above reduction rule can create parallel edges (or self loops).

Reduction Rule 5. *If there exists an edge $(u, v) \in E(G)$ whose multiplicity is more than two, then reduce its multiplicity to two.*

Now, our algorithm proceeds as follows. We guess a subset $S' \subseteq S$ and a subset $C' \in \binom{C}{\geq |C|-2}$ that intersect the feedback vertex set of G we are looking for. Note that $|C'| \geq |C| - 2$ as C is a clique. If $G[(S \backslash S') \cup (C \backslash C')]$ is not a forest, then clearly the guess is wrong, we move on to a different subset. So, $G[(S \backslash S') \cup (C \backslash C')]$ is a forest. Now, we have to find a subset I' from I with at most $\ell' = \ell - |S'| - |C'| \leq |S| + |C| - 1 - |C| + 2 - |S'| = |S| - |S'| + 1$ vertices such that $G \backslash (S' \cup I' \cup C')$ becomes a forest. So, now our goal is to solve DISJOINT FEEDBACK VERTEX SET problem defined below on the instance $(G', (S \backslash S') \cup (C \backslash C'), I, |S| - |S'| + 1)$.

DISJOINT FEEDBACK VERTEX SET   **Parameter:** $|S_1|$
Input: An undirected graph $G = (V, E)$, $S_1 \cup S_2 = V(G)$, $S_1 \cap S_2 = \emptyset$, $G[S_1]$ is a forest, S_2 is an independent set and an integer ℓ'.
Question: Is there a feedback vertex set W of G such that $W \cap S_1 = \emptyset$ and $|W| \leq \ell'$?

Note that for our case, $\ell \leq |S_1| - 1$ as we are looking for a set of at most $|S| - |S'| + 1$ vertices from I that kills all cycles in $G[(S \backslash S') \cup (C \backslash C') \cup I]$. Also for our case I is an independent set. Let $C'' = C \backslash C'$ and $S'' = S \backslash S'$. After deletion of $C' \cup S'$ from G, we get G'. Now we consider the DISJOINT FEEDBACK VERTEX SET instance on $(G', S'' \cup C'', I, |S''| + 1)$. We define a measure $\mu(G') = \ell' + d(G')$ where $d(G')$ is the number of components in $G'[S'' \cup C'']$. We will use μ and $\mu(G')$ interchangeably as we describe the rest of the parts for our algorithm. Let $|S'| = i$. Note that $|S'' \cup C''| = k - i + 2$. But, one (or two vertices) from C also is part of $S'' \cup C''$. So, $d \leq k - i + 1$. Then $\ell' \leq k - i + 1$ and $d \leq k - i + 1$, we have that $\mu \leq 2(k - i) + 2$. Now, we apply Reduction Rules 3, 6, 7 in sequence. Correctness of first reduction rule is easy to see. Reduction Rule 7 is a slight modification to the rule 4, but we need to be careful as we have made our guesses from S. For correctness of these rules, refer to [12].

Reduction Rule 6. *If there exists a vertex $u \in I$ such that $G'[S'' \cup C'' \cup \{u\}]$ contains a cycle, then delete u and reduce ℓ' by one.*

Note that Reduction Rule 6 works also when u has exactly two neighbors in the same component of $G'[S'' \cup C'']$.

Reduction Rule 7. *If there exists $u \in I$ such that u has exactly two neighbors that are in different components of $G'[S'' \cup C'']$, then move u from I to $S'' \cup C''$.*

Lemma 1. *Reduction rule 7 is correct.*

Proof. Reduction Rule 7 is applied only when Reduction Rule 3, 6 are not applicable. Then $G'[S'' \cup C'' \cup \{u\}]$ does not create a cycle for any vertex $u \in I$. The intuition behind this reduction rule is that there exists an optimal solution that does not contain u. Suppose, any optimal solution contains u. Take one such feedback vertex set D. As D is a minimal feedback vertex set as well, $G' \backslash (D \backslash \{u\})$ has a cycle. Let t_1, t_2 be the two neighbors of u in $G'[S'' \cup C'']$. In particular, in $G' \backslash D$, there is a unique path from t_1 to t_2 (as D is minimal). Moreover, the cycle in $G' \backslash (D \backslash \{u\})$ is also unique. Now, t_1, t_2 are in different components of $G'[S'' \cup C'']$. So, the unique path from t_1 to t_2 in $G' \backslash D$ must pass through a vertex u' in I. So, $(D \cup \{u'\}) \backslash \{u\}$ is also a minimum feedback vertex set of G'. □

Now, we argue that the measure does not increase when Reduction Rules 3, 6, 7 are applied.

Lemma 2. *Application of Reduction Rules 3, 6, 7 does not increase $\mu(G')$.*

Proof. Reduction Rule 3 does not increase ℓ' or $d(G')$. Such a vertex $u \in I$ on which this rule has been applied can be adjacent to only one vertex of $S'' \cup C''$. Such a vertex does not increase $d(G')$ also. So, $\mu(G')$ does not increase.

Reduction Rule 6 deletes a vertex from I, reduces ℓ', but does not increase $d(G')$. So, $\mu(G')$ does not increase. Reduction Rule 7 does not increase ℓ', but it decreases $d(G')$ as two components in $G'[S'' \cup C'']$ merges into one single component after pushing such a vertex u from I to $S'' \cup C''$. So, again μ does not increase. □

Now, when Reduction Rules 3, 6, 7 are not applicable, then every vertex of I has at least three neighbors that are in $S'' \cup C''$ and all these three neighbors are in different components. If for every vertex $u \in I$, u has exactly three neighbors all in different components, then the instance is polynomial time solvable by Kocuimaka and Pilipczuk [12]. Otherwise there exists a vertex $u \in I$ that has at least four neighbors in $S'' \cup C''$ and all are in different components. In such case, we apply the following branching rule.

Branching Rule 1. *If there exists a vertex $u \in I$ such that u has at least four neighbors all of whom are in different components of $G'[S'' \cup C'']$, then in one branch, pick u into the solution (or in other words, delete u and reduce ℓ' by 1) and in another branch, we push the vertex u from I to $S'' \cup C''$.*

It is clear that when Branching Rule 1 is applied, then in one branch $\mu(G')$ drops by 1 and in other branch, $\mu(G')$ drops by at least 3 (as four components get merged into a single one). So, we get the following recurrence for this branching rule.

$$T(\mu) \leq T(\mu - 1) + T(\mu - 3)$$

Solving this recurrence, we get

$$T(\mu) \leq 1.4656^\mu \leq 2.148^{k-i+1}.$$

As this branching rule is applied over all subsets of S, the algorithm runs in time $\sum_{i=0}^{k} \binom{k}{i} 2.148^{k-i+1} \cdot n^{\mathcal{O}(1)} = \mathcal{O}(3.148^k \cdot n^{\mathcal{O}(1)})$. Given k, finding a split vertex deletion set of size at most k takes $\mathcal{O}^*(1.2738^k \cdot k^{\mathcal{O}(\log_2 k)})$ time. So, we have the following theorem.

Theorem 2. FEEDBACK VERTEX SET *parameterized by* SPLIT VERTEX DELETION SET *can be solved in* $\mathcal{O}(3.148^k \cdot n^{\mathcal{O}(1)})$ *time.*

4.2   When $c \geq 2$

For this case, first we need the assumption that the deletion set ($S \subseteq V(G)$) and the partition of $V(G) \setminus S = A_1 \dot\cup \ldots \dot\cup A_c \cup I$ are given with the input. Because, even when $c \geq 3$, the recognition of $(c, 0)$-Graph is **NP**-Complete. So, when $c \geq 3$, then we can not even hope to have an FPT algorithm that outputs an $(c, 0)$ (or $(c, 1)$) Deletion Set. When $c \leq 2$, Kolay and Panolan [13,14] provided an algorithm with runtime $\mathcal{O}^*(3.314^k)$.

XP Algorithm. An XP algorithm is known for FVS parameterized by the number of cliques in a vertex clique cover by Jansen et al. [10]. It follows that for $c \geq 2, i = 0, k = 0$, FVS DELETION TO (c, i)-GRAPH admits an algorithm with running time $\mathcal{O}(n^{\mathcal{O}(c)})$. We give an explicit algorithm and extend it to the case when $i = 1$ with running time $\mathcal{O}(3.148^{k+c} \cdot n^{2c+\mathcal{O}(1)})$ where S is a set of at most k vertices whose deletion from G results into a $(c, 1)$-Graph.

Algorithm. Let $G \backslash S = A_1 \dot\cup \dots \dot\cup A_c \dot\cup B$. As $A_1, \dots, A_c$ are all cliques, any feedback vertex set of G intersects A_j in at least $|A_j| - 2$ vertices for all $j \in [c]$. If $\ell \geq |S| + \sum_{j=1}^{c} |A_j|$, then it is an yes-instance. So, $\ell \leq |S| + \sum_{j=1}^{c} |A_j| - 1$. Now we make the minimum degree of G to three by using Reduction Rules 2, 3, 4 and 5.

Let D be a feedback vertex set of G. Then, for every $j \in [c]$, $|D \cap A_j| \geq |A_j| - 2$. So, we have $\prod_{j=1}^{c} (1 + |A_j| + \binom{|A_j|}{2})) = \mathcal{O}(n^{2c})$ many choices of intersections of a feedback vertex set of G with $A_1, \dots, A_c$. Now, fix one such choice $(A_1', \dots, A_c')$ from A_j's. And we guess a subset $S' \subseteq S$ that intersects with feedback vertex set in G. Note that we have excluded the vertices of $A_j \backslash A_j'$ in the feedback vertex set. If $G[(S \backslash S') \cup (\bigcup_{j=1}^{c} (A_j \backslash A_j'))]$ is not a forest, then we move to the next guess. Otherwise $G[(S \backslash S') \cup (\bigcup_{j=1}^{c} (A_i \backslash A_j'))]$ is a forest. We update $\ell' = |S| + \sum_{j=1}^{c} |A_j| - 1 - |S'| \sum_{j=1}^{c} |A_j'| \leq k - |S'| + 2c - 1$. Now, our goal is to identify whether there exists a feedback vertex set of size at most ℓ' contained in B, after deletion of vertices of $A_1', \dots, A_c', S'$. This is again the DISJOINT FEEDBACK VERTEX SET problem where $V(G') = S'' \dot\cup B$ such that S'' induces a forest, B is an independent set. Let $|S'| = p$. As $\ell \leq |S| + \sum_{j=1}^{c} |A_j|$, we have that $\ell' \leq |S| - |S'| + 2c - 1 \leq k - p + 2c - 1$ as at least two vertices from each of $A_1, \dots, A_c$ had been deleted. Let $S'' = (S \backslash S') \cup (\bigcup_{j=1}^{c} (A_j \backslash A_j'))$. Also, the number of connected components in $G[(S \backslash S') \cup (\bigcup_{j=1}^{c} (A_j \backslash A_j'))]$ is $c(S'')$ which is at most $|S| - |S'| + c - 1 = k + c - p - 1$ as there are c components of them having at least two vertices. We define a measure $\mu(G') = \ell' + c(S'') \leq 2k + 3c - 2p - 1$. Now, we apply Reduction Rules 3, 6 and 7 in sequence. It is easy to see that μ does not increase when a reduction rule is applied. When Reduction Rules 3, 6 and 7 are not applicable, then we have that every vertex in I has at least three neighbors in S'' and all of them are in different components of $G[S'']$. If every vertex in B has exactly three neighbors in S'' and all are in different components in $G[S'']$, then the problem is polynomial time solvable. So, we can assume that there exists some vertex $u \in B$ that has at least four neighbors in S'' and all of those neighbors are in different components of $G[S'']$. We branch on such vertices using Branching Rule 1. In one branch, we pick v into the solution. In that case μ drops by one. In other branch, we add v to S''. In this case, μ drops by at least three as at least four connected components in $G[S'']$ merges into one component. So, we have the following recurrence where α is constant not depending on k or n.

$$T(\mu) \leq T(\mu - 1) + T(\mu - 3)$$

Solving the recurrence we get that $T(\mu) \leq 1.4656^\mu = 2.148^{k-j+1.5c-1}$. The total running time of the algorithm is as follows.

$$n^{2c+\mathcal{O}(1)} \cdot \sum_{j=0}^{k} \binom{k}{j} \cdot 2.148^{k-j+1.5c} = n^{2c+\mathcal{O}(1)} 2.148^{1.5c} \cdot \sum_{j=0}^{k} \binom{k}{k-j} \cdot 2.148^{k-j}$$

$$= \mathcal{O}(3.148^{k+c} \cdot n^{2c+\mathcal{O}(1)})$$

Theorem 3. FVS DELETION TO $(c,1)$-GRAPH *admits an algorithm with running time* $\mathcal{O}(3.148^{k+c} \cdot n^{2c+\mathcal{O}(1)})$. *I.e. the problem admits an XP algorithm when* c *is part of the parameter.*

$W[1]$-hardness. Jansen et al. [10] also prove that FVS parameterized by c is $W[1]$-hard. We slightly extend this result to justify that FVS is $W[1]$-hard when it is parameterized by deletion distance to $(c,1)$-Graph when (c,k) is a parameter. We have the following lemma.

Lemma 3. *For $i \leq 1$ and $c \geq 2$ FVS-DELETION TO (c,i)-GRAPH is $W[1]$-hard when (c,k) is the parameter.*

Proof. We reduce FVS-DELETION TO $(c,0)$-GRAPH to FVS-$(c,1)$-GRAPH as follows. We assume that the partition $V(G)\backslash S = A_1 \dot\cup \ldots \dot\cup A_c$ is given with the input. Find a maximal independent set from $G\backslash S$. Let B be the set of such vertices. We delete B from $A_1, \ldots, A_c$. We get $A'_1, \ldots, A'_c$ where for all $j \in [c]$, A'_i is also a clique and B is an independent set of size at least 2. The reason we are guaranteed to get an independent set of size at least 2 from $G\backslash S$ is that otherwise we could merge all $C_1, \ldots, C_r$ into a single partition which is also a clique, that contradicts the assumption that $c \geq 2$. Note that we have not changed the graph at all and we have adjusted the partitions slightly. So, FVS-DELETION TO $(c,1)$-GRAPH is $W[1]$-hard when (c,k) is the parameter. $\square$

5   FVS Parameterized by Cluster Vertex Deletion Set

Now, we consider a special case of (c,i)-Graph where $i = 0$. Moreover, there is no edge between a vertex u in one partition and a vertex v in a different partition of $G\backslash S$. In particular, each clique partition is a connected component, and the parameter is the number of vertices in a cluster vertex deletion set. Here we can omit the assumption that the cluster vertex deletion set (S) is given along with the input as there is an algorithm by Boral et al. [3] that runs in $\mathcal{O}^*(1.9106^k \cdot (n + m))$ time and either outputs a cluster vertex deletion set of size at most k (if exists) or says that no cluster vertex deletion set of size at most k exists. Central to our algorithm is the following theorem which is what is essentially used in the proof of Theorem 4.5 of Bodlaender et al. [2].

Theorem 4. *Let G be a graph given with a nicer path decomposition $\mathcal{P} = (X_1, \ldots, X_q)$ such that the intersection of any feedback vertex set with any bag X_i has at most $2^k \cdot (|X_i| - k)^{\mathcal{O}(1)}$ possibilities. Then, FEEDBACK VERTEX SET can be solved in $\mathcal{O}(5^k \cdot n^{\mathcal{O}(1)})$ time.*

Now, we explain nicer path decomposition and justify how G has such a nicer path decomposition as required by the Theorem 4. We define path decomposition and pathwidth of a graph.

Definition 4 (Path Decomposition). *Let $G = (V, E)$ be an undirected graph. Path decomposition of G is a sequence of bags $\mathcal{P} = (X_1, \ldots, X_q)$ where $\forall i \in [q], X_i \subseteq V(G)$ such that the following properties are satisfied.*

1. *For every vertex $u \in V(G)$, there exists $i \in [q]$ such that $u \in X_i$.*
2. *For every edge $(u, v) \in E(G)$, there exists $i \in [q]$ such that $u, v \in X_i$.*
3. *For any vertex $u \in V(G)$, if $u \in X_i \cap X_k$ for some $i \leq k$, then $u \in X_j$ for all $i \leq j \leq k$.*

Definition 5 (Pathwidth). *Let $\mathcal{P} = (X_1, \ldots, X_q)$ be a path-decomposition of G. Then, width of $\mathcal{P}$ is denoted as $pw(\mathcal{P}) = max_{i \in [q]}\{|X_i| - 1\}$. And pathwidth of G is the minimum over the width of all possible path-decompositions of G. More specifically, if $\mathcal{\hat{A}}$ be the set of path decompositions of G, then $pw(G) = min_{\mathcal{P} \in A} pw(\mathcal{P})$.*

A path decomposition is called *nice* if all of its bags are one of the following types.

- Introduce Bag: A bag X_{i+1} is called an *introduce vertex bag* if $X_{i+1} = X_i \cup \{u\}$ where $u \notin X_i$.
- Forget Bag: A bag X_{i+1} is called a *forget bag* if $X_{i+1} = X_i \setminus \{u\}$ where $u \in X_i$.
- $X_1 = X_q = \emptyset$.

We also have the following lemma which is Lemma 7.2 of [7].

Lemma 4. *If a graph G admits a path decomposition of width p, then it also admits a nice path decomposition of width p. Moreover, given a path decomposition $\mathcal{P} = (X_1, \ldots, X_q)$ of G of width at most p, one can in time $\mathcal{O}(p^2 max(q, |V(G)|))$ compute a nice path decomposition of G of width p.*

In particular, we can convert a nice path decomposition into a *nicer path decomposition* in polynomial time where the bags are of three types (see again [7]).

- Introduce Vertex Bag: A bag X_{i+1} is called an *introduce vertex bag* if $X_{i+1} = X_i \cup \{u\}$ where $u \notin X_i$.
- Introduce Edge Bag: We say a bag X_i is *introduce edge bag* if it is labeled by an edge (u, v) and $X_i = X_{i-1}$. Note that in such case $u, v \in X_i$. Also note that an edge is introduced exactly once in the entire decomposition.
- Forget Bag: A bag X_{i+1} is called a *forget bag* if $X_{i+1} = X_i \setminus \{u\}$ where $u \in X_i$.

We consider $V(G_i) = (\bigcup_{j=1}^{i} X_i)$ and $E(G_i) = \{(u, v) \in E(G) | (u, v)$ is introduced in one of the bags X_j for some $j \in [i]\}$. In particular, it is easy to see that a given a nice path decomposition can be converted into a *nicer path decomposition* also in $\mathcal{O}(p^2 max(q^2, |V(G)|^2))$ time.

Now, we get back to our problem FVS-CVD. Let $C_1, \ldots, C_c$ be the set of connected components of $G \backslash S$. Consider the following subsets of vertices $\forall i \in [c], S \cup C_i \subseteq V(G)$. Let us denote these subsets in the following order. $S \cup C_1, \ldots, S \cup C_c$. We argue that $(S \cup C_1, \ldots, S \cup C_c)$ is a path-decomposition.

Lemma 5. *For the problem FVS-CVD, $(S \cup C_1, \ldots, S \cup C_c)$ forms a path-decomposition of G.*

Proof. Consider any vertex $u \in V(G)$. If $u \in S$, then $u \in S \cup C_i$ for all $i \in [q]$. So, both property 1 and 3 are satisfied for some $u \in S$. If $u \in C_i$ for some $i \in [c]$, then $u \in S \cup C_i$ and there is exactly one $i \in [c]$ in which u exists. So, property 1 and 3 are satisfied for all $u \in V(G)$. Consider any edge $(u,v) \in E(G)$. If $u, v \in S$, then $u, v \in S \cup C_1$. If $u \in S, v \in C_i$ for some $i \in [c]$, then $u, v \in S \cup C_i$. If $u, v \in C_i$, then $u, v \in S \cup C_i$. So, property 2 is satisfied. $\square$

Now, notice that in this path decomposition, a feedback vertex set of G can intersect any bag $(S \cup C_i)$ in at most $2^{|S|} \cdot |C_i|^2$ possibilities. It can be easily shown that this path decomposition can be converted into a nicer path decomposition with similar intersection behavior with a feasible feedback vertex set of G. Now, from Lemma 5 and Theorem 4 we get the following theorem.

Theorem 5. *FVS-CVD admits an algorithm that runs in $\mathcal{O}(5^k \cdot n^{\mathcal{O}(1)})$ time.*

6   Conclusion

We have considered structural parameterization of FEEDBACK VERTEX SET where the parameter is the deletion distance to a graph whose vertex set is partitioned into cliques and independent sets. When parameterized by the size of the deletion set to split graphs, our FPT algorithm is faster than the best known FPT algorithm for FVS when parameterized by the solution size, though the split deletion set size is incomparable to the solution size. A clear open problem is to improve the runtime of the FPT algorithms we considered in this paper. Improving the FPT runtime of FVS parameterized by solution size remains as an open problem by itself.

References

1. Bafna, V., Berman, P., Fujito, T.: A 2-approximation algorithm for the undirected feedback vertex set problem. SIAM J. Discret. Math. **12**(3), 289–297 (1999)
2. Bodlaender, H.L., Cygan, M., Kratsch, S., Nederlof, J.: Deterministic single exponential time algorithms for connectivity problems parameterized by treewidth. Inf. Comput. **243**, 86–111 (2015)
3. Boral, A., Cygan, M., Kociumaka, T., Pilipczuk, M.: A fast branching algorithm for cluster vertex deletion. Theory Comput. Syst. **58**(2), 357–376 (2016)
4. Cao, Y., Chen, J., Liu, Y.: On feedback vertex set: new measure and new structures. Algorithmica **73**(1), 63–86 (2015)

5. Chen, J., Fomin, F.V., Liu, Y., Lu, S., Villanger, Y.: Improved algorithms for feedback vertex set problems. J. Comput. Syst. Sci. **74**(7), 1188–1198 (2008)
6. Chen, J., Kanj, I.A., Jia, W.: Vertex cover: further observations and further improvements. J. Algorithms **41**(2), 280–301 (2001)
7. Cygan, M., Fomin, F.V., Kowalik, L., Lokshtanov, D., Marx, D., Pilipczuk, M., Pilipczuk, M., Saurabh, S.: Parameterized Algorithms. Springer, Heidelberg (2015)
8. Cygan, M., Pilipczuk, M.: Split Vertex Deletion meets Vertex Cover: new fixed-parameter and exact exponential-time algorithms. Inf. Process. Lett. **113**(5–6), 179–182 (2013)
9. Diestel, R.: Graph Theory. Graduate Texts in Mathematics, vol. 173, 4th edn. Springer, Heidelberg (2012)
10. Jansen, B., Raman, V., Vatshelle, M.: Parameter ecology for feedback vertex set. Tsinghua Sci. Technol. **19**(4), 387–409 (2014)
11. Jansen, B.M.P., Fellows, M.R., Rosamond, F.A.: Towards fully multivariate algorithmics: parameter ecology and the deconstruction of computational complexity. Eur. J. Comb. **34**(3), 541–566 (2013)
12. Kociumaka, T., Pilipczuk, M.: Faster deterministic feedback vertex set. Inf. Process. Lett. **114**(10), 556–560 (2014)
13. Kolay, S., Panolan, F.: Parameterized algorithms for deletion to (r, ell)-graphs. In: Proceedings of FSTTCS, pp. 420–433 (2015)
14. Kolay, S., Panolan, F.: Parameterized algorithms for deletion to (r, l)-graphs. CoRR, abs/1504.08120 (2015)
15. Lokshtanov, D., Narayanaswamy, N.S., Raman, V., Ramanujan, M.S., Saurabh, S.: Faster parameterized algorithms using linear programming. ACM Trans. Algorithms **11**(2), 15:1–15:31 (2014)
16. Majumdar, D.: Structural parameterizations of feedback vertex set. In: IPEC, Aarhus, Denmark, pp. 21:1–21:16 (2016)
17. Raman, V., Saurabh, S., Subramanian, C.R.: Faster fixed-parameter tractable algorithms for finding feedback vertex sets. ACM Trans. Algorithms **2**(3), 403–415 (2006)
18. Thomassé, S.: A $4k^2$ kernel for feedback vertex set. ACM Trans. Algorithms **6**(2), 1–8 (2010)

A Constant Amortized Time Algorithm for Generating Left-Child Sequences in Lexicographic Order

Kung-Jui Pai[1], Jou-Ming Chang[2($\boxtimes$)], and Ro-Yu Wu[3]

[1] Department of Industrial Engineering and Management,
Ming Chi University of Technology, New Taipei City, Taiwan
poter@mail.mcut.edu.tw
[2] Institute of Information and Decision Sciences,
National Taipei University of Business, Taipei, Taiwan
spade@ntub.edu.tw
[3] Department of Industrial Management,
Lunghwa University of Science and Technology, Taoyuan, Taiwan
eric@mail.lhu.edu.tw

Abstract. Wu et al. (Theoret. Comput. Sci. 556:25–33, 2014) recently introduced a new type of sequences, called left-child sequences (LC-sequences for short), for representing binary trees. They pointed out that such sequences have a natural interpretation from the view point of data structure and gave a characterization of them. Based on this characterization, Pai et al. (International conference on combinatorial optimization and applications. Springer, Cham, pp. 505–518, 2016) showed that there is an easily implementing algorithm that uses generate-and-test approach to filter all LC-sequences of binary trees with n internal nodes in lexicographic order, while in general this algorithm is not efficient at all. In this paper, we design two novel rotations that allow us to drastically alter the shape of binary trees (and thus their corresponding LC-sequences). As an application, these operations can be employed to generate all LC-sequences in lexicographic order. Accordingly, we present a more efficient algorithm associated with the new types of rotations for generating all LC-sequences and show that it takes only constant amortized running cost.

Keywords: Constant amortized time algorithm · Binary trees · Left-child sequences · Lexicographic order · Generation algorithms · Amortized cost

1 Introduction

Binary trees are one of the most fundamental data structures in computer science and have been widely studied over half a century. Usually, binary trees are encoded by using integer sequences and many types of integer sequences have been introduced (e.g., see [9,11] for surveys). For convenience, hereafter the terms

© Springer International Publishing AG 2017
M. Xiao and F. Rosamond (Eds.): FAW 2017, LNCS 10336, pp. 221–232, 2017.
DOI: 10.1007/978-3-319-59605-1_20

of binary trees and their corresponding sequences are often used interchangeably. Due to many practical applications in computer science, such as combinatorial object search or algorithm performance analysis, exhaustively generating all binary tree sequences is an important issue in research topic. Generation algorithms customarily produce sequences in a specific ordering, such as lexicographic order [16, 22, 23] or Gray-code order [14, 17]. For algorithmic efficiency, in general, sequences generated in lexicographic order is demanded to run in constant amortized time [2, 3, 18]. By contrast, sequences generated in Gray-code order is in need of taking a constant time for each generation (e.g., the so-called loopless algorithms proposed by Ehrlich [4]). For more references of binary tree sequences generation, we refer to [10, 13–16, 19, 20].

Recently, Wu et al. [20] proposed a loopless algorithm associated with the usual tree rotations (i.e., left rotation and right rotation for AVL-trees) to generate four types of binary tree sequences simultaneously. In particular, the generation includes two new types of sequences called *left-child sequences* (LC-sequences for short) and their mirror images called *right-child sequences* (RC-sequences for short), as defined later in Sect. 2. It is well-known that the practice in implementing binary trees usually adopted the so-called structure-pointer representation so that the spaces of nodes in a tree are dynamically allocated by structured memory and children of nodes are accessed via pointers. Wu et al. [20] thereby claimed that LC- and RC-sequences are inspired by such a natural structure representation. Moreover, they gave characterizations of the two types of sequences (see Theorem 1). However, both LC- and RC-sequences generated in [20] are not in lexicographic order or Gray-code order. In fact, the difference between two successive LC-sequences (resp., RC-sequences) in the generated list is either one or two digits.

Later on, based on the characterization of LC-sequences provided in [20], Pai et al. [12] showed that there is an algorithm using generate-and-test approach that allows developers to easily implement for generating all LC-sequences of binary trees with n internal nodes in lexicographic order (see Procedure Lex-Gen-Tree in Sect. 3), while this algorithm is quite not efficient. Indeed, the purpose of [12] is to develop efficient ranking algorithm (i.e., a function that determines the rank of a given sequence in the generated list) and unranking algorithm (i.e., a function that produces the sequence corresponding to a given rank) of LC-sequences in lexicographic order. As expected, their ranking and unranking algorithms can be run in amortized cost of $\mathcal{O}(n)$ time and space. Since the difference between two consecutive LC-sequences in the lexicographic order may vary widely, it means that the shapes of corresponding binary trees are possibly changed drastically. To adapt to this unavoidable situation, in this paper we design two massive rotations which can deal with a great variety of changes to assist our generation. As a result, we develop an algorithm called Refined-Lex-Gen-Tree that associates with these tree rotations to generate all LC-sequences in lexicographic order. Moreover, we show that this algorithm is more efficient and has constant amortized running cost. By symmetry, generation of RC-sequences can be developed by a similar way.

The rest of this paper is organized as follows. In Sect. 2, we formally give the definitions of LC- and RC-sequences, and introduces a coding trees structure for representing all LC-sequences in lexicographic order. In Sect. 3, we define two new types of rotations for binary trees, and then propose a constant amortized-time algorithm associated with these rotations for generating all LC-sequences in lexicographic order. Finally, concluding remarks are given in the last section.

2   Preliminaries

An *extended binary tree* is a rooted and ordered tree such that every internal node has exactly two children called the *left child* and the *right child* [9]. Let T be an extended binary tree with n internal nodes numbered from 1 to n in inorder (i.e., visit recursively the left subtree, the root and then the right subtree of T). Henceforth, we shall not distinguish the terms between a node and its inorder number. For a node $i \in T$, the subtree rooted at i is denoted by T_i. Also, the sub-tree rooted at the left child (resp., right child) of i is called the *left subtree* (resp., *right subtree*) of i and is denoted by L_i (resp., R_i). The *left arm* (resp., *right arm*) of T is the path from the root to its leftmost leaf (resp., rightmost leaf).

2.1   Left-Child Sequences

Recently, Wu et al. [20] introduced new types of sequences called *left-child sequence* (LC-sequence for short) and *right-child sequence* (RC-sequence for short) to represent binary trees. Given a binary T with n internal nodes labeled by $1, 2, \ldots, n$ in inorder, the LC-sequence of T, denoted by $\ell(T) = (\ell[1], \ell[2], \ldots, \ell[n])$, is an integer sequence so that the term $\ell[i]$, $1 \leqslant i \leqslant n$, is defined as follows:

$$\ell[i] = \begin{cases} 0 & \text{if the left child of } i \text{ is a leaf;} \\ j & \text{if } j \text{ is the left child of } i \text{ in } T. \end{cases} \tag{1}$$

Similarly, $r(T) = (r[1], r[2], \ldots, r[n])$ denotes the RC-sequence of T, where we use the right child instead of the left child in Eq. (1) to define the term $r[i]$. For instance, the LC-sequence and RC-sequence of the binary tree T shown in Fig. 1 are $\ell(T) = (0, 1, 0, 3, 0, 2, 0, 0, 7)$ and $r(T) = (0, 4, 0, 5, 0, 9, 8, 0, 0)$, respectively.

Wu et al. [20] showed that the two types of binary tree sequences can transform to each other in linear time. Moreover, they characterized the two types of sequences as follows.

Theorem 1 (Wu et al. [20]). *Let $c = (c_1, c_2, \ldots, c_n)$ be an integer sequence. Then,*

(a) *c is the LC-sequence of a binary tree T with n internal nodes if and only if the following conditions are fulfilled for all $i \in \{1, 2, \ldots, n\}$: (1) $0 \leqslant c_i < i$ and (2) $c_j = 0$ or $c_j > c_i$ for all $c_i < j < i$.*

(b) *c is the RC-sequence of a binary tree T with n internal nodes if and only if the following conditions are fulfilled for all $i \in \{1, 2, \ldots, n\}$: (1) $c_i > i$ or $c_i = 0$ and (2) $c_j = 0$ or $c_j < c_i$ for all $i < j < c_i$.*

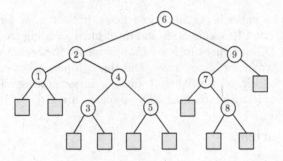

Fig. 1. A binary tree T with LC-sequence $\ell(T) = (0, 1, 0, 3, 0, 2, 0, 0, 7)$ and RC-sequence $r(T) = (0, 4, 0, 5, 0, 9, 8, 0, 0)$.

2.2   Coding Tree Structure

Let $\mathscr{T}_n$ be the set of binary trees with n internal nodes. It is well-known that $|\mathscr{T}_n| = \frac{1}{n+1}\binom{2n}{n}$ (i.e., the Catalan number). To depict all binary tree sequences, a systematic way by using coding trees was suggested in [10]. For a rooted tree, a path from the root to a leaf is called a *full path*. A *coding tree* $\mathbb{T}_n$ is a rooted tree consisting of n levels of nodes such that every node is associated with a label and the labels along a full path in $\mathbb{T}_n$ represent the sequence of a binary tree with n internal nodes. Figure 2 demonstrates the coding tree $\mathbb{T}_5$ for representing LC-sequences, where each node x_i in a full path $(x_1, x_2, \ldots, x_n)$ is labeled by $\ell[i]$. For notational convenience, we also write $\ell(x_i) = \ell[i]$ when we provide the full path $(x_1, x_2, \ldots, x_n)$ corresponding to a binary tree T.

For instance, in $\mathbb{T}_5$, labels in the left arm represent the right-skewed tree with LC-sequence $(0, 0, 0, 0, 0)$, and labels in the right arm represent the left-skewed tree with LC-sequence $(0, 1, 2, 3, 4)$. Hereafter, we consider a specific coding tree $\mathbb{T}_n$ in which all LC-sequences of binary trees are emerged from left to right in lexicographic order.

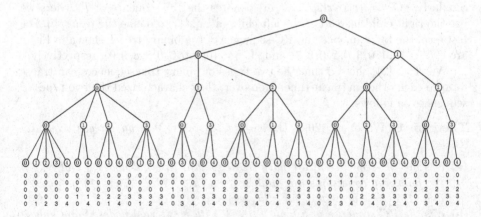

Fig. 2. A coding tree $\mathbb{T}_4$ for representing LC-sequences and LD-sequences.

3   Generating LC-sequences in Lexicographic Order

Based on the characterization of LC-sequences described in Theorem 1, Pai et al. [12] showed that there is an easy implementing algorithm (see Fig. 3) to generate all LC-sequences of binary trees with n internal nodes in lexicographic order. In this algorithm, the outer loop specifies the range of condition (a.1) in Theorem 1, and the **if** $\cdots$ **then** statement in the inner loop is the testing of condition (a.2) in Theorem 1. Initially, we set $\ell[1] = 0$, and then perform a procedure call Lex-Gen-Tree(2) to start the generation.

Procedure Lex-Gen-Tree(i)

```
begin
    if i = n + 1 then   Print(ℓ[1], ℓ[2], ..., ℓ[n]);
    else
        for ℓ[i] ← 0 to i − 1 do      // The range of condition (a.1) in Theorem 1.
            flag ← 1;   j ← ℓ[i] + 1;
            while flag = 1 and j < i do
                //, The testing of the condition (a.2) in Theorem 1.
                if ℓ[j] ≠ 0 and ℓ[j] ⩽ ℓ[i] then
                    flag ← 0;
                j ← j + 1;
            if flag = 1 then   call Lex-Gen-Tree(i + 1);
```

Fig. 3. An procedure for generating LC-sequences in lexicographic order.

The above algorithm uses generate-and-test approach for filtering out all non-valid LC-sequences. Although the algorithm is correct, in general it is not efficient at all. For instance, we reveal some non-efficient evidences as follows. Suppose that $c = (c_1, c_2, \ldots, c_n)$ is a non-valid LC-sequence satisfying $c_j \neq 0$ and $c_j \leqslant c_i$ for some integers $i \in \{3, 4, \ldots, n\}$ and $c_i < j < i$. By Theorem 1, all subsequent sequences $(c_1, \ldots, c_{i-1}, c_i', c_{i+1}, \ldots, c_n)$ for $c_i' \in \{c_i + 1, \ldots, i - 2\}$ are also non-valid. However, Lex-Gen-Tree does not detect this aspect and it performs the sequence generation and testing continuously.

A *rotation* is a simple operation that reconstructs a binary tree into another tree with the same number of nodes and preserves its inorder to be unchanged. In what follows, we design two new types of rotations for binary trees, where one is adjustable and the other is non-adjustable. Then, we present a more efficient algorithm, called Refined-Lex-Gen-Tree, that associates with these rotations for generating all LC-sequences in lexicographic order. Particularly, both rotations in a binary tree T are performed at node n (i.e., the parent of the rightmost leaf in T). Also, we imagine that T is the right subtree of a dummy node numbered by 0. The first one is called the *flip-on-site rotation*, denoted by FOS(k), which is an operation that flips the node n and its k immediate descendants in the left arm of the subtree T_n. Note that this operation is adjustable because the number of all flipped nodes is dependent on k. In fact, the degenerate case of this operation when $k = 1$ is the usual right rotation (for AVL trees) at node n in T. See Fig. 4(a) for an illustration.

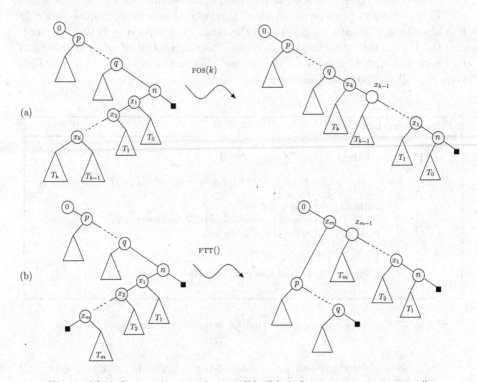

Fig. 4. (a) A flip-on-site rotation FOS(k); (b) A flip-to-top rotation FTT().

The second operation is called the *flip-to-top rotation*, denoted by FTT(), which is an operation that flips the node n and its all descendants in the left arm of T_n, and then moves this flipped list to the position between the root 0 and its right child. See Fig. 4(b) for an illustration. In general, the two types of rotations consist of a sequence of usual right rotations except for the last move of the flipped list in FTT().

Initially, the improved algorithm, called Refined-Lex-Gen-Tree, generates the first tree (i.e., the right-skewed tree with n internal nodes), and then repeatedly call a procedure Next-Tree() to generate all subsequent trees by using rotations defined above until a certain condition (i.e., CARRY $\leqslant n$, as explained later in the list of algorithm) is fulfilled. In each generation, a boolean variable FLIP_TO_TOP can determine which of FOS(k) and FTT() is the current operation. To preserve the inorder of binary trees to be unchanged after a rotation, we need the following three arrays $\ell[0..n]$, $r[0..n]$ and $p[0..n]$, where the first two are used for LC- and RC-sequences, and the last one is used for storing the parent information of nodes. For each usual right rotation in the flipped list, the node where the rotation acts on is indicated by the variable "this", and its parent and left child are indicated by variables "prev" and "next", respectively. The detail of the refined algorithm is as shown in Fig. 5.

Algorithm   Refined-Lex-Gen-Tree

begin
    for $i \leftarrow 0$ **to** n **do**   // Generate the first tree.
        $\ell[i] \leftarrow 0$; $r[i] \leftarrow i+1$; $p[i] \leftarrow i-1$;
    $r[n] \leftarrow 0$;
    Print$(\ell[1], \ell[2], \ldots, \ell[n])$;   // Print the first tree sequence;
    CARRY $\leftarrow 2$;
    while CARRY $\leqslant n$ **do**   // Generate the next tree sequence
        Next-Tree();

Procedure Next-Tree()

```
1   begin
2       this ← n;    prev ← p[this];    next ← ℓ[this];
3       if ℓ[n] = 0 then FLIP_TO_TOP ← true;
4       else if r[ℓ[n]] ≠ 0 then FLIP_TO_TOP ← false;
5       else
6           repeat                                    // Flip a sequence of nodes.
7               r[next] ← this;    p[this] ← next;    ℓ[this] ← 0;
8               p[next] ← 0;       this ← next;       next ← ℓ[this];
9           until ℓ[this] ≠ this − 1;
10          if ℓ[this] = 0 then FLIP_TO_TOP ← true;
11          else FLIP_TO_TOP ← false;

12      if FLIP_TO_TOP then                           // Perform a flip-to-top rotation.
13          r[prev] ← 0;    ℓ[this] ← r[0];    p[r[0]] ← this;
14          r[0] ← this;    p[this] ← 0;

15      else                                          // Perform a usual right rotation.
16          ℓ[this] ← r[next];    p[r[next]] ← this;    r[next] ← this;
17          p[this] ← next;       r[prev] ← next;       p[next] ← prev;

18      Print(ℓ[1], ℓ[2], …, ℓ[n]);    // Print the current tree sequence;
19      if ℓ[CARRY] = CARRY − 1 then CARRY ← CARRY + 1;
```

Fig. 5. A refined algorithm for generating LC-sequences in lexicographic order.

In the above algorithm, we use a global variable CARRY to control the generation in progress. Since the last tree in the generated list is the left-skewed tree satisfying $\ell[i] = i - 1$ for all $1 \leqslant i \leqslant n$, it is the errand of CARRY for completing this setting. Since $\ell[1] = 0$ is never changed, we set CARRY $= 2$ at the beginning. Once the setting in the current position is accomplished, CARRY is increased by one and goes ahead to the next position (see Line 19). The algorithm terminates when the condition CARRY $= n + 1$ is achieved.

In Next-Tree(), the variable FLIP_TO_TOP determines which type of rotations will be invoked. The decision is relied on the following rules:

(R1) if $\ell[n] = 0$, call FTT(); (Lines 13–14)

(R2) if $\ell[n] \neq 0$ and $r[\ell[n]] \neq 0$, call FOS(1); (Lines 16–17)

(R3) if $\ell[n] \neq 0$ and $r[\ell[n]] = 0$, flip a sequence of nodes by using usual right rotations until a rotation is performed at a node satisfying $\ell[\text{this}] \neq \text{this}-1$; (Lines 6–9)

Note that, for our Next-Tree() procedure, all flipped nodes triggered by the rule (R3) indeed have no right child because these nodes in the sequence meet with the condition $\ell[\text{this}] = \text{this} - 1$. After flipping the list of nodes in (R3), there are two statuses of the node where the last rotation acted on (i.e., the node indicated by the variable "this"). Accordingly, two kinds of subsequent processes are as follows:

(R4) if $\ell[\text{this}] = 0$, call FTT(); (Lines 13–14)

(R5) if $\ell[\text{this}] \neq 0$, call FOS(1); (Lines 16–17)

Actually, a flip-on-site rotation FOS(k) is an operation integrated with rules (R3) and (R5) for flipping the node n and partial descendants with $k - 1$ nodes in the prefix segment of the left arm of T_n, as shown in Fig. 4(a). By contrast, a flip-to-top rotation FTT() is an operation integrated with rules (R3) and (R4) for flipping the node n and all descendants in the left arm of T_n, and then moving the flipped list to the top of the tree, as shown in Fig. 4(b). Obviously, FOS(k) operation requires $\mathcal{O}(k)$ time, and the complexity of FTT() operation is dependent on the length of the left arm of T_n. Since a usual right rotation requires only constant time, the complexity of each operation is indeed equal to the number of different digits between two consecutive LC-sequences, i.e., before operating and after operating.

We now at a position to show that the above rules can correctly generate the next sequence in lexicographic order by using the two types of rotations.

Lemma 1. *Let T be a binary tree and suppose that $\ell(T) = (\ell[1], \ell[2], \ldots, \ell[n])$ satisfies the condition of (R1). If $\widetilde{T}$ is the binary tree obtained from T by taking a FTT() rotation at the node n, then $\ell(\widetilde{T})$ is the immediately succeeding sequence of $\ell(T)$ in lexicographic order.*

Proof. Let p be the right child of the root (i.e., the dummy node 0) in T. Since the node n has no left child, after performing FTT(), the whole sequence $\ell(T)$ keeps unchanged except the last position. In fact, the difference between $\ell(T)$ and $\ell(\widetilde{T})$ only occurs at $\ell[n]$, which is changed from 0 to p. We suppose to the contrary that the immediately succeeding sequence of $\ell(T)$ is $\ell(T') = (\ell[1], \ell[2], \ldots, \ell[n] = i)$ where $0 < i < p$ and T' is the corresponding binary tree. Since every number in an LC-sequence can appear at most once except the number 0, it implies that $\ell[j] \neq i$ for $1 \leqslant i \leqslant n - 1$, and thus $\ell(T)$ does not contain i as an element. Moreover, since $i < p$, it follows that i must be the right child of some node in the subtree L_p of T. Let q be the least ancestor of i in T for which some node $k \in L_p$ takes q as its left child (i.e., $\ell[k] = q$). Clearly,

$q < i < k < p$. For T', since i is the left child of n, it implies $k \in R_i$ and $q \notin R_i$. However, this contradicts the fact that $\ell[k] = q$ remains unchanged in T' because $k \neq n$.   □

Lemma 2. *Let T be a binary tree and suppose that $\ell(T) = (\ell[1], \ell[2], \ldots, \ell[n])$ satisfies the condition of (R2). If $\widetilde{T}$ is the binary tree obtained from T by taking a usual right rotation at the node n, then $\ell(\widetilde{T})$ is the immediately succeeding sequence of $\ell(T)$ in lexicographic order.*

Proof. Suppose that x is the left child of n and let y be the right chile of x in T, i.e., $\ell[n] = x$ and $r[\ell[n]] = y$. Clearly, after performing FOS(1), the sequence $\ell(T)$ keeps unchanged except the last position $\ell[n]$, which is changed from x to y. Suppose to the contrary that the immediately succeeding sequence of $\ell(T)$ is $\ell(T') = (\ell[1], \ell[2], \ldots, \ell[n] = i)$ where $x < i < y$ and T' is the corresponding binary tree. Since $x < i < y$ and $\ell[j] \neq i$ for $1 \leqslant i \leqslant n - 1$, i must be the right child of some node in the subtree L_y of T. Let q be the least ancestor of i in T for which some node $k \in L_y$ takes q as its left child (i.e., $\ell[k] = q$). Clearly, $q < i < k < y$. For T', since i is the left child of n, it implies $k \in R_i$ and $q \notin R_i$. However, this contradicts the fact that $\ell[k] = q$ remains unchanged in T' because $k \neq n$.   □

Lemma 3. *Suppose that T and $\widetilde{T}$ are two binary trees generated by the procedure Next-Tree() such that $\widetilde{T}$ is obtained from T. Then, $\ell(\widetilde{T})$ is the immediately succeeding sequence of $\ell(T)$ in lexicographic order.*

Proof. By Lemmas 1 and 2, we have proved the correctness of generation produced by rotations without a flipped list of nodes, i.e., the status meets with the condition of (R1) or (R2). In general, to show the correctness when the current status meets with the condition of (R3), we may imagine that nodes in the flipped list are contracted to form a single node and it is indicated by the variable "this". This is due to the fact that every node in the flipped list contains no right child. As a result, the rule (R4) is in keeping with the rule (R1) if we treat the variable "this" as n. Also, we note that if the current status meets with the condition of (R5), it guarantees $r[\ell[\text{this}]] \neq 0$. Otherwise, we have $\ell[\text{this}] = \text{this} - 1$ and the loop (Lines 6–9) goes ahead to the next round. Similarly, if we treat the variable "this" as n, the rule (R5) is again in keeping with the rule (R2). Therefore, using arguments similar to Lemmas 1 and 2, we can prove the correctness if the status meets with the condition of (R4) or (R5), and thus the lemma follows.   □

Theorem 2. *The algorithm Refined-Lex-Gen-Tree can correctly generates all LC-sequences of binary trees with n internal nodes in lexicographic order. In particular, each generation requires only constant amortized time with no more than 2.*

Proof. The correctness of the algorithm Refined-Lex-Gen-Tree directly follows from Lemma 3. We now give the complexity analysis as follows. Recall that we use $\mathcal{T}_n$ to denote the set of binary trees with n internal nodes. Let EC_n be the expected cost of generating an LC-sequences of length n in Refined-Lex-Gen-Tree. In [10], Lucas et al. showed that several coding trees for representing binary tree sequences are isomorphic. Actually, a coding tree for representing a certain type of binary tree sequences may come from an old one by changing the sequence representation and rearranging sequence order. Thus, the number of nodes in each level of the coding tree does not be changed, i.e., a Catalan number $|\mathcal{T}_k|$ for $k \in \{1 \ldots n\}$. Let N_k be the number of pairs of two consecutive LC-sequences of length n with k different digits in the lexicographic order. It is easy to observe that for each $k \in \{1 \ldots n-1\}$,

$$N_k = |\mathcal{T}_{n-k+1}| - |\mathcal{T}_{n-k}|.$$

Since $|\mathcal{T}_1| = 1$, the total complexity of generating all LC-sequences of length n is

$$\sum_{k=1}^{n-1} kN_k = \left(|\mathcal{T}_n| - |\mathcal{T}_{n-1}|\right) + 2\left(|\mathcal{T}_{n-1}| - |\mathcal{T}_{n-2}|\right) + 3\left(|\mathcal{T}_{n-2}| - |\mathcal{T}_{n-3}|\right) + \cdots$$
$$+ (n-2)\left(|\mathcal{T}_3| - |\mathcal{T}_2|\right) + (n-1)\left(|\mathcal{T}_2| - |\mathcal{T}_1|\right)$$
$$= S_n - n,$$

where S_n denote the sum of the first n Catalan numbers [1]. In fact, it has been pointed out in [24] that $S_n < |\mathcal{T}_{n+1}|$. Thus, we have

$$EC_n = \frac{\displaystyle\sum_{k=1}^{n-1} kN_k}{|\mathcal{T}_n|} = \frac{|\mathcal{T}_n| + S_{n-1} - n}{|\mathcal{T}_n|} < \frac{2|\mathcal{T}_n| - n}{|\mathcal{T}_n|} < 2.$$

In particular, the expected cost $EC_n = \frac{4}{3}$ when n tends to infinite. This completes the proof.   □

4   Concluding Remarks

In this paper, we propose a constant amortized-time algorithm for generating all LC-sequences of binary trees with n internal nodes in lexicographic order. It is especially interested that the proposed algorithm is associated with two new types of rotations called flip-on-site and flip-to-top. As we know that a rotation can be viewed as a transformation that changes the shape of a binary tree and usually preserves some desired property, such as keeping the inorder unchanged or adjusting to be a balanced tree, this leads to that tree transformation has many applications [5–8]. Thus, the design of efficient way for tree transformation is an important issue. However, up to now there are many discussions related to the usual rotations, and only a few attention has been focused on the design of massive rotations (e.g., see [21] as an instance). Since both flip-on-site and flip-to-top are massive rotations, we expect to find more applications that can be dealt with by these rotations in the near future.

Acknowledgments. This research was partially supported by MOST grants MOST 105-2221-E-131-027 (Kung-Jui Pai), 104-2221-E-141-002-MY3 (Jou-Ming Chang) and 104-2221-E-262-005 (Ro-Yu Wu) from the Ministry of Science and Technology, Taiwan.

References

1. Adamchuk, A.: A014138. The On-Line Encyclopedia of Integer Sequences (2006). http://oeis.org/A014138
2. Boyer, J.M.: Simple constant amortized time generation of fixed length numeric partitions. J. Algorithms **54**, 31–39 (2005)
3. Effler, S., Ruskey, F.: A CAT algorithm for generating permutations with a fixed number of inversions. Inform. Process. Lett. **86**, 107–112 (2003)
4. Ehrlich, G.: Loopless algorithms for generating permutations, combinations, and other combinatorial configurations. J. ACM **20**, 500–513 (1973)
5. Gibbons, A., Sant, P.: Rotation sequences and edge-colouring of binary tree pairs. Theor. Comput. Sci. **326**, 409–418 (2004)
6. Guibas, L., Hershberger, J., Suri, S.: Morphing simple polygons. Discret. Comput. Geometry **24**, 1–34 (2000)
7. Hershberger, J., Suri, S.: Morphing binary trees. In: Proceedings of the ACM-SIAM Sixth Annual Symposium on Discrete Algorithms (SODA), pp. 396–404 (1995)
8. Kensler, A.: Tree rotations for improving bounding volume hierarchies. In: IEEE Symposium on Interactive Ray Tracing, pp. 73–76. IEEE Computer Society, Washington (2008)
9. Knuth, D.E.: The Art of Computer Programming. Fascicle 4A - Generating All Trees, vol. 4. Addison-Wesley, Boston (2005)
10. Lucas, J.M., van Baronaigien, D.R., Ruskey, F.: On rotations and the generation of binary trees. J. Algorithms **15**, 343–366 (1993)
11. Mäkinen, E.: A survey on binary tree codings. Comput. J. **34**, 438–443 (1991)
12. Pai, K.-J., Wu, R.-Y., Chang, J.-M., Chang, S.-C.: Amortized efficiency of ranking and unranking left-child sequences in lexicographic order. In: Chan, T.H., Li, M., Wang, L. (eds.) COCOA 2016. LNCS, vol. 10043, pp. 505–518. Springer, Cham (2016). doi:10.1007/978-3-319-48749-6_37
13. Pallo, J.: Enumerating, ranking and unranking binary trees. Comput. J. **29**, 171–175 (1986)
14. Proskurowski, A., Ruskey, F.: Binary tree Gray codes. J. Algorithms **6**, 225–238 (1985)
15. van Baronaigien, D.R.: A loopless algorithm for generating binary tree sequences. Inform. Process. Lett. **39**, 189–194 (1991)
16. Ruskey, F., Hu, T.C.: Generating binary trees lexicographically. SIAM J. Comput. **6**, 745–758 (1977)
17. Savage, C.D.: A survey of combinatorial Gray codes. SIAM Rev. **39**, 605–629 (1997)
18. Sawada, J.: Generating bracelets in constant amortized time. SIAM Comput. **31**, 259–268 (2001)
19. Vajnovszki, V.: On the loopless generation of binary tree sequences. Inform. Process. Lett. **68**, 113–117 (1998)
20. Wu, R.-Y., Chang, J.-M., Chan, H.-C., Pai, K.-J.: A loopless algorithm for generating multiple binary tree sequences simultaneously. Theoret. Comput. Sci. **556**, 25–33 (2014)

21. Wu, R.-Y., Chang, J.-M., Wang, Y.-L.: A linear time algorithm for binary tree sequences transformation using left-arm and right-arm rotations. Theoret. Comput. Sci. **355**, 303–314 (2006)
22. Zaks, S.: Lexicographic generation of ordered trees. Theoret. Comput. Sci. **10**, 63–82 (1980)
23. Zaks, S., Richards, D.: Generating trees and other combinatorial objects lexicographically. SIAM J. Comput. **8**, 73–81 (1979)
24. Zumkeller, R.: A014138. The On-Line Encyclopedia of Integer Sequences (2010). http://oeis.org/A014138

Geodetic Contraction Games on Trees

Yue-Li Wang[✉]

Department of Digital Multimedia Design,
National Taipei University of Business, Taipei, Taiwan
ylwang@ntub.edu.tw

Abstract. The geodetic contraction game was introduced by Fraenkel and Harary (Int. J. Game Theor. 18:327–338, 1989). They showed that the problem on trees can be solved by using the algorithm for solving the Hackendot game. However, if we use the algorithm for solving the Hackendot game directly, then it will take $O(n^3)$ time for solving the geodetic contraction game on trees, where n is the number of vertices in a tree. They also posed the following open question: Is there a more efficient strategy to solve the geodetic contraction game on trees? In this paper, we show that the geodetic contraction game on trees can be solved in $O(n \log n)$ time.

1 Introduction

In [3], Fraenkel and Harary introduced the geodetic contraction game on graphs which is defined as follows:

Definition 1. Let $G = (V, E)$ be a finite graph. The set L of labeled vertices is initially empty. Two players Alice and Bob move alternately (Alice first), by choosing an unlabeled vertex $u \in V \backslash L$; then u itself and all vertices on shortest paths between u and any vertex of L are adjoined to L. When $L = V$, the game is over and the first player unable to move loses and the other player wins.

Example 1. In this example, the geodetic contraction game is played on the graph as shown in Fig. 1(a). Initially, the set L is empty and Alice moves first. If Alice selects vertex a, then vertex a is labeled and is put to L (see Fig. 1(b)). In the turn by Bob, if he selects vertex e, then e is labeled (see Fig. 1(c)). Since vertices b and d are on the shortest paths from e to a, vertices b and d are also labeled. Thus $L = \{a, b, d, e\}$ and vertices $a, b, d,$ and e are contracted to one vertex (see Fig. 1(d)). After that, no matter what vertex is selected by Alice, Bob will select the last unlabeled vertex which makes $L = V$ and wins.

In [3], Fraenkel and Harary showed that playing the geodetic contraction game on a tree after a vertex is labeled is equivalent to playing the Hackendot game on the rooted tree with the labeled vertex as the chopped root. The Hackendot game was invented by J. Von Neumann. The definition of the Hackendot game is as follows:

Y.-L. Wang—This work is supported under grants MOST 104–2221–E-011–023–MY3 and MOST 105–2221–E-011–086 .

© Springer International Publishing AG 2017
M. Xiao and F. Rosamond (Eds.): FAW 2017, LNCS 10336, pp. 233–240, 2017.
DOI: 10.1007/978-3-319-59605-1_21

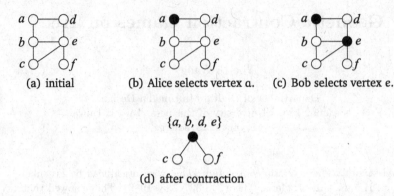

(a) initial   (b) Alice selects vertex a.   (c) Bob selects vertex e.

$\{a, b, d, e\}$

c ◯   ◯ f

(d) after contraction

Fig. 1. An example of geodetic contraction games.

Definition 2. The initial position of the Hackendot game is a directed tree T with root r. Two players Alice and Bob move alternately (Alice first). In the first move, Alice can select any vertex, say a, from T. Then every vertex on the path from the root r to a is deleted. This results in T to be a forest, denoted by T^{-a}. If T^{-a} is not empty, Bob can select a vertex, say b, from one of the rooted trees in T^{-a}. Similarly, all vertices from b to the root of the tree containing b are deleted. This leaves a forest T^{-ab}. Provided that the resulting forest T^{-ab} is nonempty, Alice selects a further vertex c, and forms T^{-abc}, and so on, with the players moving alternately. The first player unable to move loses.

Example 2. Assume that we want to play the Hackendot game on the rooted tree as shown in Fig. 2(a). If Alice selects vertex a, then the vertices in the path from r to a are deleted. Thus T^{-a} is a forest (see Fig. 2(b)). In the turn by Bob, if he selects vertex b, then the vertices in the path from f to b are deleted. The forest becomes T^{-ab} (see Fig. 2(c)). Figure 2(d) depicts T^{-abc}. In this way, Alice will win the game since there exist only four isolated vertices and now is Bob's turn.

Von Neumann proved that the first player has a winning strategy in the Hackendot game. However, he did not produce any construction for such a strategy. In [5], Úlehla provided such a strategy by computing the nim-value of a rooted tree which can be computed in $O(n)$ time, where n is the number of vertices in a tree. Furthermore, we can determine which vertex will be selected to win the game after the nim-value of the rooted tree is obtained. By applying the same procedure on the subtrees repeatedly, the winning way can be constructed in $O(n^2)$ time. By trying every vertex of an unrooted tree as the root of the tree and using the algorithm in [5], the geodetic contraction game on trees can be solved in $O(n^3)$ time. In [3], Fraenkel and Harary posed the following open question: Is there a more efficient strategy for solving the geodetic contraction game on trees? In this paper, we show that the geodetic contraction game on trees can be solved in $O(n \log n)$ time.

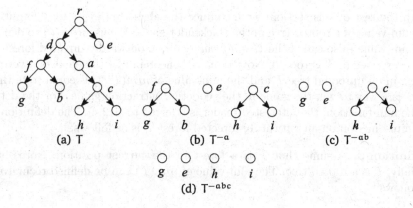

(a) T   (b) T^{-a} (c) T^{-ab}

(d) T^{-abc}

Fig. 2. An example of the Hackendot game.

This paper is organized as follows. In Sect. 2, we introduce the algorithm in [5] for computing the nim-value of a rooted tree. In Sect. 3, we show an efficient algorithm for solving the geodetic contraction game on trees.

2 Preliminaries

The following theorem states that the geodetic contraction game is closely related to the Hackendot game.

Theorem 1 [3]. *Playing the geodetic contraction game on a tree T after a vertex, say u, is labeled is equivalent to playing the Hackendot game on T^{-u}, i.e., u is a chopped root.*

We use an example to illustrate Theorem 1. In Fig. 3(a), there is an unrooted tree. If we select vertex u as a labeled vertex (see Fig. 3(b)) in playing the geodetic contraction game, then it is equivalent to playing the Hackendot game on T^{-u} (see Fig. 3(c)).

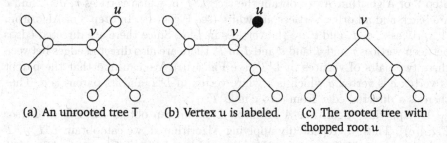

(a) An unrooted tree T   (b) Vertex u is labeled.   (c) The rooted tree with chopped root u

Fig. 3. An illustration for Theorem 1.

In the rest of this section, we introduce the algorithm in [5] for computing the nim-value of a rooted tree in the Hackendot game. We use $nim(T)$ to denote the nim-value (also called the *Grundy value*) on an outwardly directed forest T, i.e., every tree in T is rooted. Note that, by Theorem 1, if there exists a vertex, say u, in an unrooted tree T and the nim-value of $nim(T^{-u})$ is equal to 0, then there is a way to win the game in the geodetic contraction game. For the detail of the nim-function, the interested reader is referred to [1,2,4]. The definition of the nim-function on an outwardly directed forest T is as follows:

Definition 3. Assume that T is a forest in the current position. Note that, initially, T is a rooted tree. The nim-value of $nim(T)$ can be defined recursively as follows:

$$nim(T) = \begin{cases} 0 & \text{if } T \text{ is empty,} \\ nim(T_1) \oplus nim(T_2) \oplus \cdots \oplus nim(T_s) & \text{if } T \text{ has more than one tree,} \\ mex(\{nim(F_1), nim(F_2), \ldots, nim(F_t)\}) & \text{if } T \text{ is exactly a tree,} \end{cases}$$
(1)

where $T_1, T_2, \ldots$, and T_s are the trees in T with $s > 1$, the operation $\oplus$ is the bitwise-exclusive-or of the corresponding numbers, $F_1, F_2, \ldots$, and F_t are all of the possible resulting forests after a move is taken on T, and $mex(\{nim(F_1), nim(F_2), \ldots, nim(F_t)\})$ is equal to the smallest nonnegative integer which does not appear in the set $\{nim(F_1), nim(F_2), \ldots, nim(F_t)\}$.

First, we explain the bitwise-exclusive-or operation and the *mex* function. Assume that we want to compute the bitwise-exclusive-or of two numbers $x = 13$ and $y = 6$, i.e., $13 \oplus 6$. The binary representations of 13 and 6 are 1101_2 and 0110_2, respectively. Thus $13 \oplus 6 = 1101_2 \oplus 0110_2 = 1011_2$. Note that $0 \oplus 0 = 0, 0 \oplus 1 = 1, 1 \oplus 0 = 1$, and $1 \oplus 1 = 0$. Now let us use an example to explain the *mex* function. We have $mex(\{5, 3, 1, 0\}) = 2$ since 2 is the smallest nonnegative integer which does not appear in the set $\{5, 3, 1, 0\}$.

If we apply Eq. (1) to compute $nim(T)$ directly, then it will be very inefficient. In [5], Úlehla gave an $O(n)$-time algorithm for computing the nim-value of a rooted tree. In Úlehla's algorithm, there is an operation to transform a tree T to another tree L^1T which is described in Algorithm A.

We use the tree in Fig. 2(a) as an example to illustrate Algorithm A. After Step 2 of Algorithm A, we obtain the tree L^0T in which vertices r, d, f, and c are black and all other vertices are white (see Fig. 4(a)). In Step 3 of Algorithm A, vertices r, d, f, and c are the vertices in L^1T. Since there are directed edges between vertices r and d, and d and f in T, there are also directed edges between these two pairs of vertices in L^1T (see Fig. 4(b)). We can find that the parent of vertex c is vertex a which is a white vertex in L^0T and its parent is d. Thus there is a directed edge from d to c in L^1T.

By applying Algorithm A again on L^1T, we can obtain $LL^1T = L^2T$ (see Fig. 4(c)). Thus, by repeatedly applying Algorithm A, we can obtain L^2T, L^3T and so on until L^mT is empty. Note that $L^xT = LL^{x-1}T$ for $x \geqslant 1$. After L^iT for $0 \leqslant i \leqslant m - 1$ are obtained, the binary representation of $nim(T)$ can be computed as follows: Bit i for $0 \leqslant i \leqslant m - 1$ is 1 if the number of the rooted

Algorithm A

Input: A tree T.

Output: L^1T.

Step 1. /* Initialization. */

Label the vertices of T from 1 to n by the breadth-first search and set every vertex to be a white vertex.

Step 2. /* Finding L^0T. */

For $i = n$ down to 1,

if the vertex with label i is not a leaf and one of its children is a white vertex, then set this vertex to be a black vertex.

Step 3. /* Constructing L^1T. */

Select all black vertices from T to be the vertices of L^1T.

Add a directed edge between vertices x and y in L^1T if one of the following conditions is satisfied:

(1) There is a directed edge between x and y in T.

(2) The parent of y is a white vertex and x is a grand parent of y in L^0T.

Step 4. /* Output. */

Output L^1T.

white vertices in the forest L^iT is odd; otherwise, bit i is 0. Thus the binary representation of $nim(T)$ for the tree in Fig. 2(a) is 110_2. Therefore, it follows that $nim(T) = 6$ for the tree in Fig. 2(a).

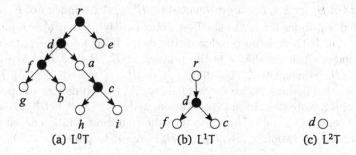

Fig. 4. An example of L^iT for $0 \leqslant i \leqslant 2$.

It is easy to show that, by using Algorithm A, the nim-value of a forest can be computed in $O(n)$ time. If the nim-value of a forest is not equal to zero, then the player who is going to make a move on the forest wins the game; otherwise, he/she loses. In the example above, to make $nim(T) = 6$ to be zero, we have to chop vertex d in T which is the only vertex in L^2T. Note that, after we obtain the nim-value of a forest, we can only determine the next move. After the opponent played, we have to recompute the nim-value of the resulting forest. Therefore, if we apply Algorithm A for each move in the Hackendot game, then it takes $O(n^2)$-time for constructing a winning way. Note that, in [6], Deuber and Thomassé showed that the Grundy value of N-free posets of order n can be computed in $O(n^4)$ time, where N-free posets are a larger class of rooted trees.

3   Main Results

In this section, we show that constructing a winning way for the Hackendot game can be done in $O(n \log n)$ time. Furthermore, we also show that determining whether Alice can win the game in the geodetic contraction game or not can also be done in $O(n \log n)$ time. Our main idea is to use the dynamic programming technique to compute the nim-value for every subtree in a rooted tree in the Hackendot game. Then, by using the depth-first search on a tree, we can compute the nim-value for every vertex in an unrooted tree T as a root of T for the geodetic contraction game.

Before we introduce our algorithm for computing $nim(T)$, we want to make a minor modification on Úlehla's algorithm when computing the ith bit for the binary representation of $nim(T)$. That is, if bit i is 0 in the binary representation of a forest T and the corresponding L^iT has more than one rooted white vertex, then we use 0^+ instead. In this case, we say that 0^+ is an *impure* 0; otherwise, we say that the zero is a *pure* 0. Moreover, we want to define an operation on a set of binary numbers which is defined as follows:

Definition 4. Let $B_1, B_2, \ldots, B_n$ for some integer $n \geqslant 1$ be a set of nonnegative integers and $b_{m-1}^t b_{m-2}^{t'} \ldots b_0^t$ be the binary representation of B_t for $1 \leqslant t \leqslant n$. Note that b_i^t for some $1 \leqslant t \leqslant n$ and $1 \leqslant i \leqslant m-1$ may contain impure 0. The leftmost k bits of B_t for $1 \leqslant t \leqslant n$ are denoted by $B_{m-1,k}^t$. An index i of $B_1, B_2, \ldots,$ B_n is called a *p-index* if i is the smallest index satisfying that b_i^t is a pure 0 for $t = 1, 2, \ldots, n$. If there is no p-index in the range from 1 to $m - 1$, then let m be the p-index. Assume that s is the p-index of $B_1, B_2, \ldots, B_n$. A *p-value* of $B_1, B_2, \ldots, B_n$ is equal to $2^s(1 + 2(B_{m-1,k}^1 \oplus B_{m-1,k}^2 \oplus \cdots \oplus B_{m-1,k}^n))$, where $k = m - s - 1$. For brevity, we use $\sigma_s(B_1, B_2, \ldots, B_n)$ to denote the p-value of $B_1, B_2, \ldots, B_n$. Note that, in the $\oplus$ operation, we have $0^+ \oplus 0^+ = 0^+$, $0^+ \oplus 0 = 0^+$, $0^+ \oplus 1 = 1$, and $1 \oplus 1 = 0^+$. Note also that the rightmost s bits, i.e., bit 0 to bit $s - 1$, after the operation $\sigma_s(B_1, B_2, \ldots, B_n)$ are set to pure 0.

Now we are at a position to describe our dynamic programming algorithm for computing $nim(T)$. Clearly, if a rooted tree T contains only one vertex; then the vertex is a white vertex in L^0T and $nim(T) = 1$. Now assume that v is the root of T and $v_1, v_2, \ldots, v_t$ are the children of v. The subtrees rooted at $v_1, v_2, \ldots, v_t$ are denoted by $T_1, T_2, \ldots, T_t$, respectively. In the following, we assume that $nim(T_1), nim(T_2), \ldots, nim(T_t)$ have already computed. Note that all those values of $nim(T_1), nim(T_2), \ldots, nim(T_t)$ are in the binary representation.

$$nim(T) = \begin{cases} 1 & \text{if } T \text{ contains only one vertex,} \\ \sigma_s(nim(T_1), nim(T_2), \ldots, nim(T_t)) & \text{otherwise,} \end{cases}$$

$$(2)$$

where s is the p-index of $nim(T_1), nim(T_2), \ldots,$ and $nim(T_t)$.

The meaning of the equation $nim(T) = \sigma_s(nim(T_1), nim(T_2), \ldots, nim(T_t))$ in Eq. (2) is that vertex v will be a black vertex in L^iT for $0 \leqslant i \leqslant s-1$ and a white vertex in L^sT. After that, vertex v will not appear in L^iT for $i > s$. We use the following example to illustrate Eq. (2).

Example 3. Assume that we have $nim(T_1) = 10100011$, $nim(T_2) = 101010^+00$, and $nim(T_3) = 10001010$. It can be found that the p-index of $nim(T_1), nim(T_2)$, and $nim(T_3)$ is 4. Thus, by Eq. (2), it follows that $\sigma_4(nim(T_1), nim(T_2), nim(T_3)) = 100^+10000 = 2^4(1 + 2(5 \oplus 5 \oplus 4)) = 144$. Figure 5 shows the computations of the nim-values for all subtrees in Fig. 2(a) by using Eq. (2).

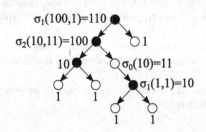

Fig. 5. An example to illustrate Eq. (2).

Theorem 2. *Constructing a winning way for the Hackendot game can be done in $O(n \log n)$ time, where n is the number of vertices in a tree.*

Proof. Note that the binary representation of the nim-value of each subtree has at most $\log n$ bits. By using Eq. (2), the nim-value of each subtree rooted at vertex v can be computed in $O(\deg(v) \log n)$ time. Therefore, the total time-complexity for computing the nim-values of all subtrees is $\sum_{v \in V} \deg(v) \log n = O(n \log n)$. □

Let T be an unrooted tree and u and v be two adjacent vertices in T. Let T^u and T^v denote the rooted trees of T with roots u and v, respectively. In T^u (respectively, T^v), the subtree with root v (respectively, u) is denoted by T_v^u (respectively, T_u^v). Comparing T^u with T^v, we can find that all subtrees are the same in these two rooted trees except those two subtrees T^v and T_u^v which are different from T_v^u and T^u, respectively. Thus, when we have all the nim-values of the subtrees in T^u, we only need to compute $nim(T_u^v)$ and $nim(T^v)$ to obtain all the nim-values of the subtrees in T^v. Accordingly, we have the following theorem.

Theorem 3. *Determining whether there is a winning way in the geodetic contraction game on an unrooted tree can be done in $O(n \log n)$ time, where n is the number of vertices in the tree.*

Proof. To solve the geodetic contraction game on trees T, we only need to compute the nim-value of each tree rooted at a vertex of T. To find the nim-value of each vertex, we can select any vertex in T as the root and compute the nim-values of all subtrees in the rooted tree. By using the depth-first search, we can compute the nim-value of each vertex as the root of T in $2\deg(v)\log n$ time. Thus the total time-complexity is still $O(n\log n)$. If there exists a rooted tree such that the nim-value of the chopped root is equal to 0, then Alice can win the game; otherwise, Bob wins. This completes the proof.   □

References

1. Bouton, C.: Nim, a game with a complete mathematical theory. Ann. Math. **3**, 35–39 (1902)
2. Conway, J.H.: On Numbers and Games. CRC Press, London (1976)
3. Fraenkel, A., Harary, F.: Geodetic contraction games on graphs. Int. J. Game Theory **18**, 327–338 (1989)
4. Gale, D., Neymann, A.: Nim-type games. Int. J. Game Theory **11**, 17–20 (1982)
5. Úlehla, J.: A complete analysis of Von Neumann's Hackendot. Int. J. Game Theory **9**, 107–113 (1980)
6. Deuber, W., Thomassé, S.: Grundy sets of partial orders. http://citeseer.nj.nec.com/19302.html (1996, preprint)

On Approximation Algorithms
for Two-Stage Scheduling Problems

Guangwei Wu[1,2], Jianer Chen[1,3], and Jianxin Wang[1(✉)]

[1] School of Information Science and Engineering, Central South University,
Changsha, People's Republic of China
jxwang@csu.edu.cn
[2] College of Computer and Information Engineering,
Central South University of Forestry and Technology,
Changsha, People's Republic of China
[3] Department of Computer Science and Engineering,
Texas A&M University, College Station, USA

Abstract. We study scheduling on parallel two-stage flowshops in which
each job has to pass through two operations: an R-operation and a T-
operation. Motivated by the current research in data centers, we consider
two restricted versions of the problem: one restricts that for each job, the
R-operation consumes no less time than the T-operation, while the other
assumes that the T-operation takes more time than the R-operation for
each job. For the first case, we present an online 2-competitive algorithm
and an offline 11/6-approximation algorithm. For the second case, we give
an online 5/2-competitive algorithm, and prove, for the offline setting,
that the problem can be reduced to the problem in the first case.

1 Introduction

A job is a *two-stage job* if it consists of an R-operation and a T-operation.
Correspondingly, a flowshop is called a *two-stage flowshop* if it contains an R-
processor and a T-processor. If a job is assigned to a flowshop, its T-operation
cannot start on the T-processor of this flowshop unless the R-operation is finished
on the R-processor of the same flowshop. Scheduling a set of two-stage jobs on
multiple two-stage flowshops is to assign the jobs to the flowshops and for each
flowshop to determine the execution order of the R- and T-operations of the jobs
assigned to it. In this paper, we will focus on schedulings whose objective is to
minimize the makespan, i.e., the completion time of the last job.

We will consider two restricted versions of the problem. The first assumes that
for each job the R-operation consumes no less time than the T-operation, while
the second assumes that the T-operation takes more time than the R-operation
for each job. These scheduling models are motivated by our current research in

This work is supported by the National Natural Science Foundation of China under
grants 61420106009, 61672536, 61232001 and 61472449, Scientific Research Fund of
Hunan Provincial Education Department under grant 16C1660.

M. Xiao and F. Rosamond (Eds.): FAW 2017, LNCS 10336, pp. 241–253, 2017.
DOI: 10.1007/978-3-319-59605-1_22

data centers. A modern data center contains hundreds of thousands of servers, and each server contains processors, network interface, and local I/O, etc. [1]. Softwares and data are stored as resources in the servers. Clients dynamically request the resources, and the cloud sends the resources to the clients over Internet. When a request arrives at a server, the server needs to read the resource from the secondary memory to the main memory first, and then sends the resource to the client over Internet. A server can read resource for one request while sending resources for other requests. Correspondingly, each request needs to pass through two stage operations, i.e., disk-read (the R-operation) and network-transmission (the T-operation), and the network-transmission cannot start unless the disk-read is finished. Thus scheduling such requests on servers in data centers to minimize the makespan is exactly the scheduling model studied by this paper.

The setting of servers in data centers is chosen based on the requested services [3]. For certain services that require high reliability of data, disk I/O transfer rate may be much lower than that of network, i.e., the R-operation costs more than the T-operation. On the other hand, services such as many web services have high I/O rate requirement to tackle with the huge requests from clients. In this circumstances, the R-operation is less expensive than the T-operation. As a result, if the servers in a cloud are divided into clusters based on the services they provide, then the requests to the servers in each cluster will be likely to be all with more expensive R-operations or all with more expensive T-operations.

It is easy to see that our scheduling models are generalizations of the classical Makespan problem, which schedules n (one-stage) jobs on m (one-stage) machines to minimize the makespan. The Makespan problem is NP-hard even when m is a constant, and becomes strongly NP-hard when m is an input [8]. As a consequence, both the versions of the parallel two-stage flowshop problem studied in this paper are strongly NP-hard.

We focus on approximation algorithms and competitive algorithms for the problem. A scheduling algorithm is an α-*approximation algorithm*, with an *approximation ratio* α, if for all instances, the algorithm delivers a schedule whose makespan is bounded by α times the minimum makespan of the instance [16]. Scheduling problems where the input is received online and the output must be given online are called *online scheduling problems*, and algorithms solving the problems are called *online algorithms*. An online algorithm is α-*competitive*, with a *competitive ratio* α, if for all finite input sequences I, the algorithm constructs an online schedule whose makespan is within α times the minimum makespan of the online schedules for the sequence I.

For classical Makespan problem, *ListRanking* is a well-known algorithm for both online and offline settings, which will also impact our algorithms for the new scheduling models. The *ListRanking* algorithm was proposed by Graham [9], who proved that the algorithm achieves $(2 - 1/m)$-approximation ratio. Later he improved the approximation ratio to $(4 - 1/m)/3$ by sorting the jobs before *ListRanking* [10]. *ListRanking* is also the first online algorithm for the problem. The competitive ratio $(2 - 1/m)$ stood for a long time, until Galambos and Woeginger provided a $(2 - \frac{1}{m} - \epsilon_m)$-competitive algorithm, where ϵ is a positive constant for $m \geq 4$, but tends to be 0 with the growth of m [7].

Bartal *et al.* presented a 1.986-competitive algorithm for $m > 70$, that first achieves a competitive ratio $2 - \epsilon$ for a positive constant ϵ [4]. Albers presented a 1.923-competitive algorithm based on a new strategy for $m \geq 2$ [2].

The paralled two-stage flowshop problem had not been studied throughly until very recently, and most of the studies focused on the problem when the number m of flowshops is a fixed constant. He *et al.* [12] first proposed this problem, motivated by applications in glass manufacturing. Vairaktarakis *et al.* [15] studied the problem in their work on hybrid flowshops, and proposed a formulation that leads to a pseudo-polynomial time algorithm when $m = 2$. Zhang *et al.* [18] presented constant ratio approximation algorithms for scheduling on two and three two-stage flowshops. The approximation ratio of these algorithms can reach $\frac{3}{2}$ for $m = 2$ and $\frac{12}{7}$ for $m = 3$. Using the formulation similar to that of [15], Dong *et al.* [6] presented a pseudo-polynomial time algorithm and developed a fully polynomial-time approximation scheme for the problem. Recently, Wu *et al.* [17] proposed a new formulation for the problem that leads to improvements over the algorithm complexity. They also studied algorithms for the cases where the costs of the two stages are different significantly. Approximation algorithms for k-stage jobs on multiple k-stage flowshops for general k had also been studied [14].

To the best of our knowledge, no approximation results for the paralled two-stage flowshop problem have been known when the number m of flowshops is not a fixed constant, which makes the problem become strongly NP-hard. This paper considers two restricted versions of the problem: one restricts that the R-operation consumes no less time than the T-operation for each job, while the other assumes that the T-operation takes more time than the R-operation for each job. For the first case, we present an online 2-competitive algorithm and an offline 11/6-approximation algorithm, for arbitrary m. For the second case, we give an online 5/2-competitive algorithm for arbitrary m, and prove, for the offline setting, that the problem can be reduced to the problem in the first case.

2   Scheduling on a Single Two-Stage Flowshop

Let $G = \{J_1, \ldots, J_n\}$ be a set of two-stage jobs to be scheduled in a system $\{M_1, \ldots, M_m\}$ of m identical two-stage flowshops. We assume that (1) each job $J_i = (r_i, t_i)$ consists of an R-operation of cost r_i (i.e. the R-*time*) and a T-operation of cost t_i (i.e. the T-*time*); (2) each flowshop has an R-processor and a T-processor that can run in parallel and can process the R-operations and the T-operations, respectively, of the assigned jobs; (3) the R- and T-operation of a job must be executed in the R- and T-processor, respectively, of the same flowshop, in a way that the T-operation cannot start unless the R-operation is completed; (4) there are no precedence constraints among the jobs; and (5) preemption is not allowed.

Under the above assumptions, each job J_i is represented by a pair (r_i, t_i) of integers. A *schedule* S consists of an *assignment* that assigns all jobs in G to flowshops, and for each flowshop, the execution orders of the R- and T-operations of the jobs assigned to that flowshop. The *completion time* of a flowshop M

under the schedule $\mathcal{S}$ is the time when M finishes the last T-operation of the jobs assigned to M. The *makespan* $C_{\max}$ of $\mathcal{S}$ is the largest flowshop completion time over all flowshops. In this paper, the number of flowshops is an input. Following the three-field notation $\alpha|\beta|\gamma$ suggested by Graham *et al.* [11], the general parallel two-stage flowshop problem can be written as $P|2\mathrm{FL}|C_{\max}$, and as $P|2\mathrm{FL}_{R\geq T}|C_{\max}$ for the first version, as $P|2\mathrm{FL}_{R<T}|C_{\max}$ for the second version.

Without loss of generality, we say an ordered job sequence $\langle J_1, \ldots, J_n \rangle$ is a schedule on a flowshop if the R- and T-operations of the jobs are executed by the R- and T-processor of the flowshop in the way which strictly follows the given order. Let $\bar{\rho}_i$ and $\bar{\tau}_i$, respectively, be the times when the R-operation and T-operation of job J_i are started. The following lemma holds if the objective of scheduling is to minimize the makespan.

Lemma 1 [17]. *Let $\mathcal{S} = \langle J_1, J_2, \ldots, J_n \rangle$ be a two-stage job sequence scheduled on a single two-stage flowshop, where $J_i = (r_i, t_i)$, for $1 \leq i \leq n$. Then for all i, $1 \leq i \leq n$, we can assume: $\bar{\rho}_i = \sum_{j=1}^{i-1} r_j$; and $\bar{\tau}_i = \max\{\bar{\rho}_i + r_i, \bar{\tau}_{i-1} + t_{i-1}\}$.*

Note that the execution of the R-processor of flowshop is continuous, the R-operation of a job starts as soon as the R-operations of previous jobs scheduled on the same flowshop are finished. On the other hand, the way the T-processor executes is different. For a job, its T-operation can not start unless not only the previous T-operations are finished but also its own R-operation is finished. Thus the execution of T-processor may not be continuous, and each "gap" in the execution means that there is a job whose T-operation waits for its R-operation being completed.

Let ρ_q, τ_q be the finishing times of the R- and T-processor of the flowshop M_q, respectively, when the jobs assigned to M_q are finished. From Lemma 1, we have Algorithm 1 that assigns a new job J_i to the flowshop M_q. The scheduling algorithms presented in the subsequent sections will call it whenever they need to assign a new job to a flowshop.

Algorithm 1. Assign job J_i to flowshop M_q

1: $\rho_q = \rho_q + r_i$;
2: **if** $\rho_q > \tau_q$ **then** $\tau_q = \rho_q + t_i$ **else** $\tau_q = \tau_q + t_i$

For a two-stage job $J_i = (r_i, t_i)$, define $J_i^d = (t_i, r_i)$ to be the *dual job* of J_i (i.e., the dual job J_i^d is obtained from the job J_i by swapping its R- and T-time). Let $\mathcal{S} = \langle J_1, J_2, \ldots, J_n \rangle$ be a schedule of two-stage jobs on a two-stage flowshop. Denote by $\mathcal{S}^d = \langle J_n^d, \ldots, J_2^d, J_1^d \rangle$ the *dual schedule* of $\mathcal{S}$ on the dual jobs, where J_i^d is the dual job of J_i for $1 \leq i \leq n$.

Theorem 1 [17]. *On a single two-stage flowshop, the optimal schedule of the job set $G = \{J_1, J_2, \ldots, J_n\}$ and the optimal schedule of the dual job set $G^d = \{J_1^d, J_2^d, \ldots, J_n^d\}$, where J_i^d is the dual job of J_i for $1 \leq i \leq n$, have the same completion time. Moreover, if a schedule $\mathcal{S}$ is optimal for the job set G then its dual schedule $\mathcal{S}^d$ is optimal for the dual job set G^d.*

Suppose that S is a schedule of the job set $G = \{J_1, J_2, \ldots, J_n\}$ on m two-stage flowshops, where for each i, S assigns a subset G_i of jobs in G to the i-th flowshop. By replacing the schedule of G_i on the i-th flowshop by its dual schedule for the dual job set G_i^d, we can get a schedule S^d on the m flowshops for the dual job set G^d. Based on the above discussion, it is easy to see that the completion times of each flowshop of the two schedules are the same. Furthermore, the following theorem holds [17].

Theorem 2. *On multiple two-stage flowshops, the optimal schedule of the job set G and the optimal schedule of the dual job set G^d have the same makespan. Moreover, an optimal schedule for the job set G can be easily obtained from an optimal schedule for the dual job set G^d.*

3   The Case $P|2\mathrm{FL}_{R \geq T}|C_{\max}$

In this section, we consider the problem $P|2\mathrm{FL}_{R \geq T}|C_{\max}$, which assumes that the R-operation consumes no less time than the T-operation for each job, i.e., $r_i \geq t_i$ for each job $J_i = (r_i, t_i)$. This case is suitable for the situation where the major time consumed by each job comes from data reading from server disks, rather than data transmission over Internet.

3.1   An Online Algorithm

We first consider the online version of the model, where jobs come in sequence and a scheduling algorithm has to decide which flowshop to assign a job to when the job arrives without any information about the following jobs.

Let ρ and τ denote the finishing times of the R-processor and T-processor of a flowshop, respectively. Specially, as defined in the last section, ρ_q, τ_q represent the finishing times of the R- and T-processor of the flowshop M_q, $1 \leq q \leq m$, when the assigned jobs on it are finished. The online algorithm is given as Algorithm 2. The main idea of the algorithm is that: for a coming job J_i, the algorithm provides the flowshop with the minimum ρ over all flowshops. Note that after assigning a new job, the algorithm needs to permute the flowshops (i.e., to reorganize flowshops) in non-decreasing order by their ρ value. Thus ρ_1 is always the minimum value and M_1 is always the flowshop for the coming job. Using a heap structure to hold the flowshops, sorted by their ρ value, the permuting operation takes time $O(\log m)$, while keeping the decision time $O(1)$. Therefore the algorithm can be regarded as "effective" [5].

Algorithm 2. The online algorithm for $P|2\mathrm{FL}_{R \geq T}|C_{\max}$.

1: When a job J_i arrives, assign it to the flowshop M_1 using Algorithm 1.
2: permute the list of flowshops in non-decreasing order by their ρ.

We introduce some notations for analysis. Let $OPT(G)$ be the minimum makespan of scheduling a two-stage job sequence G in a given system of m

two-stage flowshops. Without loss of generality, it is supposed that the algorithm achieves its makespan on the flowshop M_h. The completion time of M_h is denoted by τ^* instead of $C_{\max}$ for convenience. Let the job sequence $\langle J_1, \ldots, J_c \rangle$ be the schedule on M_h, and k $(1 \leq k \leq c)$ be the minimum job index from which the T-operations of the following jobs on M_h are executed continuously by M_h.

Let $a_0 = \sum_{i=1}^{k-1} r_i$ be the sum of the R-times of the jobs from J_1 to J_{k-1}, $b_0 = \sum_{i=1}^{k-1} t_i$ be the sum of the T-times of these jobs correspondingly. Similarly, $a_1 = \sum_{i=k+1}^{c-1} r_i$ and $b_1 = \sum_{i=k+1}^{c-1} t_i$. Figure 1 illustrates the state of M_h after scheduling. Note that we denote by b_0 the sum of the T-times of the jobs before J_k, though these T-operations may not be executed continuously. It is easy to see that such action will not affect the following analysis.

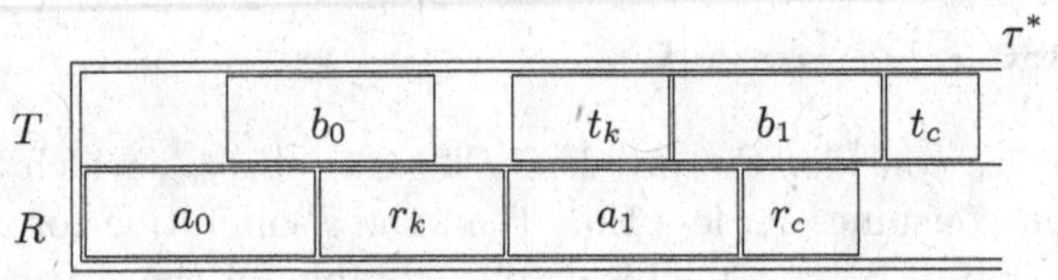

Fig. 1. The state of M_h after scheduling by the online algorithm in $P|2FL_{R \geq T}|C_{\max}$

We start the competitive analysis of Algorithm 2 with some lemmas.

Lemma 2. $OPT(G)$ is no smaller than r_i plus t_i for every job J_i. Moreover, $OPT(G)$ is no smaller than $2t_i$ for every job J_i.

Lemma 3. b_1 is no larger than a_1.

Lemma 4. $OPT(G)$ is no smaller than $a_0 + r_k + a_1$.

Proof. J_c is the last job scheduled on the flowshop M_h. According to Algorithm 2, M_h must be in the first place in the list of the flowshops when scheduling the coming job J_c, thus has the minimum ρ over all flowshops at that time. Therefore, the completing times of R-processors of all flowshops must be no smaller than the minimum ρ, which equals $a_0 + r_k + a_1$. Combining the fact that by Lemma 1, the R-operations are executed continuously in any flowshop, it is obvious that $OPT(G)$ is no smaller than the minimum completing time of R-processor over all flowshops, that is $OPT(G) \geq a_0 + r_k + a_1$.   □

Based on the above lemmas, we give a competitive analysis of this online algorithm as the following theorem.

Theorem 3. Algorithm 2 is 2-competitive for all $m \geq 2$.

Proof. The makespan achieved on M_h by Algorithm 2 is expressed as:

$$\tau^* = a_0 + r_k + t_k + b_1 + t_c \leq a_0 + r_k + t_k + a_1 + t_c \tag{1}$$

$$\leq OPT(G) + \frac{1}{2}OPT(G) + \frac{1}{2}OPT(G) = 2OPT(G). \tag{2}$$

We explain the derivations in (1) and (2). As discussed above, the execution of R-processor of any flowshop is continuous. By the definition of the job index k, the T-operations of the jobs from J_k to J_c are executed continuously. Moreover, the T-operation of J_k starts once its own R-operation is finished, otherwise by Lemma 1, the T-operation of J_k must be waiting for the previous T-operations on M_h being completed at that time. Thus there exists a job $J_{k'}$, $k' < k$, from which the T-operations of the following jobs on M_h are executed continuously, which is a contradiction against the definition of k. As a result, the equation in (1) holds. Lemma 3, $b_1 \leq a_1$, explains the inequality in (1). By Lemma 4, $a_0 + r_k + a_1 \leq OPT(G)$. We also have $t_i \leq \frac{1}{2}OPT(G)$ for each job J_i by Lemma 2. These explain the inequality in (2). The equation in (2) completes the proof that Algorithm 2 can construct a schedule with makespan bounded by $2OPT(G)$.   □

This competitive ratio almost matches that of the best online algorithm for the classical scheduling problem $P||C_{\max}$ [13], which can be regarded as a simpler version of the $P|2FL|C_{\max}$ problem in which all jobs are one-stage jobs and all machines are one-stage flowshops.

3.2   An Offline Algorithm

In this section, we discuss the same problem $P|2FL_{R \geq T}|C_{\max}$ in the situation where all the jobs are given offline, i.e., every job $J_i = (r_i, t_i)$, where $1 \leq i \leq n$, is known at the beginning. Offline version is also meaningful for scheduling problem [5]. Unlike online algorithm which makes decision when a job comes, offline algorithm outputs a schedule for all jobs at once. An offline algorithm for scheduling m multiple two-stage flowshops is given as Algorithm 3.

Algorithm 3. The offline algorithm for $P|2FL_{R \geq T}|C_{\max}$.

1: Sort all jobs into job sequence $G = \langle J_1, \ldots, J_n \rangle$ in non-increasing order by R-time;
2: **for** $i = 1$ to n **do**
3:     select a flowshop M_q which has the minimum ρ_q over all q, $1 \leq q \leq m$;
4:     assign the job J_i to the flowshop M_q using Algorithm 1;
5: **return** $\max_{1 \leq q \leq m}\{\tau_q\}$ as *makespan*

We give more detailed descriptions about step 3. There may be more than one flowshops which have the same minimum finishing time of R-processor when scheduling a job. In such case, the algorithm always picks the flowshop whose flowshop index is minimum over them. The time complexity of the offline algorithm is shown in the following theorem.

Theorem 4. *Algorithm 3 runs in time $O(n \log n + nm)$.*

The main idea of Algorithm 3 is similar to that of Algorithm 2. When assigning a job J_i, Algorithm 3 always picks the flowshop M_q with the minimum ρ over all flowshops to process J_i. The difference between them is that: in the offline algorithm, all the jobs are sorted into non-increasing order by their R-time at the first step, while such sorting can not be done in online setting where all jobs are coming online and algorithm does not have any information about the following jobs. As a consequence, after scheduling by Algorithm 3, the schedule on each flowshop must be a subsequence of the job sequence G, thus also follows non-increasing order by R-time.

Before analysis, we introduce some further notations used in the following part of this section. For a two-stage job $J_i = (r_i, t_i)$ in G, define by $J_i^r = (r_i, 0)$ the R-partial job of J_i, i.e., the job J_i^r is constructed from the original job J_i by setting its T-time to 0. Given a two-stage job set $G = \{J_1, \ldots, J_n\}$, its R-partial job set is $G^r = \{J_1^r, \ldots, J_n^r\}$, where J_i^r is the R-partial job of J_i for $1 \leq i \leq n$. Let $OPT(G)$ and $OPT(G^r)$ be the minimum makespan of scheduling G and G^r respectively on m two-stage flowshops. It is straightforward that $OPT(G) \geq OPT(G^r)$. We also have the following observation.

Theorem 5. *On m two-stage flowshops by Algorithm 3, a two-stage job J_i is assigned to the flowshop M_q when scheduling a two-stage job set G if its R-partial job J_i^r is assigned to M_q when scheduling the corresponding R-partial job set G^r, and vice versa.*

We use the same notations supposed in Sect. 3.1. M_h is the flowshop achieving the makespan τ^*. The job sequence $\langle J_1, \ldots, J_c \rangle$ is the schedule on M_h, and k $(1 \leq k \leq c)$ is the minimum job index from which the T-operations of the following jobs on M_h are executed continuously. For the sake of analysis, we remove the jobs following J_c in the job sequence G, which is sorted by Algorithm 3 in non-increasing order by R-time. This removal does not affect our analysis of the approximation ratio for this algorithm for two reasons: first, τ^* can not decrease, because all the removed jobs are scheduled after J_c thus are assigned to other flowshops. Second, $OPT(G)$ can not increase by removing jobs in G. After this removal, J_c is also the last job in the job sequence G, and hence has the minimum R-time over all jobs.

With these notations and observation, we start the analysis of the approximation ratio for Algorithm 3, which is divided into two cases based on whether the R-time r_c of the job J_c is larger than $\frac{1}{3}OPT(G)$. Now we consider the first case where $r_c > \frac{1}{3}OPT(G)$.

Lemma 5. *After scheduling a two-stage job set G on m flowshops by Algorithm 3, if $r_c > \frac{1}{3}OPT(G)$, then the value $\max_{1 \leq q \leq m}\{\rho_q\}$ is no larger than $OPT(G)$.*

Based on Lemma 5, the approximation ratio of the algorithm when $r_c > \frac{1}{3}OPT(G)$ is given as the following lemma.

Lemma 6. *If $r_c > \frac{1}{3}OPT(G)$, Algorithm 3 is $\frac{3}{2}$-approximation for all $m \geq 2$.*

Then we consider the second case where the last job J_c on M_h has $r_c \leq \frac{1}{3}OPT(G)$. We use the same notations in Sect. 3.1, such as the costs a_0, b_0,

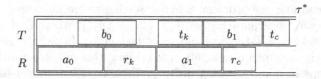

Fig. 2. The state of M_h after scheduling by the offline algorithm in $P|2\mathrm{FL}_{R \geq T}|C_{\max}$ when $r_c \leq \frac{1}{3}OPT(G)$

a_1 and b_1. The state of M_h after scheduling is shown in Fig. 2. The difference between Figs. 1 and 2 is that: since Algorithm 3 sorts the jobs in non-increasing order by R-time before scheduling and J_c is the last job in this sorted sequence, it follows that r_c is no larger than r_k in Fig. 2, while the relationship between r_c and r_k is uncertain in Fig. 1. Note that Lemmas 2 and 3, which are valid in arbitrary job sequence in this problem, still hold here. Lemma 4 is also valid because the property, M_h has the minimum ρ over all flowshops, still holds when scheduling J_c by Algorithm 3. The case is discussed in the following lemma.

Lemma 7. If $r_c \leq \frac{1}{3}OPT(G)$, Algorithm 3 is $\frac{11}{6}$-approximation for all $m \geq 2$.

Proof. The makespan is expressed as:

$$\tau^* = a_0 + r_k + t_k + b_1 + t_c \leq a_0 + r_k + t_k + a_1 + r_c \tag{3}$$

$$\leq OPT(G) + \frac{1}{2}OPT(G) + \frac{1}{3}OPT(G) = \frac{11}{6}OPT(G). \tag{4}$$

As discussed in Sect. 3.1, the equation in (3) holds obviously. b_1 is no larger than a_1 by Lemma 3, and t_c is no larger than r_c in this section. These explain the inequality in (3). By the same reasons for the inequality in (2), we have $a_0 + r_k + a_1 \leq OPT(G)$ and $t_k \leq \frac{1}{2}OPT(G)$ in this problem. Combining the assumption that r_c is no larger than $\frac{1}{3}OPT(G)$, the inequality in (4) is proved. Finally we give the conclusion that the makespan, which is achieved on the flowshop M_h by the offline algorithm, is bounded by $\frac{11}{6}OPT(G)$. □

By Lemmas 6 and 7 about the two cases based on whether the R-time of the last job scheduled on M_h is larger than $\frac{1}{3}OPT(G)$, the approximation ratio of the offline algorithm follows immediately as the following theorem.

Theorem 6. Algorithm 3 is $\frac{11}{6}$-approximation for all $m \geq 2$.

4 The Case $P|2\mathrm{FL}_{R<T}|C_{\max}$

In this section, the other problem $P|2\mathrm{FL}_{R<T}|C_{\max}$ is considered, which restricts that the R-time r_i is smaller than the T-time t_i for each job J_i where $1 \leq i \leq n$. This version is considered for the situation where the jobs on any servers in a cluster are all with more time-consuming T-operations.

We first consider the problem in online setting where jobs come online. For convenience, we define an array set $\psi = \{\psi_1, \ldots, \psi_m\}$, where ψ_q $(1 \leq q \leq m)$ records the sum of the T-times of jobs scheduled on M_q. In order to decide which flowshop to choose when a new job J_i is coming, this online algorithm always sorts flowshops in non-decreasing order by their ψ each time it assigns a job, and then picks M_1 as the flowshop to assign the job J_i to, i.e., the flowshop with the minimum sum of T-times of the jobs scheduled on it. This online algorithm is shown as Algorithm 4.

Algorithm 4. The online algorithm for $P|2\mathrm{FL}_{R<T}|C_{\max}$.

1: When a job J_i arrives, assign it to flowshop M_1 using Algorithm 1;
2: $\psi_1 = \psi_1 + t_i$; permute the list of flowshops in non-decreasing order by their ψ;

The only difference between Algorithms 2 and 4 is that: Algorithm 4 bases its scheduling strategy on the sum of the T-times instead of the sum of the R-times in Algorithm 2, of the jobs on each flowshop. The permuting operation and the decision also take time $O(\log m)$ and $O(1)$ respectively here, which makes Algorithm 4 "effective". Let $OPT(G)$ be the minimum makespan of scheduling a two-stage job sequence G in a given system of m flowshops. Without loss of generality, suppose that Algorithm 4 achieves makespan τ^* on the flowshop M_h. Let the job sequence $\langle J_1, \ldots, J_c \rangle$ denote the schedule on M_h, and k be the minimum job index from which the T-operations of the following jobs on M_h are executed continuously. We define the following notations similar to that in Sect. 3.1. Denote by $a_0 = \sum_{i=1}^{k-1} r_i$ and $b_0 = \sum_{i=1}^{k-1} t_i$ the sum of the R and T-times of the jobs $J_1, \ldots, J_{k-1}$ respectively. Similarly, $a_1 = \sum_{i=k+1}^{c-1} r_i$ and $b_1 = \sum_{i=k+1}^{c-1} t_i$. The state of M_h after scheduling is illustrated in Fig. 3. Note that the execution of the T-operations of the jobs on M_h before J_k may not be continuous. As Lemmas 2, 3 and 4, we have the following similar lemmas.

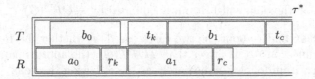

Fig. 3. The state of M_h after scheduling by the online algorithm in $P|2\mathrm{FL}_{R<T}|C_{\max}$

Lemma 8. *$OPT(G)$ is no smaller than r_i plus t_i for every job J_i. Moreover, $OPT(G)$ is larger than $2r_i$ for every job J_i.*

Lemma 9. *a_0 is smaller than b_0.*

Lemma 10. *$OPT(G)$ is no smaller than $b_0 + t_k + b_1$.*

The analysis for the online algorithm is shown as Theorem 7.

Theorem 7. *Algorithm 4 is $\frac{5}{2}$-competitive for all $m \geq 2$.*

Proof. The makespan achieved on the flowshop M_h is expressed as:

$$\tau^* = a_0 + r_k + t_k + b_1 + t_c < b_0 + r_k + t_k + b_1 + t_c \tag{5}$$
$$< OPT(G) + \frac{1}{2}OPT(G) + OPT(G) = \frac{5}{2}OPT(G). \tag{6}$$

It is obvious from Fig. 3 that the equation in (5) holds, whose reason is the same as that for the equation in (1). The reason for the inequality in (5) is also straightforward: by Lemma 9, a_0 is smaller than b_0. By Lemma 10, $b_0 + t_k + b_1 \leq OPT(G)$. We have that $r_k < \frac{1}{2}OPT$ from Lemma 8. By the same lemma, $OPT(G)$ is no smaller than $r_i + t_i$ for every job J_i, thus is no smaller than t_c. These explain the inequality in (6). The equation in (6) completes the proof. □

Now we describe how to deal with the problem $P|2\mathrm{FL}_{R<T}|C_{\max}$ in offline setting, where all the jobs are given from the beginning. Given a job set G of n two-stage jobs and m flowshops, we first construct the dual job set G^d of G, where J_i^d in G^d is obtained from J_i in G by swapping its R-time and T-time for $1 \leq i \leq n$. It is easy to see that for each job J_i^d in G^d, its R-time is larger than its T-time. Thus by Theorems 4 and 6, applying offline Algorithm 3 on the job set G^d can generate a schedule for G^d with makespan bounded by $\frac{11}{6}OPT(G^d)$ in time $O(n \log n + nm)$. By Theorem 2, $OPT(G^d)$ equals $OPT(G)$, and moreover, according to the description of Theorem 1, a schedule for the job set G can be easily constructed from a schedule for its dual job set G^d with no change in makespan in time $O(n)$. Using this method, we can get a schedule for the original job set G whose makespan is also bounded by $\frac{11}{6}OPT(G)$. The offline algorithm is given as Algorithm 5.

Algorithm 5. The offline algorithm for $P|2\mathrm{FL}_{R<T}|C_{\max}$.

1: **for** $i = 1$ to n **do** $r_i' = t_i$, $t_i' = r_i$;
2: let $G^d = \{J_1^d, \ldots, J_n^d\}$, where for each i, $J_i^d = (r_i', t_i')$;
3: Call Algorithm 3 on G^d, which returns *makespan* for G^d;
4: **return** *makespan* which is also the makespan for the job set G.

Compared to Algorithm 3, the additional step in Algorithm 5 of constructing the dual job set G^d of G costs time $O(n)$. Therefore the time complexity of Algorithm 5 is still $O(n \log n + nm)$.

5   Conclusion

Motivated by the current research in data centers and cloud computing, we studied the problem, that schedules two-stage jobs on m multiple two-stage flowshops to minimize the makespan. Especially, we considered this problem in the situation where the number m of flowshops is a part of input, which makes the problem become strongly NP-hard. To the best of our knowledge, there are no approximation results for the paralled two-stage flowshop problem when m is not a fixed constant. To meet the practice demand, this paper studied two restricted versions of the problem based on whether the R-time is smaller than the T-time for each job.

Online and offline algorithms for both versions are provided. For the first case which assumes that the R-time is no smaller than the T-time for each job, an online 2-competitive algorithm for all $m \geq 2$ is provided, whose competitive ratio almost matches that of the best online algorithm for the classical scheduling problem, which can be regarded as a simpler case of our problem. An offline $\frac{11}{6}$-approximation algorithm is also given for all $m \geq 2$. Then for the second case which constricts that the R-time is smaller than the T-time for each job, we give an online $\frac{5}{2}$-competitive algorithm for all $m \geq 2$. Using the concept and the theorems of dual job built in chapter 2, we state that by applying the offline algorithm in the first case on the dual job set, we can get an offline $\frac{11}{6}$-approximation algorithm for the original job set in this case.

References

1. Abts, D., Felderman, B.: A guided tour through data-center networking. Queue **10**(5), 10–23 (2012)
2. Albers, S.: Better bounds for online scheduling. SIAM J. Comput. **29**(2), 459–473 (1999)
3. Barroso, L.A., Clidaras, J., Hölzle, U.: The datacenter as a computer: an introduction to the design of warehouse-scale machines. Synth. Lect. Comput. Archit. **8**(3), 1–154 (2013)
4. Bartal, Y., Fiat, A., Karloff, H., Vohra, R.: New algorithms for an ancient scheduling problem. In: Proceedings of the Twenty-Fourth Annual ACM Symposium on Theory of Computing, pp. 51–58. ACM (1992)
5. Borodin, A., El-Yaniv, R.: Online Computation and Competitive Analysis. Cambridge University Press, New York (2005)
6. Dong, J., Tong, W., Luo, T., Wang, X., Hu, J., Xu, Y., Lin, G.: An FPTAS for the parallel two-stage flowshop problem. Theoret. Comput. Sci. **657**, 64–72 (2017)
7. Galambos, G., Woeginger, G.J.: An on-line scheduling heuristic with better worst-case ratio than Graham's list scheduling. SIAM J. Comput. **22**(2), 349–355 (1993)
8. Garey, M.R., Johnson, D.S.: Computers and Intractability: A Guide to the Theory of NP-Completeness. W. H. Freeman and Company, New York (1979)
9. Graham, R.L.: Bounds for certain multiprocessing anomalies. Bell Syst. Tech. J. **45**(9), 1563–1581 (1966)
10. Graham, R.L.: Bounds on multiprocessing timing anomalies. SIAM J. Appl. Math. **17**(2), 416–429 (1969)
11. Graham, R.L., Lawler, E.L., Lenstra, J.K., Kan, A.H.G.R.: Optimization and approximation in deterministic sequencing and scheduling: a survey. Ann. Discret. Math. **5**(1), 287–326 (1979)
12. He, D.W., Kusiak, A., Artiba, A.: A scheduling problem in glass manufacturing. IIE Trans. **28**(2), 129–139 (1996)

13. Pinedo, M.L.: Scheduling: Theory, Algorithms, and Systems. Springer Science, New York (2016)
14. Tong, W., Miyano, E., Goebel, R., Lin, G.: A PTAS for the multiple parallel identical multi-stage flow-shops to minimize the makespan. In: Zhu, D., Bereg, S. (eds.) FAW 2016. LNCS, vol. 9711, pp. 227–237. Springer, Cham (2016). doi:10. 1007/978-3-319-39817-4_22
15. Vairaktarakis, G., Elhafsi, M.: The use of flowlines to simplify routing complexity in two-stage flowshops. IIE Trans. **32**(8), 687–699 (2000)
16. Williamson, D.P., Shmoys, D.B.: The Design of Approximation Algorithms. Cambridge University Press, New York (2011)
17. Wu, G., Chen, J., Wang, J.: On scheduling two-stage jobs on multiple two-stage flowshops. Technical report, School of Information Science and Engineering Central South University (2016)
18. Zhang, X., van de Velde, S.: Approximation algorithms for the parallel flow shop problem. Eur. J. Oper. Res. **216**(3), 544–552 (2012)

A New Lower Bound for Positive Zero Forcing

Boting Yang[✉]

Department of Computer Science, University of Regina, Regina, SK, Canada
boting.yang@uregina.ca

Abstract. The positive zero forcing number is a variant of the zero forcing number, which is an important parameter in the study of minimum
rank/maximum nullity problems. In this paper, we first introduce the
propagation decomposition of graphs; then we use this decomposition
to prove a lower bound for the positive zero forcing number of a graph.
We apply this lower bound to find the positive zero forcing number of
matching-chain graphs. We prove that the positive zero forcing number
of a matching-chain graph is equal to its zero forcing number. As a consequence, we prove the conjecture about the positive zero forcing number
of the Cartesian product of two paths, and partially prove the conjecture about the positive zero forcing number of the Cartesian product of
a cycle and a path. We also show that the positive zero forcing number
and the zero forcing number agree for claw-free graphs. We prove that it
is NP-complete to find the positive zero forcing number of line graphs.

1 Introduction

This paper studies the positive zero forcing number and introduces the propagation decomposition of graphs which is used to prove a lower bound for the
positive zero forcing number of a graph. The zero forcing number was introduced
in [1]. The motivation is to bound the maximum nullity of a graph (and hence
the minimum rank) because the zero forcing process is a method of forcing zeros
in a null vector of a symmetric matrix described by a graph. The maximum
nullity problem for a graph G is to determine the largest possible nullity over all
real symmetric matrices described by the graph (the edges of G correspond to
nonzero off-diagonal entries), which is equivalent to the minimum rank problem.
Independently, the zero forcing number was introduced in physics as the graph
infection number [6], and it was also introduced in graph searching as the fast-
mixed search number [19,20]. The zero forcing number has been studied as an
interesting graph parameter by many researchers, see for example, [2,5,9,13].

The positive zero forcing number, also called positive semidefinite zero forcing
number, was introduced in [3] as a variant of the zero forcing number. The maximum positive semidefinite nullity problem (equivalently, the minimum positive
semidefinite rank problem) for a graph G is to determine the largest possible nullity (smallest possible rank) over all real positive semidefinite matrices described

---

Research supported in part by an NSERC Discovery Research Grant, Application
No.: RGPIN-2013-261290.

M. Xiao and F. Rosamond (Eds.): FAW 2017, LNCS 10336, pp. 254–266, 2017.
DOI: 10.1007/978-3-319-59605-1_23

by the graph. The motivation of the positive zero forcing is to bound the maximum positive semidefinite nullity. Details of the positive zero forcing process and the zero forcing process will be given in Sect. 2.

Barioli et al. [4] showed that for an outerplanar multigraph G, its maximum positive semidefinite nullity is equal to its tree cover number, which is the minimum number of vertex disjoint induced trees that cover all vertices of G. Ekstrand et al. [7] showed that every vertex of G is in some minimum positive zero forcing set. They also showed that the tree cover number of G is bounded by the positive zero forcing number. Ekstrand et al. [8] extended the definition of the positive zero forcing number for simple graphs to multigraphs. They showed that for a multigraph G of tree-width at most 2, its positive zero forcing number is equal to its maximum positive semidefinite nullity, and is also equal to its tree cover number. Fallat et al. [10] established a relation between the zero forcing and the fast-mixed searching [19], which implies some NP-completeness results for the zero forcing problem. They also gave a linear time algorithm for computing the positive zero forcing number of chordal graphs. Fallat et al. [11] investigated positive zero forcing within the context of certain edge clique coverings. They introduced the compressed cliques graph and studied a number of properties associated with the compressed cliques graph, including: uniqueness, forbidden subgraphs, connections to Johnson graphs, and positive zero forcing. Yang [21] proved that it is NP-complete to find the positive zero forcing number of a graph, and it remains NP-complete for graphs with maximum vertex degree seven. A linear time algorithm is presented in [21] for computing the positive zero forcing number of generalized series-parallel graphs.

In [16] (Table 3.1, p. 828), Peters conjectured that for the Cartesian product of a path of m vertices and another path of n ($n \geq m$) vertices, the positive zero forcing number is equal to m. He proved a special case of the conjecture for $m = 2$. Warnberg [17] proved another special case of the conjecture for $m = 3$. We will prove the general case of this conjecture in Sect. 5. Peters also conjectured that for the Cartesian product of a cycle of m vertices and a path of n vertices, the positive zero forcing number is equal to $\min\{m, 2n\}$. He proved two special cases, one for $m \geq 4$ and $n = 2$ and the other for $m = 4$ and $n \geq 2$. We will prove the conjecture for the case when $m \leq 2n$.

The main contributions of this paper are the following:

- We introduce the propagation decomposition of graphs (Sect. 3).
- We introduce the matrix form of graphs and use it to prove a lower bound for the positive zero forcing number of a graph (Sect. 4).
- We introduce the matching-chain graphs. We show that the positive zero forcing number of a matching-chain graph is equal to its zero forcing number, which can be used to prove the conjectures mentioned above (Sect. 5).
- We prove that the positive zero forcing number of a claw-free graph is equal to its zero forcing number (Sect. 6). We give an NP-completeness result for the positive zero forcing number of line graphs.

2   Preliminaries

In this paper, we only consider finite undirected graphs with no loops or multiple edges. Let $G = (V, E)$ denote a graph with vertex set V and edge set E. We also use $V(G)$ for the vertex set of G and $E(G)$ for the edge set. For vertices $u, v \in V$, we use $\{u, v\}$ to denote an edge with endpoints u and v; if $\{u, v\} \in E$, we say that u and v are adjacent. The *degree* of v, denoted $\deg_G(v)$, is the number of edges incident to v. The *neighborhood* of v, denoted $N_G(v)$, is the set of all vertices adjacent to v. For a subset $V' \subseteq V$, let $G[V']$ denote the subgraph induced by V', which is a subgraph whose vertex set is V' and the edge set is $\{\{u, v\} \in E(G) : u, v \in V'\}$. We use $G - V'$ to denote the induced subgraph $G[V \setminus V']$. Definitions omitted here can be found in [18].

We now define the positive zero forcing, and the associated positive zero forcing number. Let G be a graph and B_0 be a subset of vertices of G such that each vertex of B_0 is colored blue and all other vertices of G are uncolored. Let B be the current set of blue vertices of G (initially, $B = B_0$). Let $U_1, \ldots, U_k$ be the sets of uncolored vertices in each of the connected components of $G - B$ (note that it is possible that $k = 1$). *Positive zero forcing* (also called *Positive semidefinite zero forcing* [5]) is based on the following coloring rule: If v is a blue vertex in B and it has exactly one uncolored neighbor u in $G[U_i \cup B]$, then color the vertex u blue. In this case, we say that the vertex v *forces* u. We repeatedly applying this coloring rule (adding the new blue vertices into B at the end of each iteration) until no more vertices can be forced. If all uncolored vertices of G are forced to blue, then the set of initial blue vertices (i.e., B_0) is called a *positive zero forcing set*; it is called a *minimum positive zero forcing set* if the positive zero forcing set has the minimum cardinality; this minimum cardinality is called the *positive zero forcing number* of G, denoted $Z_+(G)$. That is,

$$Z_+(G) = \min\{|B_0| : B_0 \text{ is a positive zero forcing set of } G\}.$$

The procedure of coloring a graph using the coloring rule of the positive zero forcing is called a *positive zero forcing process*.

The positive zero forcing is a variant of the zero forcing, which has a different coloring rule: If v is a blue vertex in G and it has only one uncolored neighbor, say u, then color the vertex u blue. Similar to the positive zero forcing set/number, we can define the zero forcing set and the zero forcing number (see e.g., [1]).

Given two graphs H_1 and H_2, the *Cartesian product* of H_1 and H_2, denoted $H_1 \square H_2$, is the graph such that its vertex set is the Cartesian product $V(H_1) \times V(H_2)$ and two vertices $(u, v), (u', v') \in V(H_1) \times V(H_2)$ are adjacent if and only if $u = u'$ and $\{v, v'\} \in E(H_2)$, or $v = v'$ and $\{u, u'\} \in E(H_1)$.

Let $\mathcal{S}_n(\mathbb{R})$ denote the set of $n \times n$ real symmetric matrices. The *graph of* $A = [a_{ij}] \in \mathcal{S}_n(\mathbb{R})$, denoted $\mathcal{G}(A)$, is the graph with vertex set $\{1, 2, \ldots, n\}$ and edge set $\{\{i, j\} : a_{ij} \neq 0, 1 \leq i < j \leq n\}$. Note that the diagonal of A is ignored in determining $\mathcal{G}(A)$. For a given graph G, the set of symmetric matrices described by G is defined to be $\mathcal{S}(G) = \{A \in \mathcal{S}_n(\mathbb{R}) : \mathcal{G}(A) = G\}$. Let $\mathcal{S}_+(G)$ denote the subset of positive semidefinite matrices in $\mathcal{S}(G)$. The nullity of a matrix B is denoted by null(B). The *maximum nullity* of G is defined to be $\mathrm{M}(G) = \max\{\mathrm{null}(B) : B \in \mathcal{S}(G)\}$, and, similarly, the *maximum positive semidefinite nullity* of G is defined to be

$$\mathrm{M}_+(G) = \max\{\mathrm{null}(B) : B \in \mathcal{S}_+(G)\}.$$

It is shown in [3] that $\mathrm{M}_+(G) \le Z_+(G) \le Z(G)$.

A connected graph G is said to be k-*connected* (also called k-vertex-connected) if it contains at least $k+1$ vertices and it remains connected whenever fewer than k vertices are removed. The vertex connectivity $\kappa(G)$ is defined as the largest k such that G is k-connected. A connected graph G is k-*edge-connected* if it remains connected whenever fewer than k edges are removed. A *bridge* of a connected graph is an edge whose removal disconnects the graph.

A *path* is a sequence $(v_0, v_1, \ldots, v_k)$ of *distinct* vertices of G with $\{v_{i-1}, v_i\} \in E(G)$ for all i $(1 \le i \le k)$. The *length* of a path is the number of edges on it. The *distance* between two vertices u and v of G, denoted $\mathrm{dist}_G(u, v)$, is the length of the shortest path between them.

3 Propagation Tree Decomposition

Definition 1. Given a connected graph $G = (V, E)$, a *propagation tree decomposition* of G is a pair (T, W), where $T = (I, F)$ is a rooted tree with vertex set $I = \{1, 2, \ldots, m\}$ and edge set F such that each directed edge $(i, j) \in F$ satisfies $i < j$ (the vertex j is called a *child* of the vertex i), and $W = \{W_i \subseteq V : i = 1, 2, \ldots, m\}$ is a family of distinct non-empty subsets (called *bags*) associated with the vertices of T such that the following properties are satisfied:

1. For each vertex $v \in V \setminus W_1$, there is a unique directed edge $(i, j) \in F$ such that $W_j \setminus W_i = \{v\}$.
2. For each edge $\{u, v\} \in E$, either there is at least one vertex $i \in I$ such that $u, v \in W_i$, or there is a directed edge $(i, j) \in F$ such that $\{u\} = W_i \setminus W_j$ and $\{v\} = W_j \setminus W_i$.
3. For each directed edge $(i, j) \in F$, there is an edge $\{u, v\} \in E$ such that $\{u\} = W_i \setminus W_j$ and $\{v\} = W_j \setminus W_i$.
4. For all $i, j, k \in I$, if i is an ancestor of j and j is an ancestor of k on the directed path from i to k in T, then $W_i \cap W_k \subseteq W_j$.

The *size* of a propagation tree decomposition (T, W) is the number of directed edges in T (i.e., $|F|$). The *maximum propagation tree size* of G, denoted pts(G), is the maximum size over all propagation tree decompositions of G.

Theorem 1. *Let G be a connected graph that contains at least one edge. Then*

$$Z_+(G) + \mathrm{pts}(G) = |V(G)|.$$

Proof. Let $|V(G)| = n$. We first show that $Z_+(G) + \mathrm{pts}(G) \ge n$. Let $Z_+(G) = k$ and $B^1 = \{b_1, \ldots, b_k\}$ be a minimum positive zero forcing set in which every vertex is blue initially, and let $\mathcal{S}$ be a corresponding positive zero forcing process whose positive zero forcing set is B^1. We construct a propagation tree decomposition (T, W) with a rooted tree $T = (I, F)$ and a set of distinct non-empty bags $W = \{W_i : i \in I\}$ in the following way: Let $W_1 = B^1$. Let $U_1^1, \ldots, U_{j_1}^1$ be the sets of uncolored vertices in each of the connected components of $G - B^1$. Let the root of T be the vertex 1. For two vertices $i, j \in I$, i is a descendant of j if the unique directed path from the root 1 to i passes through j. The vertex 1 will have j_1 children and each child will be the root of the rooted subtree consisting of this child and its descendants after the construction of T is finished. Each U_i^1, $1 \le i \le j_1$, will be equal to the union of the bags associated to the vertices of the rooted subtree whose root is the i-th child of the vertex 1 after the construction of T is finished. Suppose that b^1 is the blue vertex in W_1 that forces v^1 to blue in the first forcing action of $\mathcal{S}$. Then let $W_2 = (W_1 \setminus \{b^1\}) \cup \{v^1\}$ and create the vertex 2 and edge $(1, 2)$ for T. Let $B^2 = B^1 \cup \{v^1\}$ and $U_1^2, \ldots, U_{j_2}^2$ be the sets of uncolored vertices in each of the connected components of $G - B^2$. Then after the construction of T is finished, there will be j_2 rooted subtrees in the forest $T - \{1, 2\}$, and each $U_i^2, 1 \le i \le j_2$, will be equal to the union of the bags associated to the vertices of a rooted subtree whose root is a child of the vertex 1 or 2. Suppose b^2 is the blue vertex in $W_{k'}$ $(1 \le k' \le 2)$ that forces v^2 to blue in the second forcing action of $\mathcal{S}$. Then let $W_3 = (W_{k'} \setminus \{b^2\}) \cup \{v^2\}$ and create the vertex 3 and edge $(k', 3)$ for T. In general, repeat the following until the last forcing action of $\mathcal{S}$ is considered: let $B^\ell = B^{\ell-1} \cup \{v^{\ell-1}\}$ and $U_1^\ell, \ldots, U_{j_\ell}^\ell$ be the sets of uncolored vertices in each of the connected components of $G - B^\ell$. Then after the construction of T is finished, there will be j_ℓ rooted subtrees in the forest $T - \{1, 2, \ldots, \ell\}$, and each U_i^ℓ, $1 \le i \le j_\ell$, will be equal to the union of the bags associated to the vertices of a rooted subtree whose root is a child of a vertex in $\{1, 2, \ldots, \ell\}$. Suppose b^ℓ is a blue vertex in $W_{k'}$ $(1 \le k' \le \ell)$ that forces v^ℓ to blue in the ℓ-th forcing action of $\mathcal{S}$. Then let $W_{\ell+1} = (W_{k'} \setminus \{b^\ell\}) \cup \{v^\ell\}$ and create the vertex $\ell + 1$ and edge $(k', \ell + 1)$ for T.

Since $|B^1| = k$, we know that $\mathcal{S}$ contains $n - k$ forcing actions. Thus T is a rooted tree with vertex set $I = \{1, 2, \ldots, n-k+1\}$ and edge set F such that each directed edge $(i, j) \in F$ satisfies $i < j$. We now show that W satisfies the four properties in Definition 1. The first property is easy to verify. Let $\{u, v\}$ be an edge of G. If u and v are blue initially, then $u, v \in W_1$, otherwise, suppose that v is uncolored when u is colored (u could be blue initially or forced by another blue vertex). If v is forced to blue by a vertex u', then there is a directed edge $(k', \ell') \in F$ such that $u' \in W_{k'}$ and $v \in W_{\ell'} = (W_{k'} \setminus \{u'\}) \cup \{v\}$. If $u = u'$ or $u \in W_{\ell'}$, then property 2 is satisfied. Suppose that $u \ne u'$ and $u \notin W_{\ell'}$: Let $(j', j + 1) \in F$ be a directed edge on the directed path of T from the root 1 to ℓ' such that $u \in W_{j'}$, $u \notin W_{j+1}$ and u forces u''. Since u'' and v belong to the same connected component of $G - B^j$ and both of them are uncolored before u forces u'', u cannot force u'' to blue by the coloring rule of the positive zero forcing (because $v, u'' \in N_G(u)$). This is a contradiction. Thus property 2 is held. For each directed edge $(k', \ell + 1)$ in F, where $1 \le k' \le \ell$, there is an edge $\{b^\ell, v^\ell\} \in E$ such that $\{b^\ell\} = W_{k'} \setminus W_{\ell+1}$ and $\{v^\ell\} = W_{\ell+1} \setminus W_{k'}$. Thus, the third property is satisfied. For any $i, j, \ell \in I$, where i is an ancestor of j and j is an ancestor of ℓ on the directed path from i to ℓ in T, suppose that u is a vertex in $W_i \cap W_\ell$. If u forces a vertex v to blue such that $v \in W_{j'}$ and j' is on the directed path from i to ℓ in T, then $u \notin W_{j'}$, and furthermore, $u \notin W_\ell$. This is a contradiction. Thus $u \in W_j$. Hence the fourth property is satisfied.

From the above, we know that (T, W) is a propagation tree decomposition of size $n - k$. Therefore $\text{pts}(G) \geq n - Z_+(G)$.

We next show that $Z_+(G) + \text{pts}(G) \leq n$. Let $\text{pts}(G) = p$. Given a propagation tree decomposition (T, W) of size p, we will describe a positive zero forcing process such that the corresponding positive zero forcing set has size $n - p$. Let $T = (I, F)$. Since $\text{pts}(G) = p$, from Definition 1, $|F| = p$ and $|I| = p + 1$. Let $I = \{1, 2, \ldots, p + 1\}$ and let $W = \{W_i \subseteq V(G) : 1 \leq i \leq p + 1\}$ be a family of distinct non-empty bags satisfying the properties in Definition 1. Color all vertices of W_1 blue and leave all other vertices of $V(G) \setminus W_1$ uncolored initially. Assume that all vertices in $W_1, \ldots, W_j$ are blue just before the j-th forcing action of the positive zero forcing process. We will show that all vertices of W_{j+1} are blue just after the j-th forcing action of the positive zero forcing process. Suppose $(i, j + 1) \in F$ is a directed edge of T such that the vertex $j + 1$ is a child of i. Then there is an edge $\{u, v\} \in E(G)$ such that $\{u\} = W_i \setminus W_{j+1}$ and $\{v\} = W_{j+1} \setminus W_i$. Since $i \leq j$, we know that v is the only uncolored vertex in W_{j+1} just before the j-th forcing action of the positive zero forcing process. Let B be the set of blue vertices just before the j-th forcing action, and let U be the set of uncolored vertices in a connected component of $G - B$ that contains v. Note that $\{v\} = U \cap W_{j+1}$. Let $H = G[U \cup B]$. Suppose that $N_H(u)$ contains at least two uncolored vertices, say v and v'. Let $W_{j'}$ be a bag in W which contains v'. Since $\{v\} = U \cap W_{j+1}$, from the fourth property of W, we have $j' > j + 1$. So $u \notin W_{j'}$. Thus, the second property of W is violated for the edge $\{u, v'\} \in E(G)$. This is a contradiction. Hence v is the only uncolored vertex in $N_H(u)$, and therefore, we can let "u forcing v" to be the j-th forcing action of the positive zero forcing process. Thus, all vertices of W_{j+1} are blue just after the j-th forcing action of the positive zero forcing process.

Since W contains $p + 1$ bags and each bag W_i, $2 \leq i \leq p + 1$, contains an uncolored vertex that is forced to blue by another vertex, we must have $n - p$ blue vertices initially. Thus $Z_+(G) \leq n - p = n - \text{pts}(G)$.

From the first part of the proof of Theorem 1, we obtain the following result.

Corollary 1. *For a connected graph G with n vertices, if $Z_+(G) = k$, then G has a propagation tree decomposition (T, W) of size $n - k$ such that $W = \{W_i \subseteq V(G) : 1 \leq i \leq n - k + 1\}$ satisfies the following*

$$|W_1| = |W_2| = \cdots = |W_{n-k+1}| = k.$$

4   Lower Bound

In this section, we first introduce the matrix form and then apply the propagation tree decomposition to show a lower bound on positive zero forcing numbers. We will use this lower bound to find the positive zero forcing number of matching-chain graphs in Sect. 5.

Definition 2. We say that a graph G has an $m \times n$ *matrix form* if all vertices of G can be arranged in an $m \times n$ matrix

$$\begin{bmatrix} v_{11} & v_{12} & \cdots & v_{1n} \\ v_{21} & v_{22} & \cdots & v_{2n} \\ \vdots & \vdots & \ddots & \vdots \\ v_{m1} & v_{m2} & \cdots & v_{mn} \end{bmatrix}$$

such that each *row* $A_i = \{v_{i1}, v_{i2}, \ldots, v_{in}\}$, $1 \le i \le m$, induces a connected subgraph $G[A_i]$, and each *column* $B_j = \{v_{1j}, v_{2j}, \ldots, v_{mj}\}$, $1 \le j \le n$, also induces a connected subgraph $G[B_j]$.

From Definition 2, it is easy to see that every connected graph G has a $1 \times |V(G)|$ matrix form. If $|V(G)|$ (as a number) has more than two factors, then G may have two or more different matrix forms.

Lemma 1. *Let G be a connected graph and (T, W) be a propagation tree decomposition of G such that $T = (I, F)$ is a rooted tree with $I = \{1, 2, \ldots, m\}$ and $W = \{W_i \subseteq V(G) : 1 \le i \le m\}$ is a family of distinct non-empty bags satisfying the properties in Definition 1. Let H be a connected subgraph of G. If there is a $W_j \in W$ such that $V(H) \cap W_j = \emptyset$, then there is a connected component T' in the forest $T - \{j\}$ such that $V(H) \subseteq \cup_{\ell \in V(T')} W_\ell$.*

By Lemma 1, we can prove the main result of this section.

Theorem 2. *Let G be a connected graph that has an $m \times n$ matrix form with m rows $A_1, \ldots, A_m$, and n columns $B_1, \ldots, B_n$.*

(i) *If $\sum_{j=1}^n \kappa(G[B_j]) \ge m \ge 2$, then $Z_+(G) \ge m$; and*
(ii) *If $\sum_{i=1}^m \kappa(G[A_i]) \ge n \ge 2$, then $Z_+(G) \ge n$.*

Proof. (i) Let $Z_+(G) = k$. From Corollary 1, G has a propagation tree decomposition (T, W) such that $T = (I, F)$ is a rooted tree with $I = \{1, 2, \ldots, mn-k+1\}$ and $W = \{W_i \subseteq V(G) : 1 \le i \le mn - k + 1\}$ is a family of distinct non-empty bags satisfying the properties in Definition 1. It follows from Corollary 1 that $|W_1| = |W_2| = \cdots = |W_{mn-k+1}| = k$.

For the sake of contradiction, we assume that $k < m$. Let $A_i = \{v_{i1}, v_{i2}, \ldots, v_{in}\}$, $1 \le i \le m$, and $B_j = \{v_{1j}, v_{2j}, \ldots, v_{mj}\}$, $1 \le j \le n$. For any W_ℓ, $1 \le \ell \le mn - k + 1$, since $k < m$ and $|W_\ell| = k$, at least one of the intersection sets $W_\ell \cap A_i$, $1 \le i \le m$, must be empty. Let $\ell_1 = 1$ be the root of T. Then the induced subgraph $G[V \setminus W_{\ell_1}]$ contains a connected component that contains all vertices of some row, say A_{h_1}, because $G[A_{h_1}]$ is connected. By Lemma 1, there is a connected component T_{ℓ_1} in the forest $T - \{\ell_1\}$ such that $A_{h_1} \subseteq \cup_{\ell \in V(T_{\ell_1})} W_\ell$. Suppose that vertex ℓ_2 is a child of ℓ_1 on T such that $\ell_2 \in V(T_{\ell_1})$. Similarly, $G[V \setminus W_{\ell_2}]$ contains a connected component that contains all vertices of a row, say A_{h_2} (note that it is possible that $A_{h_2} = A_{h_1}$). From Lemma 1, there is a connected component T_{ℓ_2} in the forest $T - \{\ell_2\}$ such that $A_{h_2} \subseteq \cup_{\ell \in V(T_{\ell_2})} W_\ell$. Continue in this way, we will find a path $(\ell_1, \ell_2, \ldots, \ell_t, \ell_{t+1})$ on T, where each vertex is a parent of the next vertex along the path, such that the following properties are satisfied:

1. For each $s \in \{1, \ldots, t\}$, $G[V \setminus W_{\ell_s}]$ contains a connected component that contains all vertices of a row, say A_{h_s}, and there is a connected component T_{ℓ_s} in the forest $T - \{\ell_s\}$ such that $\ell_{s+1} \in V(T_{\ell_s})$ and $A_{h_s} \subseteq \cup_{\ell \in V(T_{\ell_s})} W_\ell$; and

2. $G[V \setminus W_{\ell_{t+1}}]$ contains a connected component that contains all vertices of a row, say $A_{h_{t+1}}$, and there is a connected component $T_{\ell_{t+1}}$ in the forest $T - \{\ell_{t+1}\}$ such that $\ell_t \in V(T_{\ell_{t+1}})$ and $A_{h_{t+1}} \subseteq \cup_{\ell \in V(T_{\ell_{t+1}})} W_\ell$.

To simplify the notation, let $A_{h_t} = A_x = \{v_{x1}, v_{x2}, \ldots, v_{xn}\}$ and $A_{h_{t+1}} = A_y = \{v_{y1}, v_{y2}, \ldots, v_{yn}\}$. From the properties of the path $(\ell_1, \ell_2, \ldots, \ell_t, \ell_{t+1})$, we know that $A_x \neq A_y$. Consider an arbitrary $j \in \{1, \ldots, n\}$. Suppose W_ℓ contains v_{xj} and $W_{\ell'}$ contains v_{yj}. Let $\kappa(G[B_j]) = k'$. Note that $m \geq 2$. So $G[B_j]$ is a connected graph with at least two vertices, and thus, $k' \geq 1$. Since $G[B_j]$ is k'-connected, from Menger's theorem [18], there are k' pairwise internally disjoint paths from v_{xj} to v_{yj} on $G[B_j]$. Let P_G be one of the k' paths from v_{xj} to v_{yj} on $G[B_j]$ and P_T be a path from ℓ to ℓ' on T. From the properties of the path $(\ell_1, \ell_2, \ldots, \ell_t, \ell_{t+1})$, we know that ℓ_t is a vertex on the path P_T and ℓ' is a descendant of ℓ_t. Note that ℓ_t is a descendant or sibling of ℓ. By properties 2 and 4 in Definition 1, W_{ℓ_t} must contain at least one vertex of P_G because W_ℓ contains v_{xj} and $W_{\ell'}$ contains v_{yj}. Let v be the closest vertex to v_{yj} among the vertices in $W_{\ell_t} \cap V(P_G)$. From property 3 in Definition 1, if P_G contains at least one internal vertex, then v must be an internal vertex of P_G; otherwise, $v = v_{xj}$ and $V(P_G) = \{v_{xj}, v_{yj}\}$, and thus $v_{xj} \in W_{\ell_t}$. Note that there is at most one edge between v_{xj} and v_{yj} because G does not contain multiple edges. Therefore, W_{ℓ_t} must contain at least k' vertices of $G[B_j]$. Since j is an arbitrary number in $\{1, \ldots, n\}$ and the k' paths from v_{xj} to v_{yj} on $G[B_j]$ are pairwise internally disjoint, W_{ℓ_t} must contain at least $\sum_{j=1}^{n} \kappa(G[B_j])$ distinct vertices. Hence $|W_{\ell_t}| \geq \sum_{j=1}^{n} \kappa(G[B_j]) \geq m > k$, which is a contradiction.

(ii) The result follows from (i) if we consider the transpose of the matrix form of G.

Corollary 2. *Let G be a connected graph that has an $m \times n$ matrix form. Then*

$$Z_+(G) \geq \min\{m, n\}.$$

5   Matching-Chain Graphs

In this section, we introduce a class of graphs, called matching-chain graphs, on which the zero forcing number equals the positive zero forcing number.

Definition 3. A graph $G = (V, E)$ is an $m \times n$ *matching-chain graph*, denoted by $M_{m,n}$, if its vertex set V can be partitioned into n disjoint subsets $V_1, V_2, \ldots, V_n$ such that each subset contains m vertices and they satisfy the following properties:

1. for each $j \in \{1, 2, \ldots, n\}$, the induced subgraph $G[V_j]$ is connected;
2. for each edge $\{u, v\} \in E$, there is a $j \in \{1, 2, \ldots, n\}$ such that either both u and v belong to V_j or one belongs to V_j and the other belongs to V_{j+1}; and
3. for each $j \in \{1, 2, \ldots, n-1\}$, the bipartite graph $(V_j, V_{j+1}; E_{j,j+1})$, where $E_{j,j+1} = \{\{u, v\} \in E : u \in V_j \text{ and } v \in V_{j+1}\}$, has a unique perfect matching.

Lemma 2 [15]. *Let $G = (U, V; E)$ be a bipartite graph with two parts U and V. If G has a unique perfect matching, then there are vertices $u \in U$ and $v \in V$ such that $\deg_G(u) = \deg_G(v) = 1$.*

Theorem 3. *Let $M_{m,n} = (V, E)$ and $(V_1, V_2, \ldots, V_n)$ be a partition of the vertex set V such that each V_i, $1 \leq i \leq n$, contains m vertices and the properties in Definition 3 are satisfied. If $\sum_{i=1}^{n} \kappa(G[V_i]) \geq m \geq 2$, then*

$$Z_+(M_{m,n}) = Z(M_{m,n}) = m.$$

Proof. From [3] we know that $Z(M_{m,n}) \geq Z_+(M_{m,n})$. We now show that $Z_+(M_{m,n}) \geq m$. For each $j \in \{1, 2, \ldots, n-1\}$, consider the bipartite graph $H_j = (V_j, V_{j+1}; E_{j,j+1})$, where $E_{j,j+1} = \{\{u, v\} \in E : u \in V_j \text{ and } v \in V_{j+1}\}$. Let $M_j = \{\{v_{i,j}, v_{i,j+1}\} \in E : 1 \leq i \leq m, v_{i,j} \in V_j \text{ and } v_{i,j+1} \in V_{j+1}\}$ be the unique perfect matching for H_j. For each $i \in \{1, 2, \ldots, m\}$, we have a path $(v_{i,1}, v_{i,2}, \ldots, v_{i,n})$ such that $\{v_{i,j}, v_{i,j+1}\} \in M_j, j \in \{1, 2, \ldots, n-1\}$. Thus $M_{m,n}$ has an $m \times n$ matrix form with rows $\{v_{i,1}, v_{i,2}, \ldots, v_{i,n}\}, 1 \leq i \leq m$, and columns $\{v_{1,j}, v_{2,j}, \ldots, v_{m,j}\}, 1 \leq j \leq n$. By Theorem 2, we have $Z_+(M_{m,n}) \geq m$.

We next describe a zero forcing process for $M_{m,n}$, which has m blue vertices initially. Color all vertices of $\{v_{1,1}, v_{2,1}, \ldots, v_{m,1}\}$ blue and leave all other vertices of $M_{m,n}$ uncolored. Since M_1 is the unique perfect matching for H_1, from Lemma 2, there is a vertex in V_1, say $v_{1,1}$, such that $\deg_{H_1}(v_{1,1}) = 1$. Since $v_{1,1}$ has only one uncolored neighbor (i.e., $v_{1,2}$) in $M_{m,n}$, it can force $v_{1,2}$ to blue. Notice that $M_1 \setminus \{\{v_{1,1}, v_{1,2}\}\}$ is the unique perfect matching for the bipartite graph $H_1 - \{v_{1,1}, v_{1,2}\}$. Thus, there is a vertex in $V_1 \setminus \{v_{1,1}\}$, say $v_{2,1}$, whose degree is one in $H_1 - \{v_{1,1}, v_{1,2}\}$. Note that $v_{2,1}$ currently has only one uncolored neighbor (i.e., $v_{2,2}$) in $M_{m,n}$. Hence $v_{2,1}$ can force $v_{2,2}$ to blue. Continue like this, all vertices of V_2 can be blue. Similarly, each vertex of V_j, $3 \leq j \leq n$, can be forced to blue by the corresponding vertex in V_{j-1}. So $Z(M_{m,n}) \leq m$, and therefore, $Z(M_{m,n}) = Z_+(M_{m,n}) = m$.

From Theorem 3, we have the following result.

Corollary 3. *Let G be a connected graph and P be a path. If $|V(G)| \leq |V(P)|$, then*

$$Z_+(G \square P) = Z(G \square P) = |V(G)|.$$

Let P_n denote a path with n vertices. As a consequence of Corollary 3, we can prove the conjecture that $Z_+(P_m \square P_n) = \min\{m, n\}$ (see p. 828 in [16]).

Corollary 4. $Z_+(P_m \square P_n) = Z(P_m \square P_n) = \min\{m, n\}$.

Let C_m denote a cycle with m vertices. Peters conjectured that $Z_+(C_m \square P_n) = \min\{m, 2n\}$ ([16]; p. 828). We next prove this conjecture partially by using Theorem 3.

Corollary 5. If $m \leq 2n$, then $Z_+(C_m \square P_n) = Z(C_m \square P_n) = m$.

Corollary 6. Let G be a k-connected graph. If $k \geq |V(G)|/2 \geq 1$ and $n \geq 2$, then

$$Z_+(G \square P_n) = Z(G \square P_n) = |V(G)|.$$

Note that $\kappa(K_m) = m - 1$ for a complete graph K_m with $m \geq 2$ vertices. By Corollary 6 we can obtain the following result.

Corollary 7. For any $m \geq 2$ and $n \geq 2$, $Z_+(K_m \square P_n) = Z(K_m \square P_n) = m$.

6   Claw-Free Graphs

In this section, we first show that the zero forcing number of a claw-free graph is equal to its positive zero forcing number. We then prove that finding the positive zero forcing number of line graphs (which is a subclass of claw-free graphs) is NP-complete.

6.1 $Z(G) = Z_+(G)$ for Claw-Free Graphs

In a positive zero forcing process, let B be a positive zero forcing set of G and $U_1, \ldots, U_k$ be the sets of uncolored vertices in each of the connected components of $G - B$. When the coloring rule of the positive zero forcing is applied, a blue vertex in B can force an uncolored vertex in some connected component U_i to blue. Then we update the blue vertex set and apply the coloring rule again. We repeat this process until all vertices are blue. For each initial blue vertex v, the forces determine an induced rooted-tree T, referred to as a *forcing tree* with *root* v, and write $v \Rightarrow T$. If B is a positive zero forcing set of G, then the set of rooted-trees $\{T : v \in B, v \Rightarrow T\}$ is called a *positive zero forcing tree cover* of G; furthermore, if B is a minimum positive zero forcing set of G, then this set is called a *minimum positive zero forcing tree cover* of G. Since every vertex not in the positive zero forcing set B is forced by exactly one vertex, it is easy to see that $\{T : v \in B, v \Rightarrow T\}$ is a set of vertex-disjoint induced rooted-trees that partition $V(G)$.

In [12], Fallat and Soltani showed that the zero forcing number of a line graph is equal to its positive zero forcing number. We extend this result to claw-free graphs.

Theorem 4. *If a graph G is claw-free, then $Z_+(G) = Z(G)$ and each forcing tree in a minimum positive zero forcing tree cover of G is a path.*

Proof. Suppose there is a forcing tree T in a minimum positive zero forcing tree cover of G which contains a vertex v with $\deg_T(v) \geq 3$. Let $a, b, c \in N_T(v)$. If v is the root of T, then v forces a, b, c to blue. Since G is claw-free, at least two of a, b, c must be adjacent. Without loss of generality, suppose a and b are adjacent. When v forces a or b to blue, both a and b belong to the same connected component of the graph induced by the uncolored vertices. This violates the coloring rule of the positive zero forcing. Assume that v is not the root of T. Then a neighbor of v, say a, must force v to blue and v forces all other vertices in $N_T(v)$ to blue. So there is no edge between a and $N_T(v) \setminus \{a\}$. Since G is claw-free, we know that $G[N_T(v) \setminus \{a\}]$ is a clique in G. Hence, v cannot force b and c to blue. This is a contradiction.

From the above, we know that each tree in a minimum positive zero forcing tree cover of G must be a path, and thus $Z_+(G) \geq Z(G)$. On the other hand, we have $Z_+(G) \leq Z(G)$ [3]. Therefore $Z_+(G) = Z(G)$. $\blacksquare$

The reverse of Theorem 4 is not always true. For example, consider the Cartesian product of an edge and a path with at least three vertices. This graph is not claw-free, but it has a minimum positive zero forcing tree cover that contains two paths.

6.2 Complexity

In [19], Yang proved that finding the fast-mixed search number of a given graph is NP-complete and it remains NP-complete even for biconnected graphs with maximum vertex degree 4. Since the fast-mixed search number of a graph is equal to its zero forcing number, this NP-completeness result is also held for the zero forcing problem [10]. Recently, Yang [21] proved that finding the positive zero forcing number of a given graph is NP-complete and it remains NP-complete for graphs with maximum vertex degree 7. In this subsection, we consider the complexity of computing the positive zero forcing number of another class of graphs – the line graphs. We will show that finding the positive zero forcing number (or zero forcing number) of line graphs is still NP-complete. This result enhances the NP-completeness results in [10, 21].

Theorem 5. *Given a graph G and an integer k, the problem of determining whether $Z_+(L(G)) \leq k$ is NP-complete. The problem remains NP-complete even when G is a 2-edge-connected planar bipartite graph with maximum vertex degree 3.*

Since line graphs are claw-free, from Theorems 4 and 5, we obtain the following result.

Corollary 8. *Given a graph G and an integer k, the problem of determining whether $Z(L(G)) \leq k$ is NP-complete for 2-edge-connected planar bipartite graphs G with maximum vertex degree 3.*

Fallat et al. [10] considered the structure of trees in a zero forcing tree cover of a graph. For a given graph G and a positive integer ℓ, among all the positive zero forcing tree covers of G with size ℓ, they wanted to minimize the number of positive zero forcing trees that are non-trivial trees (a tree is non-trivial if it contains at least one edge). We now consider a strong version of this problem, that is, we want to decide if the number of non-trivial positive zero forcing trees is one. The following result is derived from the proof of Theorem 5.

Corollary 9. *Given a graph $L(G)$ that is the line graph of a graph G and an integer k, the problem of determining whether $L(G)$ has a positive zero forcing tree cover of size at most k such that there is only one positive zero forcing tree that is non-trivial is NP-complete. The problem remains NP-complete even when G is a 2-edge-connected planar bipartite graph with maximum vertex degree 3.*

6.3 Line Graphs

Recall that a path in a graph G is a sequence $(v_0, v_1, \ldots, v_k)$ of distinct vertices with $\{v_{i-1}, v_i\} \in E(G)$ for all $i \in \{1, \ldots, k\}$, and its length is the number of edges on it.

Theorem 6. *Let G be a connected graph and $\mathcal{P} = \{P_1, \ldots, P_k\}$ be a set of vertex-disjoint paths in G such that each P_i contains at least two edges. Suppose the sum of the lengths of all paths in $\mathcal{P}$ is k'. Then $Z_+(L(G)) = Z(L(G)) \leq |E(G)| - k' + k$.*

The upper bound in Theorem 6 is tight when G is a path or cycle. The next result is from [21]. We state it here for easier referencing.

Lemma 3 [21]. *Let $G = (V, E)$ be a connected graph and $G_i = (V_i, E_i), 1 \leq i \leq k$, be maximal cliques of G satisfying the following conditions: (1) $\cup_{i=1}^{k} V_i = V$ and $\cup_{i=1}^{k} E_i = E$, (2) for any $1 \leq i < j \leq k, |V_i \cap V_j| \leq 1$, and (3) each $G_i, 1 \leq i \leq k$, contains a vertex v_i that is not in any $G_j, j \neq i$. Then*

$$Z_+(G) = \mathrm{M}_+(G) = |V| - k.$$

Theorem 7. *Let G be a connected graph with minimum vertex degree 2 and G' be a graph obtained from G by adding a pendant edge to each vertex of G. Then*

$$Z(L(G')) = \mathrm{M}(L(G')) = Z_+(L(G')) = \mathrm{M}_+(L(G')) = |E(G)|.$$

References

1. AIM Minimum Rank-Special Graphs Work Group: Zero forcing sets and the minimum rank of graphs. Linear Algebra Appl. **428**, 1628–1648 (2008)
2. Barioli, F., Barrett, W., Fallat, S., Hall, H.T., Hogben, L., Shader, B., van den Driessche, P., van der Holst, H.: Parameters related to tree-width, zero forcing, and maximum nullity of a graph. J. Graph Theory **72**, 146–177 (2013)

3. Barioli, F., Barrett, W., Fallat, S., Hall, H.T., Hogben, L., Shader, B., van den Driessche, P., van der Holst, H.: Zero forcing parameters and minimum rank problems. Linear Algebra Appl. **433**(2), 401–411 (2010)
4. Barioli, F., Fallat, S., Mitchell, L., Narayan, S.: Minimum semidefinite rank of outerplanar graphs and the tree cover number. Electron. J. Linear Algebra **22**, 10–21 (2011)
5. Booth, M., Hackney, P., Harris, B., Johnson, C.R., Lay, M., Mitchell, L.H., Narayan, S.K., Pascoe, A., Steinmetz, K., Sutton, B.D., Wang, W.: On the minimum rank among positive semidefinite matrices with a given graph. SIAM J. Matrix Anal. Appl. **30**, 731–740 (2008)
6. Burgarth, D., Giovannetti, V.: Full control by locally induced relaxation. Phys. Rev. Lett. **99**(10), 100501 (2007)
7. Ekstrand, J., Erickson, C., Hall, H.T., Hay, D., Hogben, L., Johnson, R., Kingsley, N., Osborne, S., Peters, T., Roat, J., Ross, A., Row, D., Warnberg, N., Young, M.: Positive semidefinite zero forcing. Linear Algebra Appl. **439**, 1862–1874 (2013)
8. Ekstrand, J., Erickson, C., Hay, D., Hogben, L., Roat, J.: Note on positive semidefinite maximum nullity and positive semidefinite zero forcing number of partial 2-trees. Electron. J. Linear Algebra **23**, 79–87 (2012)
9. Fallat, S., Hogben, L.: Minimum rank, maximum nullity, and zero forcing number of graphs. In: Handbook of Linear Algebra, Discrete Mathematics and its Applications, chap. 46, pp. 775–810. CRC Press (2013)
10. Fallat, S., Meagher, K., Yang, B.: On the complexity of the positive semidefinite zero forcing number. Linear Algebra Appl. **491**, 101–122 (2016)
11. Fallat, S., Meagher, K., Soltani, A., Yang, B.: Positive zero forcing and edge clique coverings. In: Zhu, D., Bereg, S. (eds.) FAW 2016. LNCS, vol. 9711, pp. 53–64. Springer, Cham (2016). doi:10.1007/978-3-319-39817-4_6
12. Fallat, S., Soltani, A.: Line graphs: their maximum nullities and zero forcing numbers. Czechoslov. Math. J. **66**, 743–755 (2016)
13. Huang, L.-H., Chang, G.J., Yeh, H.-G.: On minimum rank and zero forcing sets of a graph. Linear Algebra Appl. **432**, 2961–2973 (2010)
14. Itai, A., Papadimitriou, C.H., Szwarcfiter, J.L.: Hamiltonian paths in grid graphs. SIAM J. Comput. **11**(4), 676–686 (1982)
15. Lovász, L., Plummer, M.D.: Matching Theory. North Holland Publishing, Amsterdam (1986)
16. Peters, T.: Positive semidefinite maximum nullity and zero forcing number. Electron. J. Linear Algebra **23**, 815–830 (2012)
17. Warnberg, N.: Positive semidefinite propagation time. Discret. Appl. Math. **198**, 274–290 (2016)
18. West, D.B.: Introduction to Graph Theory, 2nd edn. Prentice Hall, Upper Saddle River (2001)
19. Yang, B.: Fast-mixed searching and related problems on graphs. Theoret. Comput. Sci. **507**(7), 100–113 (2013)
20. Yang, B.: Fast edge-searching and fast searching on graphs. Theoret. Comput. Sci. **412**, 1208–1219 (2011)
21. Yang, B.: Lower bounds for positive semidefinite zero forcing and their applications. J. Comb. Optim. **33**, 81–105 (2017)

Phase Transition for Maximum Not-All-Equal Satisfiability

Junping Zhou[1], Shuli Hu[1], Tingting Zou[2], and Minghao Yin[1(✉)]

[1] College of Computer Science and Information Technology,
Northeast Normal University, Changchun 130117, China
ymh@nenu.edu.cn
[2] College of Information Science and Technology,
Dalian Maritime University, Dalian 116026, China

Abstract. Phase transition is a dramatic transition from one state to another state when a particular parameter varies. This paper aims to study the phase transition of maximum not-all-equal satisfiability problem (Max NAE SAT), an optimization of not-all-equal satisfiability problem (NAE SAT). Given a conjunctive normal formula (CNF) F with n variables and rn k-clauses (the clause exactly contains k literals), we use first-moment method to obtain an upper bound for $f(n, rn)$ the expectation of the maximum number of NAE-satisfied clauses of random Max NAE k-SAT. In addition, we also consider the phase transition of decision version of random Max NAE k-SAT—bounded not-all-equal satisfiability problem (NAE k-SAT(b)). We demonstrate that there is a phase transition point $r_{k,b}$ separating the region where almost all NAE k-SAT(b) instances can be solved from the region where almost all NAE k-SAT(b) instances can't be solved. Furthermore, we analyze the upper bound and lower bound for $r_{k,b}$.

1 Introduction

Propositional satisfiability (SAT) is a prototypical NP-complete problem, and consists in deciding whether there exists an assignment to all variables satisfying a given propositional logic formula in conjunctive normal formula (CNF). This problem has been well studied due to its academic and practical significance [1]. Not-all-equal satisfiability (NAE SAT) is an extension of SAT, which is a problem to decide whether there exists an assignment to all variables such that every clause in a given CNF formula contains at least one satisfied literal and at least one unsatisfied literal [2]. The optimization version of NAE SAT is Maximum Not-All-Equal-Satisfiability (Max NAE SAT), which consists in finding an assignment to all variables such that the maximum number of clauses are NAE-satisfied. Nowadays, the optimization problems have arisen in many real problems like routing, bioinformatics, scheduling, probabilistic reasoning, electronic markets [3].

The phase transition phenomenon is usually a dramatic transformation from one state to another state when a particular parameter varies. Research on phase

© Springer International Publishing AG 2017
M. Xiao and F. Rosamond (Eds.): FAW 2017, LNCS 10336, pp. 267–279, 2017.
DOI: 10.1007/978-3-319-59605-1_24

transition is significant. This is because phase transition is one of the most crucial features of these inherently intractable problems and by working on it the researchers can not only analyze the structure of intractable problems, but also understand the average-case performance of solvers. Recently, numerous theoretical analyses and experiments have provided evidences that phase transition phenomena exist in a large number of problems [4]. In the pioneering work of Kirkpatrick and Selma [5], a sharp phase transition phenomenon was proved to exist in the satisfiable probability of a random SAT formula with respect to the control parameter of the density of clauses. In [6,7], Xu and Li proposed a random CSPs model, called RB, and proved that there exists an exact phase transition point in this model. Phase transition phenomena of several real-world problems, such as traveling salesman problem and manipulation problem, have been observed [8,9]. In [10–12], we have proved that phase transition phenomena exist in intelligent planning problems, counting constraint satisfaction problems, and knowledge compilation problems.

Several results on phase transition of NAE SAT have also been proposed. Achlioptas first proved that phase transition phenomena exist in NAE SAT, and provided both an upper bound and a lower bound for phase transition point of NAE 3-SAT in [13]. After that, he further improved the upper bound and lower bound for NAE k-SAT as well [14]. Coja-Oghlan et al. proposed a survey propagation inspired second moment method to study the phase transition of NAE k-SAT, and obtained the upper bound and lower bound for NAE k-SAT [15]. Ding et al. considered the regular NAE k-SAT and established a critical value of the phase transition point [16].

Phase transition phenomena exist in not only decision problems, but also optimization problems. Achilioptas et al. presented a mathematical method for locating phase transition point for optimization problems [17]. Coppersmith et al. provided bounds for Max 2-SAT, Max k-SAT, and Max Cut [18]. Further, Xu et al. presented a tighter upper bound for Max 2-SAT by using the first moment method [19]. In [10], we provided bounds for Max CSP and Min CSP problems. Some researchers also study the phase transition for bounded satisfiability problems, a decision version of Max SAT. Zhang first showed that there exists phase transition for the bounded satisfiability problems, and obtained the phase transition points of 3-SAT(b) and 2-SAT(b) respectively [20]. Based on that, Bailey and Kolaitis provided a theoretical analysis to obtain phase transition points of 3-SAT(b) and k-SAT(b) problems in [21].

This paper also follows this line of research by studying the upper bound of the Max NAE SAT and the phase transition of the decision version of the Max NAE SAT. Until now, within our knowledge, there has been no literature analyzing the bounds for Max NAE k-SAT. Firstly, we present an upper bound for the expectation of the maximum number of NAE-satisfied clauses $f(n, rn)$ of random Max NAE k-SAT. That is, when $r > (2^k - 2)\ln 2$, $f(n, rn) \leq (1 - (1/2^{k-1}) + \sqrt{(2^k - 2)\ln 2/\sqrt{r2^{2k-2}}})rn$, where k is the length of each clause, r is the clause density, n and rn denote the number of variables and clauses respectively. Moreover, we also research the decision version of Max

NAE SAT problem—NAE k-SAT(b). We prove that there exists phase transition phenomenon in NAE k-SAT(b), regardless of the value of b. Further, we precisely locate the area of the phase transition point, i.e., the upper bound and lower bound for NAE k-SAT(b) by theoretical analyses.

This paper is organized as follows. First, we recall some basic concepts in Sect. 2. Then in Sect. 3, we present how to obtain the upper bound for random Max NAE k-SAT. In Sect. 4, we show how to acquire the upper and lower bounds for NAE k-SAT(b). Finally, conclusions are provided in Sect. 5.

2   Preliminary

Let $\{x_1, x_2, \ldots, x_n\}$ denote a set of n variables. The corresponding set of literals is $\{x_1, \bar{x}_1, x_2, \bar{x}_2, \ldots, x_n, \bar{x}_n\}$, and $\bar{x}$ denotes the negation of x. A clause is a disjunction of literals which come from distinct variables and a k-length clause is the clause with exactly k literals. A conjunctive normal formula (CNF) is a conjunction of clauses. A k-CNF formula $F(n, \lfloor rn \rfloor)$ is a CNF formula with n variables and rn k-length clauses, where r is the clause density equalling the number of clause divides the number of variables. For simplicity, we use rn in lieu of $\lfloor rn \rfloor$. And a random k-CNF $F(n, rn)$ formula can be obtained by choosing rn k-length clauses over n variables uniformly and independently at random among all $(2n)^k$ possible clauses. $F_k(n, rn)$ is the collection of all random k-CNF $F(n, rn)$ with n variables and rn clauses.

Given a CNF F, a truth assignment is a mapping that assigns to each variable contained by F either 0 or 1. If the value of a literal is 1, we say the literal is satisfied; otherwise, we say the literal is unsatisfied. We say a clause is NAE-satisfied if it contains at least one satisfied literal and at least one unsatisfied literal under the assignment; otherwise, the clause is NAE-unsatisfied. A truth assignment is a Not-All-Equal-satisfying assignment (NAE-satisfying assignment) iff every clause of F is NAE-satisfied under the assignment. If a clause is NAE-unsatisfied under an assignment, then we say the assignment violates the clause. If F has a NAE-satisfying assignment we say the formula F is Not-All-Equal-satisfied (NAE-satisfied); otherwise, F is Not-All-Equal-unsatisfied (NAE-unsatisfied).

Given a random k-CNF $F(n, rn)$, the Not-All-Equal-k-Satisfiability problem (NAE k-SAT) consists in deciding whether there exists a NAE-satisfying assignment such that all clauses of $F(n, rn)$ are NAE-satisfied. The problem Maximum Not-All-Equal-k-Satisfiability (Max NAE k-SAT) consists in finding an assignment to all variables such that the maximum number of clauses of $F(n, rn)$ are NAE-satisfied. The bounded Not-All-Equal-k-Satisfiability problem (NAE k-SAT(b)) is to decide whether there is an assignment such that does not violate more than b clauses of $F(n, rn)$. That is, NAE k-SAT(b) asks whether there is an assignment such that the number of NAE-satisfied clauses is at least $rn - b$. Therefore, NAE k-SAT(b) problem can be regarded as a decision version of Max NAE k-SAT problem.

Until now, there have been several research results about NAE k-SAT. In the following, we present two relevant theorems about bounds for NAE k-SAT problem. In [13], Achlioptas proved that the phase transition phenomenon exists in NAE 3-SAT with respect to the control parameter of the clause density r, and the upper and lower bounds for its phase transition point r_3^{NAE} is presented as follows:

Theorem 1 [13]. $1.514 < r_3^{NAE} < 2.215$, where r_3^{NAE} is the clause density of NAE 3-SAT.

In [15], Coja-Oglan further presented the upper and lower bound for the phase transition point r_k^{NAE} of NAE k-SAT as follows:

Theorem 2 [15]. There is a sequence $\varepsilon_k = 2^{-(1-o_k(1))k}$, such that

$$2^{k-1} \ln 2 - (\frac{\ln 2}{2} + \frac{1}{4}) - \varepsilon_k \leq r_k^{NAE} \leq 2^{(k-1)} \ln 2 - (\frac{\ln 2}{2} + \frac{1}{4}) + \varepsilon_k,$$

where r_k^{NAE} is the clause density of NAE k-SAT.

As can be seen in Theorem 2 when $r < 2^{k-1} \ln 2 - (\frac{\ln 2}{2} + \frac{1}{4}) - \varepsilon_k$, as $n \to \infty$, the k-CNF formulas are almost all NAE-satisfied; when $r > 2^{k-1} \ln 2 - (\frac{\ln 2}{2} + \frac{1}{4}) + \varepsilon_k$, the k-CNF formulas are nearly all NAE-unsatisfied.

3   An Upper Bound for Max NAE k-SAT

For a random k-CNF $F(n, rn)$ formula, we usually use max_F to denote the maximum number of NAE-satisfied clauses of $F(n, rn)$. And let $f(n, rn) = \mathbb{E}(max\_F)$ denote the expectation of the maximum number of NAE-satisfied clauses over the random k-CNF. In this section, we shall present the upper bound for $f(n, rn) = \mathbb{E}(max\_F)$ of random Max NAE k-SAT and give the proof subsequently.

Theorem 3. When $r > (2^k - 2) \ln 2$, $f(n, rn) = \mathbb{E}(max\_F)$, where the operation $\mathbb{E}$ reflects the expectation of maximum number of NAE-satisfied clauses over the random instances, then

$$f(n, rn) \leq (1 - \frac{1}{2^{k-1}} + \sqrt{\frac{(2^k - 2) \ln 2}{r 2^{2k-2}}}) rn. \tag{1}$$

Proof. Let σ denote a real number, $0 \leq \sigma \leq 1$. If $max\_F > (1 - \sigma) rn$, then there must exist an NAE-satisfied subformula F' such that the number of clauses is more than $(1 - \sigma) rn$. For each clause, it is NAE-satisfied by any assignment with probability $\frac{2^k - 2}{2^k}$, and NAE-unsatisfied by the assignment with probability $\frac{2}{2^k}$. Then, the following inequality holds:

$$P = \mathbb{P}(\exists NAE\text{-}satisfied \ F') \leq 2^n \sum_{i=0}^{\sigma rn} \binom{rn}{i} (\frac{2^k - 2}{2^k})^{rn-i} (\frac{2}{2^k})^i. \tag{2}$$

When $\sigma < \frac{2}{2^k}$, the last term of Eq. (2) is maximum. So we have:

$$P \le 2^n(\sigma rn + 1)\binom{rn}{\sigma rn}(\frac{2^k - 2}{2^k})^{rn-\sigma rn}(\frac{2}{2^k})^{\sigma rn}. \tag{3}$$

According to the Stirling's formula $n! \approx \sqrt{2\pi n}(n/e)^n$, we know that

$$\binom{rn}{\sigma rn} = \frac{(rn)!}{(\sigma rn)!(rn - \sigma rn)!} = \frac{\sigma^{-\sigma rn}(1 - \sigma)^{-(1-\sigma)rn}}{\sqrt{2\pi\sigma(1 - \sigma)rn}}. \tag{4}$$

By using Eq. (4) we can simplify Eq. (3)

$$P \le \frac{1}{\sqrt{2\pi\sigma(1 - \sigma)rn}}2^n(\sigma rn + 1)\left[\sigma^{-\sigma}(1 - \sigma)^{\sigma-1}(\frac{2^k - 2}{2^k})^{1-\sigma}(\frac{2}{2^k})^{\sigma}\right]^{rn}. \tag{5}$$

Thus,

$$\frac{\ln P}{rn} \le \frac{\ln 2}{r} + \frac{\ln(\sigma rn + 1)}{rn} + \ln\left[\sigma^{-\sigma}(1 - \sigma)^{\sigma-1}(\frac{2^k - 2}{2^k})^{1-\sigma}(\frac{2}{2^k})^{\sigma}\right]. \tag{6}$$

Let $\sigma = \frac{2}{2^k} - \lambda(0 < \lambda < \frac{2}{2^k})$, and note that $\ln(1 + x) = \sum_{i=1}^{\infty}(-1)^{i-1}x_i/i$. Let us consider the last term of in Eq. (6):

$$\ln\left[\sigma^{-\sigma}(1 - \sigma)^{\sigma-1}(\frac{2^k - 2}{2^k})^{1-\sigma}(\frac{2}{2^k})^{\sigma}\right]$$

$$= \ln\left[\sigma^{-\sigma}(1 - \sigma)^{\sigma-1}(\frac{2^k - 2}{2^k})(\frac{2}{2^k - 2})^{\sigma}\right]$$

$$= -\sigma\ln\sigma + (\sigma - 1)\ln(1 - \sigma) + \ln\frac{2^k - 2}{2^k} + \sigma\ln(\frac{2}{2^k - 2})$$

$$= \ln(\frac{2^k - 2}{2^k}) - (\frac{2}{2^k} - \lambda)\ln(\frac{2}{2^k} - \lambda) + (\frac{2}{2^k} - \lambda - 1)\ln\left[1 - (\frac{2}{2^k} - \lambda)\right]$$

$$+ (\frac{2}{2^k} - \lambda)\ln(\frac{2}{2^k - 2})$$

$$= (\frac{1}{2^{k-1}} - \lambda - 1)\ln(1 + \frac{2^k}{2^k - 2}\lambda) - (\frac{1}{2^{k-1}} - \lambda)\ln(1 - 2^{k-1}\lambda)$$

$$= (\frac{1}{2^{k-1}} - \lambda - 1)(x_1\lambda + x_2\lambda^2 + x_3\lambda^3 + ...) + (\lambda - \frac{1}{2^{k-1}})$$

$$\times (x_1((1 - 2^{k-1})\lambda) + x_2((1 - 2^{k-1})\lambda)^2 + x_3((1 - 2^{k-1})\lambda)^3 + ...).$$

where

$$x_i = \frac{(-1)^{i-1}}{i}(\frac{2^k}{2^k - 2})^i, \quad i \ge 1.$$

So,

$$\ln \left[\sigma^{-\sigma} (1-\sigma)^{\sigma-1} (\frac{2^k - 2}{2^k})^{1-\sigma} (\frac{2}{2^k})^{\sigma} \right] = \sum_{j=2}^{\infty} \Phi_j \lambda^j, \tag{7}$$

where $\Phi_j = \left[(-1)^{j-1} - (2^{k-1} - 1)^{j-1} \right] (\frac{2^k}{2^k - 2})^{j-1} (\frac{1}{j(j-1)}), \ j \geq 2. \tag{8}$

According to Eqs. (6), (7) and (8), we have

$$\frac{\ln P}{rn} \leq \frac{\ln 2}{r} + \frac{\ln (\sigma rn + 1)}{rn} + \sum_{j=2}^{\infty} \left[(-1)^{j-1} - (2^{k-1} - 1)^{j-1} \right] (\frac{2^k}{2^k - 2})^{j-1} (\frac{1}{j(j-1)}) \lambda^j. \tag{9}$$

Let us first consider the case $k=2$. When $\xi = 1, 2, 3, \ldots$, it is obvious that according to Eq. (8), if $j = 2\xi$, then $\Phi_j < 0$, and if $j = 2\xi + 1$ then $\Phi_j = 0$. So we can draw the conclusion that $\Phi_j \leq 0$ for any $j \geq 2$ if $k = 2$. Then we can consider the case when $k \geq 3$. According to Eq. (8), obviously $\Phi_j < 0$ for any $j \geq 2$. Thus, we can simplify inequality Eq. (9) by omitting $\sum_{j=2}^{\infty} \Phi_j \lambda^j$. If we reserve λ^2 and omit $O(\lambda^3)$, then we have

$$\frac{\ln P}{rn} \leq \frac{\ln 2}{r} + \frac{\ln (\sigma rn + 1)}{rn} + \left[(-1) - (2^{k-1} - 1) \right] \frac{2^k}{2^k - 2} \frac{1}{2} \lambda^2. \tag{10}$$

To obtain $\Pr \leq 0$ when $n \to \infty$, we must make sure $(\ln 2)/r + [(-1) - (2^{k-1} - 1)] \times [2^k/(2^k - 2)](1/2)\lambda^2 < 0$. We can solve the following equation first:

$$\frac{\ln 2}{r} + \left[(-1) - (2^{k-1} - 1) \right] (\frac{2^k}{2^k - 2}) \frac{1}{2} \lambda^2 = 0. \tag{11}$$

Solving the equation, we can gain its root as follows:

$$\lambda = \pm \sqrt{\frac{(2^k - 2) \ln 2}{r 2^{2k-2}}}. \tag{12}$$

Thus, when $\lambda > \sqrt{\frac{(2^k - 2) \ln 2}{r 2^{2k-2}}}$ then $\frac{\ln 2}{r} + \left[(-1) - (2^{k-1} - 1) \right] (\frac{2^k}{2^k - 2}) \frac{1}{2} \lambda^2 < 0$ Recalling that $\lambda < \frac{2}{2^k}$, so

$$\sqrt{\frac{(2^k - 2) \ln 2}{r 2^{2k-2}}} < \lambda < \frac{2}{2^k}. \tag{13}$$

Then we have $r > (2^k - 2) \ln 2$. Therefore, we can draw the conclusion when $r > (2^k - 2) \ln 2$ and $n \to \infty$

$$P = \mathbb{P}(\exists NAE\text{-}satisfied \ F') \leq 0. \tag{14}$$

So

$$max\_F < (1 - \sigma)rn = (1 - (\frac{2}{2^k} - \lambda))rn = (1 - \frac{1}{2^{k-1}} + \lambda)rn. \qquad (15)$$

Therefore,

$$f(n, rn) \le (1 - \frac{1}{2^{k-1}} + \sqrt{\frac{(2^k - 2)\ln 2}{r 2^{2k-2}}})rn. \qquad (16)$$

Finally, we can obtain that when $r > (2^k - 2)\ln 2$, $f(n, rn) \le (1 - \frac{1}{2^{k-1}} + \sqrt{\frac{(2^k-2)\ln 2}{r 2^{2k-2}}})rn$. $\qquad \square$

According to Theorem 3, we can attain the upper bound for Max NAE k-SAT in Table 1.

Table 1. Upper bound for Max NAE k-SAT

k	r	Upper bound for $f(n, rn)$	k	r'	Upper bound for $f(n, rn)$
2	1.386	$\frac{1}{2}rn + n\sqrt{\frac{r\ln 2}{2}}$	3	4.159	$\frac{3}{4}rn + n\sqrt{\frac{3r\ln 2}{8}}$
4	9.704	$\frac{7}{8}rn + n\sqrt{\frac{7r\ln 2}{32}}$	5	20.794	$\frac{15}{16}rn + n\sqrt{\frac{15r\ln 2}{128}}$
6	41.589	$\frac{31}{32}rn + n\sqrt{\frac{31r\ln 2}{512}}$	7	87.337	$\frac{63}{64}rn + n\sqrt{\frac{63r\ln 2}{2048}}$
8	176.059	$\frac{127}{128}rn + n\sqrt{\frac{127r\ln 2}{8192}}$	9	353.505	$\frac{255}{256}rn + n\sqrt{\frac{255r\ln 2}{32768}}$
10	708.396	$\frac{511}{512}rn + n\sqrt{\frac{511r\ln 2}{131072}}$	11	1418.179	$\frac{1023}{1024}rn + n\sqrt{\frac{1023r\ln 2}{524288}}$
12	2837.745	$\frac{2047}{2048}rn + n\sqrt{\frac{2047r\ln 2}{2097152}}$	13	5676.875	$\frac{4095}{4096}rn + n\sqrt{\frac{4095r\ln 2}{8388608}}$

Note that in Theorem 3, the estimation of the upper bound by the first moment method is invalid when $r < r' = (2^k - 2)\ln 2$ (see Fig. 1).

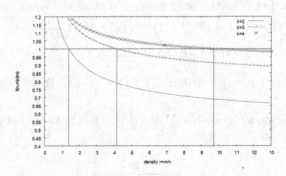

Fig. 1. The curves show the upper bound for $f(n, rn)/rn$ in Theorem 3

In Theorem 3, we only keep λ^2 and omit $O(\lambda^3)$. In order to attain a tighter upper bound for $f(n, rn)$, we can reserve all terms of $\sum\limits_{j=2}^{\infty} \Phi_j \lambda^j$, as can be seen in Theorem 4.

Theorem 4. Let $F(n, rn)$ be a random k-CNF formula, $k \geq 2$, $f(n, rn) = \mathbb{E}(max\_F)$. When $r > \dfrac{-\ln 2}{\sum\limits_{j=2}^{\infty} \Phi_j(1/2^{k-1})^j}$, then

$$f(n, rn) \leq (1 - \frac{1}{2^{k-1}} + \delta)rn, \tag{17}$$

where δ is the only positive root of the equation

$$\frac{\ln 2}{r} + \sum_{j=2}^{\infty} \Phi_j \lambda^j = 0, \text{ and} \tag{18}$$

$$\Phi_j = \left[(-1)^{j-1} - (2^{k-1} - 1)^{j-1}\right] \left(\frac{2^k}{2^k - 2}\right)^{j-1} \left(\frac{1}{j(j-1)}\right), \ j \geq 2. \tag{19}$$

Proof. Similar to Theorem 3, we have

$$\frac{\ln P}{rn} \leq \frac{\ln 2}{r} + \frac{\ln(\sigma rn + 1)}{rn} + \sum_{j=2}^{\infty} \left[(-1)^{j-1} - (2^{k-1} - 1)^{j-1}\right] \left(\frac{2^k}{2^k - 2}\right)^{j-1} \left(\frac{1}{j(j-1)}\right) \lambda^j. \tag{20}$$

To make $\frac{\ln 2}{r} + \sum\limits_{j=2}^{\infty} \left[(-1)^{j-1} - (2^{k-1} - 1)^{j-1}\right] \left(\frac{2^k}{2^k-2}\right)^{j-1} \left(\frac{1}{j(j-1)}\right) \lambda^j < 0$, we need to solve the following equation

$$\frac{\ln 2}{r} + \sum_{j=2}^{\infty} \Phi_j \lambda^j = 0, \tag{21}$$

where $\Phi_j = \left[(-1)^{j-1} - (2^{k-1} - 1)^{j-1}\right] \left(\frac{2^k}{2^k - 2}\right)^{j-1} \left(\frac{1}{j(j-1)}\right), j \geq 2$.

According to the property of this equation, there exists only one positive root of the equation. Let δ be the only positive root of the equation. Thus, when $\lambda > \delta$, $\frac{\ln 2}{r} + \sum\limits_{j=2}^{\infty} \left[(-1)^{j-1} - (2^{k-1} - 1)^{j-1}\right] \left(\frac{2^k}{2^k-2}\right)^{j-1} \left(\frac{1}{j(j-1)}\right) \lambda^j < 0$. Therefore, as $n \to \infty$, we have $P \to 0$. So we can attain

$$f(n, rn) = \mathbb{E}(max\_F) \leq (1 - \frac{1}{2^{k-1}} + \delta)rn. \tag{22}$$

Because $\delta < \lambda < \frac{2}{2^k}$, we have $\frac{\ln 2}{r} + \sum\limits_{j=2}^{\infty} \Phi_j (\frac{2}{2^k})^j < 0$. The variant of this inequality is showed as follows:

$$r > \frac{-\ln 2}{\sum\limits_{j=2}^{\infty} \Phi_j(1/2^{k-1})^j} \tag{23}$$

Let $G = \frac{\ln 2}{r} + \sum_{j=2}^{\infty} \Phi_j x^j$, then $G(\delta) = 0$. For any $x' \in [0, +\infty)$, if $G(x') < 0$, then $x' > \delta$, and if $G(x') > 0$, then $x' < \delta$.
Since,

$$G(\sqrt{\frac{(2^k - 2)\ln 2}{r2^{2k-2}}}) = \frac{\ln 2}{r} + \sum_{j=2}^{\infty} \Phi_j(\sqrt{\frac{(2^k - 2)\ln 2}{r2^{2k-2}}})^j = \sum_{j=3}^{\infty} \Phi_j(\sqrt{\frac{(2^k - 2)\ln 2}{r2^{2k-2}}})^j < 0,$$

(24)

we have $\delta < \sqrt{\frac{(2^k-2)\ln 2}{r2^{2k-2}}}$. Thus,

$$(1 - \frac{1}{2^{k-1}} + \delta)rn < (1 - \frac{1}{2^{k-1}} + \sqrt{\frac{(2^k - 2)\ln 2}{r2^{2k-2}}})rn.$$

(25)

Therefore, we gain a tighter upper bound for $f(r, rn)$. When $r > \frac{-\ln 2}{\sum_{j=2}^{\infty} \Phi_j(1/2^{k-1})^j}$,

$$f(n, rn) \leq (1 - \frac{1}{2^{k-1}} + \delta)rn,$$

where δ is the only positive root of the equation:

$$\frac{\ln 2}{r} + \sum_{j=2}^{\infty} [(-1)^{j-1} - (2^{k-1} - 1)^{j-1}] \left(\frac{2^k}{2^k - 2}\right)^{j-1} \left(\frac{1}{j(j-1)}\right) \lambda^j = 0.$$

Finally, we gain a tighter upper bound for $f(r, rn)$ by reserving all terms of $\sum_{j=2}^{\infty} \Phi_j \lambda^j$. □

4   Upper and Lower Bounds for NAE k-SAT(b)

In this section, we shall investigate the phase transition phenomena of the family of NAE k-SAT(b) problems. From the definition of NAE k-SAT(b), we can see that if $b = 0$, then NAE k-SAT(b) is equivalent to NAE k-SAT; and if b equals the result of the total number of clauses minus the optimal solution cost, then NAE k-SAT(b) is equivalent to Max NAE k-SAT. In this sense, both NAE k-SAT and Max NAE k-SAT can be regarded as special cases for NAE k-SAT(b).

Let $X_k^{n,r,b}$ be a random variable on $F_k(n, rn)$ such that $X_k^{n,r,b}(\varphi)$ is the number of assignments on n variables that violate no more than b clauses of φ and $X_k^{n,r,b} = \sum_{\varphi} X_k^{n,r,b}(\varphi)$, where φ is a random k-CNF formula in $F_k(n, rn)$.

Now, we have the following conjecture for the family of NAE k-SAT(b) problems, where $k \geq 3$ and $b \geq 0$.

Conjecture 1. For every integer $k \geq 3$ and every $b \geq 0$, there is a positive real number $r_{k,b}$ such that

$$\text{If } r > r_{k,b}, \text{ then } \lim_{n \to \infty} \mathbb{P}[X_k^{n,r,b} \geq 1] = 0.$$

$$\text{If } r < r_{k,b}, \text{ then } \lim_{n \to \infty} \mathbb{P}[X_k^{n,r,b} \geq 1] = 1.$$

We first assume there exists a phase point $r_{k,b}$ when the clause density r varies. That is, when $r > r_{k,b}$, almost all NAE k-SAT(b) instances can not be solved. And when $r < r_{k,b}$, almost all NAE k-SAT(b) instances can be solved. In order to prove the conjecture, we can establish certain analytical results, which generate the upper bound and lower bound for the value of $r_{k,b}$. In Theorem 5, we present an upper bound for $r_{k,b}$.

Theorem 5. Let $k \geq 3$ and $b \geq 0$ be two integers. If $r > \frac{1}{k-1-\log_2{(2^{k-1}-1)}}$, then $\lim_{n \to \infty} \mathbb{P}[X_k^{n,r,b} \geq 1] = 0$. It follows that if $r_{k,b}$ exists, then $r_{k,b} \leq \frac{1}{k-1-\log_2{(2^{k-1}-1)}}$, regardless of the value of b.

Proof. For each truth assignment θ on n variables, let I_θ^b be a random variable on $F_k(n, rn)$ such that $I_\theta^b(\varphi) = 1$ and $I_\theta^b = \sum_\varphi I_\theta^b(\varphi)$, if θ is an assignment that violates no more than b clauses of φ; and $I_\theta^b(\varphi) = 0$, otherwise. Obviously, there are a total of $(2^k \binom{n}{k})^{rn}$ possible formulas in $F_k(n, rn)$, and there are $\sum_{i=0}^{b} \binom{rn}{i}(2\binom{n}{k})^i((2^k - 2)\binom{n}{k})^{rn-i}$ formulas in $F_k(n, rn)$ for which NAE-satisfying assignment θ violates no more than b clauses. Therefore,

$$\mathbb{E}(I_\theta^b) = \frac{\sum_{i=0}^{\infty} \binom{rn}{i}(2\binom{n}{k})^i((2^k - 2)\binom{n}{k})^{rn-i}}{(2^k\binom{n}{k})^{rn}}. \tag{26}$$

Since $X_k^{n,r,b} = \sum_\theta I_\theta^b$, the linearity of expectation implies that $\mathbb{E}(X_k^{n,r,b}) = \sum_\theta \mathbb{E}(I_\theta^b) = 2^n \mathbb{E}(I_\theta^b)$. By Markov's inequality, we have that

$$\mathbb{P}[X_k^{n,r,b} \geq 1] \leq \mathbb{E}(X_k^{n,r,b}) = 2^n \left(\frac{\sum_{i=0}^{b} \binom{rn}{i}(2\binom{n}{k})^i((2^k - 2)\binom{n}{k})^{rn-i}}{(2^k\binom{n}{k})^{rn}} \right)$$

$$= \frac{2^n}{2^{krn}} \sum_{i=0}^{b} (2^k - 2)^{rn} \binom{rn}{i} \frac{2^i}{(2^k - 2)^i} \tag{27}$$

$$= (2(\frac{(2^k - 2)}{2^k})^r)^n \sum_{i=0}^{b} \binom{rn}{i} (\frac{2}{2^k - 2})^i$$

Thus,

$$\lim_{n\to\infty} \mathbb{P}[X_k^{n,r,b} \geq 1] \leq \lim_{n\to\infty} (2(\frac{2^k-2}{2^k})^r)^n \sum_{i=0}^{b} \binom{rn}{i} (\frac{2}{2^k-2})^i$$

$$\leq \lim_{n\to\infty} (2(\frac{2^k-2}{2^k})^r)^n O(n^b).$$

(28)

Therefore, if $r > \frac{1}{k-1-\log_2(2^{k-1}-1)}$, then $\lim_{n\to\infty} \mathbb{P}[X_k^{n,r,b} \geq 1] = 0$, regardless of the value of b. □

For the lower bound for $r_{k,b}$, obviously, if all k-CNF formulas are NAE-satisfied, then $\lim_{n\to\infty} \mathbb{P}[X_k^{n,r,b} \geq 1] = 1$, regardless of the value of b. Since [14] have proved that if $r < 2^{k-1}\ln 2 - \frac{\ln 2}{2} - \frac{1}{4} - \varepsilon_k$, then all k-CNF formulas are NAE-satifiable, where $\varepsilon_k = 2^{-(1-o_k(1))k}$, we know $r_{k,b} \geq 2^{k-1}\ln 2 - \frac{\ln 2}{2} - \frac{1}{4} - \varepsilon_k$.

In summary, we obtain the upper bound and the lower bound for NAE k-SAT(b) in the following.

Theorem 6. Let $k \geq 3$ and $b \geq 0$ be two integers.

If $r > \dfrac{1}{k-1-\log_2(2^{k-1}-1)}$, then $\lim_{n\to\infty} \mathbb{P}[X_k^{n,r,b} \geq 1] = 0$.

If $r < 2^{k-1}\ln 2 - \dfrac{\ln 2}{2} - \dfrac{1}{4} - \varepsilon_k$, $\varepsilon_k = 2^{-(1-o_k(1))k}$, then $\lim_{n\to\infty} \mathbb{P}[X_k^{n,r,b} \geq 1] = 1$.

It follows that if $r_{k,b}$ exists, then $2^{k-1}\ln 2 - \frac{\ln 2}{2} - \frac{1}{4} - \varepsilon_k \leq r_{k,b} \leq \frac{1}{k-1-\log_2(2^{k-1}-1)}$, $\varepsilon_k = 2^{-(1-o_k(1))k}$, regardless of the value b.

According to Theorem 6, for $k = 3, 4, 5, \ldots$, the upper and lower bounds for NAE k-SAT(b) can be illustrated in Table 2.

Table 2. Upper and lower bounds for NAE k-SAT(b)

k	3	4	5	7	9	10	11	12
Upper bound for $r_{k,b}$	2.4096	5.195	10.753	45.454	195.695	452.489	709.220	1419.219
Lower bound for $r_{k,b}$,	1.514	4.083	9.973	43.432	177.099	354.027	708.925	1418.712

5 Conclusion

In this paper, we systematically research the upper bound of Max NAE k-SAT and the phase transition of the decision version of Max NAE k-SAT. We use first-moment method to obtain an upper bound for $f(n, rn)$ of random Max NAE k-SAT ($k \geq 2$). Furthermore, we prove that NAE k-SAT(b) ($k \geq 3$) has phase transition at critical point $r_{k,b}$, and acquire both the upper bound and lower bound for NAE k-SAT(b).

Acknowledgement. The authors of this paper wish to extend their sincere gratitude to all anonymous reviewers for their efforts. This work was supported in part by NSFC (under Grant Nos.61503074, 61403076, 61402070, and 61403077), the Natural Science Foundation for Youths of JiLin Province (20160520104JH) and (NCET-13-0724).

References

1. Xu, K., Li, W.: The SAT phase transition. Sci. China Ser. E **42**, 494–501 (1999)
2. Sly, A., Sun, N., Zhang, Y.: The number of solutions for random regular NAE-SAT. In: FOCS (2016)
3. Larrosa, J., Heras, F., De Givry, S.: A logical approach to efficient Max-SAT solving. Artif. Intell. **172**(2–3), 204–233 (2006)
4. Monasson, R., Zecchina, R., Kirkpatrick, S., Selman, B., Troyansky, L.: Determining computational complexity from characteristic phase transitions. Nature **400**, 133–137 (1999)
5. Kirkpatrick, S., Selman, B.: Critical behavior in the satisfiability of random boolean expressions. Science **264**, 1297–1301 (1994)
6. Xu, K., Li, W.: Exact phase transitions in random constraint satisfaction problems. J. Artif. Intell. Res. **12**, 93–103 (2000)
7. Fan, Y., Shen, J., Xu, K.: A general model and thresholds for random constraint satisfaction problems original. Artif. Intell. **193**, 1–17 (2012)
8. Gent, I.P., Walsh, T.: The TSP phase transition. Artif. Intell. **88**(1–2), 349–358 (1996)
9. Walsh, T.: Where are the really hard manipulation problems? the phase transition in manipulating the veto rule. In: Proceedings of the 21st International Jont Conference on Artifical intelligence, Morgan Kaufmann Publishers Inc. (2009)
10. Huang, P., Yin, M.: An upper (lower) bound for Max (Min) CSP. Sci. China Inf. Sci. **57**(7), 1–9 (2014)
11. Gao, J., Wang, J., Yin, M.: Experimental analyses on phase transitions in compiling satisfiability problems. Sci. China Inf. Sci. **58**, 1–11 (2015)
12. Zhou, J., Yin, M., Li, X., Wang, J.: Phase transitions of EXPSPACE-complete problems: a further step. Int. J. Found. Comput. Sci. **23**(01), 173–184 (2012)
13. Achlioptas, D., Chtcherba, A., Istrate, G., Moore, C.: The phase transition in 1-in-k SAT and NAE 3-SAT. In: SODA, pp. 721–722 (2001)
14. Achlioptas, D.: The asymptotic order of the random k-SAT threshold. In: Proceedings of the 43rd Annual IEEE Symposium on Foundations of Computer Science (FOCS02), pp. 779–788 (2002)
15. Coja-Oglan, A., Panagiotou, K.: Catching the k-NAE SAT threshold. In: Proceedings of the Forty-Fourth Annual ACM Symposium on Theory of Computing, pp. 899–908 (May 2012)
16. Ding, J., Sly, A., Sun, N.: Satisfiability threshold for random regular NAE-SAT. In: Proceedings of the 46th Annual ACM Symposium on Theory of Computing, pp. 814–822 (May 2014)
17. Achlioptas, D., Naor, A., Peres, Y.: Rigorous location of phase transitions in hard optimization problems. Nature **435**, 759–764 (2005)
18. Coppersmith, D., et al.: Random MAX SAT, random MAX CUT, and their phase transitions. Random Struct. Algorithms **24**, 502–545 (2004)

19. Xu, X.L., Gao, Z.S., Xu, K.: A tighter upper bound for random MAX 2-SAT. Inf. Process. Lett. **111**, 115–119 (2011)
20. Zhang, W.: Phase transitions and backbones of 3-SAT and maximum 3-SAT. In: Walsh, T. (ed.) CP 2001. LNCS, vol. 2239, pp. 153–167. Springer, Heidelberg (2001). doi:10.1007/3-540-45578-7_11
21. Bailey, D.D., Kolaitis, P.G.: Phase transitions of bounded satisfiability problems. In: IJCAI, pp. 1187–1193 (2003)

Author Index

Printed in the United States
By Bookmasters